Handbook of Industrial Waste Treatment

Volume 1

edited by

Lawrence K. Wang
Zorex Environmental Engineers, Inc.
Pittsfield, Massachusetts

Mu Hao Sung Wang
New York State Department of Environmental Conservation
Albany, New York

Marcel Dekker, Inc.   New York • Basel • Hong Kong

Library of Congress Cataloging-in-Publication Data

Handbook of industrial waste treatment / edited by Lawrence K. Wang,
  Mu Hao Sung Wang.
    p.   cm.
  Includes bibliographical references and index.
  ISBN 0-8247-8716-1 (acid-free paper)
  1. Factory and trade waste--Management.   2. Industry-
-Environmental aspects.   I. Wang, Lawrence K.   II. Wang, Mu Hao
Sung.
TD897.5.H35   1992
628.4--dc20   92-19052
                                                                              CIP

This book is printed on acid-free paper.

Copyright © 1992 by Marcel Dekker, Inc. All Rights Reserved.

Neither this book nor any part may be reproduced or transmitted in any form or by any means, electronic or mechanical, including photocopying, microfilming, and recording, or by any information storage and retrieval system, without permission in writing from the publisher.

Marcel Dekker, Inc.
270 Madison Avenue, New York, New York 10016

Current printing (last digit):
10 9 8 7 6 5 4 3 2 1

PRINTED IN THE UNITED STATES OF AMERICA

Preface

Environmental managers, engineers, and scientists who have had experience with industrial waste management problems have noted the need for a handbook that is comprehensive in its scope, directly applicable to daily waste management problems of specific industries, and widely acceptable by practicing environmental professionals and educators.

Many standard industrial waste treatment and management texts adequately cover a few major industries, for conventional in-plant pollution control strategies, but no one book, or series of books, focuses on new developments in innovative and alternative environmental technology, design criteria, effluent standards, managerial decision methodology, and regional and global environmental conservation.

This handbook emphasizes in-depth presentation of environmental pollution sources, waste characteristics, control technologies, management strategies, facility innovations, process alternatives, costs, case histories, effluent standards, and future trends for each industrial or commercial operation, such as the metal plating and finishing industry and photographic processing industry, and in-depth presentation of methodologies, technologies, alternatives, regional effects, global effects of each important industrial pollution control practice that may be applied to all industries, such as waste minimization and stormwater management.

In a deliberate effort to complement other industrial waste treatment and management texts, this handbook covers new subjects as much as possible.

Many topics, such as waste minimization, treatment of stormwater, photographic processing wastes, and metal finishing wastes, have been presented in detail for the first time in any industrial waste treatment book. Special efforts have been made to invite experts to contribute chapters in their own areas of expertise. Since the field of industrial waste treatment is very broad, no one can claim he/she is expert in all industries. Collective contributions logically are better than a single author's presentation for a handbook of this nature.

The volumes in this handbook are intended to be used as college textbooks as well as reference books for the environmental professional. To accomplish the goal of developing a set of modern industrial waste treatment textbooks, we feature in the first three volumes the major industries and industrial practices that have significant effects on the environment. Professors, students, and researchers in environmental, civil, chemical, sanitary, mechanical, and public health engineering and science will find valuable education material here. New volumes in the handbook will be published periodically in order to cover almost all industries for reference by practicing environmental professionals who need state-of-the-art solutions and guidance for solving the pollution problems in all industrial areas. Each volume will contain extensive bibliographic references for each industrial waste treatment or practice. These should be invaluable to the environmental managers or researchers who need to trace, follow, duplicate, or even improve upon a specific industrial waste treatment practice.

A successful modern industrial waste treatment program for a particular industry will include not only traditional water pollution control, but also air pollution control, noise control, soil conservation, radiation protection, groundwater protection, hazardous waste management, solid waste disposal, and combined industrial–municipal waste treatment and management. In fact, it should be a total environmental control program. Another intention of this handbook is to provide technical and economical information on the development of the most feasible total environmental control program that can benefit both industry and local municipalities—frequently, the most economically feasible methodology is combined industrial–municipal waste treatment.

We are indebted to Dr. Constantine Yapijakis, Professor of Environmental Engineering at the Cooper Union, New York City, for his assistance in editing the first volume. Because of Dr. Yapijakis' contribution, he has been invited to be a coeditor of Volume 2.

Lawrence K. Wang
Mu Hao Sung Wang

Contents

Preface		*iii*
Contributors		*vii*
1	Waste Minimization *Eric H. Snider*	1
2	Stormwater Management and Treatment *Robert Leo Trotta and Constantine Yapijakis*	61
3	Treatment of Metal Plating and Finishing Wastes *Mark Davis and Tom Sandy*	127
4	Treatment of Photographic Processing Wastes *Thomas W. Bober, Thomas J. Dagon, and Harvey E. Fowler*	173
5	Treatment of Soap and Detergent Industry Wastes *Constantine Yapijakis*	229
6	Treatment of Acid Pickling Wastes of Metals *Veysel Eroglu and Ferruh Erturk*	293

7	Treatment of Textile Wastes *Sanjoy K. Bhattacharya*	307
8	Treatment of Phosphate Industry Wastes *Constantine Yapijakis*	323

Index *385*

Contributors

Sanjoy K. Bhattacharya   *Tulane University, New Orleans, Louisiana*

Thomas W. Bober   *Eastman Kodak Company, Rochester, New York*

Thomas J. Dagon   *Eastman Kodak Company, Rochester, New York*

Mark Davis   *CH2M HILL, Denver, Colorado*

Veysel Eroglu   *Istanbul Technical University, Ayazaga, Istanbul, Turkey*

Ferruh Erturk   *TUBITAK Mamara Research Center, Gebze, Kocaeli, Turkey*

Harvey E. Fowler   *Eastman Kodak Company, Rochester, New York*

Tom Sandy   *CH2M HILL, Denver, Colorado*

Eric H. Snider   *Engineering-Science, Inc., Atlanta, Georgia*

Robert Leo Trotta   *O'Brien & Gere Engineers, Inc., New York, New York*

Constantine Yapijakis   *The Cooper Union, New York, New York*

1

Waste Minimization

Eric H. Snider

Engineering-Science, Inc., Atlanta, Georgia

An ounce of prevention is worth a pound of cure.
American Folk Expression

1. INTRODUCTION

Waste. Every living thing produces waste. Our cities produce mountains of waste. And in this modern industrial age, our businesses, industries, and commercial activities produce waste. Waste is a reality that cannot be denied, nor can it be buried, as has often been the case in the past.

In this chapter, we shall discuss and demonstrate the concepts of waste minimization. Waste minimization is simply the philosophy and process of keeping to an absolute minimum the amount of waste that is produced, treated, stored, and disposed. We will concentrate exclusively in this discussion on the minimization of industrial wastes, since these pose the greatest threat to human health and the environment. However, the principles and practices described here by and large apply to municipal solid waste (another staggering problem) and indeed to all materials categorized as wastes.

2. WASTE GENERATION: A BRIEF HISTORY

Industrial processes of all types almost invariably produce wastes. These wastes have many forms and names—*residue, by-product, off-gas*, to name a few. Until the middle of the twentieth century, industrial wastes were considered only a casual nuisance and were handled as such by generators. Industrial plants of the time disposed of most wastes by burial in landfills, discharge into seepage basins, or by pumping directly to a body of water (a so-called receiving stream) or into a deep well. Refinements were added over the years—e.g., many wastes were drummed and the containerized wastes sent for off-site disposal. However, little if any thought was given to the fact that these wastes, once generated, ultimately ended up being released to the environment unless they were destroyed by treatment.

During the 1960s and 70s, industrial waste generators were made increasingly aware of the nature of their wastes and the problems that waste disposal imposed on our environment. Spurred on for the first time by mandates from a federal agency charged with environmental protection (the U.S. Environmental Protection Agency, EPA, and its forerunners) as well as by their own sense of corporate responsibility, industries addressed, in turn, air pollution emissions, wastewater discharges, industrial hygiene/worker safety, and a variety of related issues. However, with rare exception, the actual generation of wastes was never questioned.

During the decades of the 1970s and 80s, new information regarding industrial wastes was developed and complementary federal regulations required industry to reexamine the overall concept of waste generation. First, it was determined that many chemicals present in industrial wastes exert a permanent deleterious effect on human health. In fact, exposure to some chemicals can alter human genetic material so that the effects of exposure are passed on to future generations.

Second, industrial wastes that are not properly treated and disposed of will ultimately release constituents to the environment. For example, wastes disposed in landfills may release constituents to subsurface aquifiers that serve as drinking water supplies.

Third, testing methods have been developed to evaluate whether an industrial waste contains any constituents of concern to human health or the environment. Furthermore, the tests determine whether and at what rate a waste will release constituents into the environment. Wastes that contain any of an extensive list of hazardous constituents or that exhibit a hazard characteristic or that are generated by certain industrial processes are referred to as hazardous

Waste Minimization

wastes under the Resource Conservation and Recovery Act (RCRA). At present the United States generates more than 265 million metric tons of 71 billion gallons of hazardous wastes per year (Doyle, 1987). The treatment, storage, and disposal of these wastes are now governed by strict regulations.

3. WASTE TREATMENT AND DISPOSAL

During the 1970s and 80s, waste treatment and disposal for industrial residues have assumed growing importance. In particular, those wastes defined as *RCRA-hazardous* require meticulous attention to treatment and ultimate disposal. Furthermore, during the late 1980s, federal regulations were enacted eliminating any form of land disposal for a variety of hazardous wastes, thereby making imperative the treatment of these wastes to render them nonhazardous.

This volume is devoted to industrial waste treatment. Many treatment techniques—physical, chemical, and biological—will be presented and discussed as unit operations. Applicability of the unit operations to specific industry wastes will be covered in detail. The remaining chapters assume that a waste is generated by some industrial process and must be treated and disposed of—an *end-of-pipe* approach involving significant costs and technical complexity. By and large, this is true: In our modern society we require the benefits of industrially produced materials, and the processes used to make these materials generate wastes. However, in this chapter we shall present the framework for minimizing industrial waste generation.

4. DEFINITION AND PHILOSOPHY OF WASTE MINIMIZATION

4.1 Definitions

The definition and philosophy of waste minimization in the United States is being driven by the regulations and guidelines of the EPA. Therefore, much of the discussion in this section is taken from EPA publications.

Waste minimization means the reduction, to the extent feasible, of any solid or hazardous waste that is generated or subsequently treated, stored, or disposed of. In addition to waste regulated under RCRA, the EPA encourages the minimization of all wastes that pose risks to human health and the environment (EPA, 1987). In general, the idea underlying the promotion of waste minimization is that it makes far more sense for a generator not to produce waste rather than develop extensive (and expensive) treatment schemes to insure that the waste stream poses no threat to the quality of the environment (EPA, 1988).

The EPA has developed a hazardous waste management strategy that is applicable to all industrial wastes. The four categories in the strategy, in decreasing order of desirability, are as follows (EPA, 1988):

Source reduction. Reduce the amount of waste at the source, through changes in industrial processes.

Recycling. Reuse and recycle wastes for the original or some other purpose, such as materials recovery or energy production.

Incineration/treatment. Destroy, detoxify, and neutralize wastes into less harmful substances.

Secure land disposal. Deposit wastes on land using volume reduction, encapsulation, leachate containment, monitoring, and controlled air and surface/subsurface waste releases.

In the current working definitions used by the EPA, waste minimization (or the more recent term *pollution prevention*) includes source reduction and recycling (see Fig. 1). Source reduction is highly preferable to recycling, and both are much preferred over the latter two, treatment and disposal. Source reduction and recycling each encompass a variety of techniques, as shown in Fig. 2. These techniques will be discussed in detail later in this chapter.

Until recently, the term *waste minimization* was assumed by many to include incineration and other physical, chemical, or biological treatments that render a waste nonhazardous. In addition, the term waste minimization was often construed to include the removal of wastes from a hazardous waste listing by regulatory action without changing the form of the waste at all (the process of "delisting" covered in Volume 40 of the Code of Federal Regulations, Part 261). However, the current EPA definitions restrict waste minimization to source reduction and recycling, and the remainder of this chapter will concentrate on these two topics. Treatment techniques for industrial wastes are discussed in detail in the remaining chapters of this volume.

4.2 EPA's Philosophy on Waste Minimization

In January 1989, the EPA published a proposed policy statement outlining its philosophy on pollution prevention or waste minimization (Federal Register, 1989). The philosophy as excerpted from this policy statement is as follows:

> EPA has made substantial progress over the last 18 years in improving the quality of the environment through implementation of media-specific

Waste Minimization

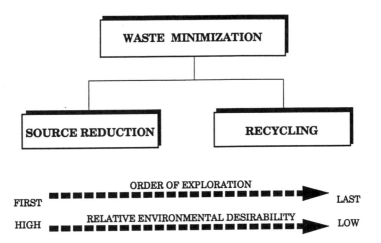

WASTE MINIMIZATION
   The reduction, to the extent feasible, of hazardous waste that is generated or subsequently treated, stored or disposed of. It includes any source reduction or recycling activity undertaken by a generator that results in either (1) the reduction of total volume or quantity of hazardous waste or (2) the reduction of toxicity of the hazardous waste, or both, so long as such reduction is consistent with the goal of minimizing present and future threats to human health and the environment (EPA's Report to Congress, 1986, EPA/530-SW-86-033).

SOURCE REDUCTION
   Any activity that reduces or eliminates the generation of hazardous waste at the source, usually within a process (op. cit.).

RECYCLING
   A material is "recycled" if it is used, reused, or reclaimed (40 CFR 261.1 (c) (7)). A material is "used or reused" if it is either (1) employed as an ingredient (including its use as an intermediate) to make a product; however a material will not satisfy this condition if distinct components of the material are recovered as separate end products (as when metals are recovered from metal containing secondary materials) or (2) employed in a particular function as an effective substitute for a commercial product (40 CFR 261.1 (c) (5)). A material is "reclaimed" if it is processed to recover a useful product or if it is regenerated. Examples include the recovery of lead values from spent batteries and the regeneration of spent solvents (40 CFR 261.1 (c) (4)).

Figure 1 Waste minimization definitions. (From EPA, 1988.)

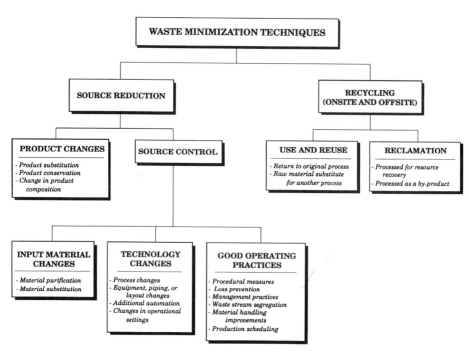

Figure 2 Waste minimization techniques. (From EPA, 1988.)

pollution control programs. Not withstanding past progress, there are economic, technological, and institutional limits on how much improvement can be achieved under these programs, which emphasize management after pollutants have been generated. As early as 1976, EPA believed the nation could not continue to reduce threats to human health and the environment while utilizing only better methods of control, treatment or disposal.

In practice, waste management activities by both the regulatory and the regulated community have largely focused on treatment, control and disposal as specified in EPA's major statutes, and to a lesser extent, on recycling. Although each of these techniques is appropriate in a comprehensive waste management strategy, government and industry are beginning to realize that end-of-pipe pollution controls alone are not enough. Significant amounts of waste containing toxic constituents continue to be released into the air, land, and water, despite stricter pollution controls and skyrocketing waste management costs.

Waste Minimization

There is increasing evidence of the economic and environmental benefits to be realized by reducing waste at the source rather than managing such waste after it is produced. Elimination of tons of pollutant discharges can be combined with cost savings estimated from the cost of pollution control facilities that did not have to be built; reduced operating costs for pollution control facilities; reduced manufacturing costs; and retained sales of products that might otherwise have been taken off the market as environmentally unacceptable.

The term, *waste minimization*, which EPA has previously used in reference to source reduction and recycling activities in its hazardous waste program, has been replaced by the phrase, *pollution prevention*. Through eliminating a term that may be perceived as closed tied to RCRA, EPA is emphasizing that the policy has applicability beyond the RCRA hazardous waste context. EPA stresses that the policy focuses primarily on the prevention of pollution through the multi-media reduction of pollutants at the source. In addition, in order to obtain additional benefits of avoiding releases to the environment, EPA's pollution prevention program secondarily promotes environmentally sound recycling.

EPA believes pollution prevention through source reduction and environmentally sound recycling is highly desirable, and that as a Nation there are many opportunities in source reduction and recycling that we have not yet pursued. However, we recognize that, while there is still much progress to be gained, the extent to which we can prevent pollution also has limitation, and that safe treatment, storage and disposal, for pollution that couldn't reasonably be reduced at the source or recycled, will continue to be important components of an environmental protection strategy. Source reduction and recycling will not totally obviate the need for or the importance of these processes. Individuals as well as industrial facilities or organizations can practice source reduction and recycling through changing their consumption or disposal habits, their driving patterns and their on-the-job practices. EPA believes that developing and implementing a new multi-media prevention strategy, focused primarily on source reduction and secondarily on environmentally sound recycling, offers enormous promise for improvements in human health protection and environmental quality and significant economic benefits.

The EPA and all 50 states of the United States are involved in waste minimization program definition, implementation, and research/demonstration funding. The EPA has published a list of state waste minimization programs (EPA, 1987), and several of these programs are producing excellent results.

5. SOURCE REDUCTION

Source reduction is by far the most preferable of all options in the waste management hierarchy. Source reduction is simply the prevention of pollution at the source. As shown in Fig. 2, source reduction consists of two broad areas: product changes and source control. We shall discuss each of these in turn with brief examples.

5.1 Product Changes

Product changes are performed by the manufacturer of a product with the intent of reducing waste resulting from a product's use (EPA, 1988). Product changes include:

Product substitution
Product conservation
Changes in product composition

The technology for implementing product changes is developed and in place for many industrial processes. However, many issues besides technical feasibility must be considered in selecting product changes as a waste minimization option. These issues include consumer acceptance, product shelf life, compatibility of the changed product in its intended use compared to the original product, and potential environmental problems of the new product.

As an example, in the paint manufacturing industry, water-based coatings are finding increasing applications where solvent-based paints were used before. These products do not contain toxic or flammable solvents that make solvent-based paints hazardous when they are disposed of. Also, cleaning the applicators with a solvent is not necessary. The use of water-based paints instead of solvent-based paints also greatly reduces volatile organic compound emissions to the atmosphere (EPA, 1988).

5.2 Source Control

In the absence of a major modification of the industrial product itself to eliminate waste, the other preferred option for reduction at the source is source control. Control of waste generation at the source while still producing the

Waste Minimization

same product can be accomplished by three broad techniques: input material changes, technology changes, and good operating practices.

Input Material Changes

Input material changes accomplish waste minimization by reducing or eliminating the hazardous materials that enter the production process. Also, changes in input materials can be made to avoid the generation of hazardous wastes within the production processes. Input material changes include material purification and material substitution (EPA, 1988).

An example involves an electronic manufacturing facility of a large diversified corporation that originally cleaned printed circuit boards with solvents. The company found that by switching from a solvent-based cleaning system to an aqueous-based system the same operating conditions and workloads could be maintained. The aqueous-based system was found to clean six times more effectively. This resulted in a lower product reject rate and eliminated hazardous waste (EPA, 1988).

Technology Changes

Technology changes are oriented toward process and equipment modifications to reduce waste, primarily in a production setting. Technology changes can range from minor changes that can be implemented in a matter of days at low cost, to the replacement of processes involving large capital costs (EPA, 1988). These changes include the following:

Changes in the production process
Equipment, layout, or piping changes
Use of automation
Changes in process operating conditions, such as flow rates, temperatures, pressures, residence times.

As an example, a manufacturer of fabricated metal products cleaned nickel and titanium wire in an alkaline chemical bath prior to using the wire in its product. The company began to experiment with a mechanical abrasive system. The wire was passed through the system that uses silk and carbide pads and pressure to brighten the metal. The system worked, but required passing the wire through the unit twice for complete cleaning. The company bought a second abrasive unit and installed it in series with the first unit. This system allowed the company to completely eliminate the need for the chemical cleaning bath (EPA, 1988).

Good Operating Practices

Good operating practices are procedural, administrative, or institutional measures that a company can use to minimize waste. Good operating practices apply to the human aspect of manufacturing operations. Many of these measures are used in industry largely as efficiency improvements and good management practices. Good operating practices can often be implemented with little cost and, therefore, have a high return on investment. These practices can be implemented in all areas of a plant, including production, maintenance operations, and in raw material and product storage. Good operating practices are often categorized broadly as "housekeeping"; estimates are that improved housekeeping can accomplish about half of the total waste minimization possible through efficiency increases and changes in process. Good operating practices include the following (EPA, 1988):

Waste minimization programs
Management and personnel practices
Material handling and inventory practices
Loss prevention
Waste segregation
Cost accounting practices
Production scheduling

Management and personnel practices include employee training, incentives and bonuses, and other programs that encourage employees to conscientiously strive to reduce waste. Material handling and inventory control practices include programs to reduce loss of shelf life of time-sensitive materials and proper storage conditions. Loss prevention minimizes waste by avoiding leaks from equipment and spills.

Waste segregation practices reduce the volume of hazardous wastes by preventing the mixing of hazardous and nonhazardous wastes. One issue that promotes serious attention to waste segregation is the "mixing rule." When a hazardous waste is mixed with a nonhazardous waste, the resulting mixture can be classified as hazardous if warranted by specific characteristics. Consequently, segregating the hazardous wastes (usually smaller quantities) from nonhazardous wastes results in a minimizing of the quantities of hazardous waste requiring further consideration (Schoenberger and Corbin, 1985). Cost accounting practices include programs to allocate waste treatment and disposal costs directly to the departments or groups that generate waste, rather than charging these costs to general company overhead accounts. In doing so, the

Waste Minimization 11

departments or groups that generate the waste become more aware of the effects of their treatment and disposal practices and have a financial incentive to minimize their waste. By the judicious scheduling of batch production runs, the frequency of equipment cleaning and the resulting waste can be reduced (EPA, 1988).

As an example of good operating practices, a large consumer product company in California adopted a corporate policy to minimize the generation of hazardous waste. In order to implement the policy, the company mobilized quality circles made up of employees representing areas within the plant that generated hazardous wastes. The company experienced a 75% reduction in the amount of wastes generated by instituting proper maintenance procedures suggested by the quality circle teams. Since the team members were also line supervisors and operators, they made sure the procedures were followed (EPA, 1988).

6. RECYCLING

Once all options for source reduction are exhausted, the most desirable of the remaining options in the waste management hierarchy is recycling. Recycling can occur entirely on-site or may involve a second facility; the two recycling techniques are use/reuse and reclamation.

6.1 Use and Reuse

Recycling *via* use and/or reuse involves the return of a waste material either to the originating process as a substitute for an input material, or to another process as an input material (EPA, 1988).

As an example, a printer of newspaper advertising in California purchased an ink recycling unit to produce black newspaper ink from its various waste inks. The unit blends the different colors of waste ink together with fresh black ink and black toner to create the black ink. This ink is then filtered to remove flakes of dried ink. This ink is used in place of fresh black ink and eliminates the need for the company to ship waste ink off-site for disposal. The price of the recycling unit was paid off in 18 months based only on the savings in fresh black ink purchases. The payback improved to nine months when the costs for disposing of ink as a hazardous waste are included (EPA, 1988).

6.2 Reclamation

Reclamation is the recovery of a valuable material from a hazardous waste. Reclamation techniques differ from use and reuse techniques in that the recovered material is not used in the facility; rather, it is sold to another company (EPA, 1988).

As an example of reclamation, a photoprocessing company used an electrolytic deposition cell to recover silver out of the rinsewater from film processing equipment. The silver is then sold to a small recycler. By removing the silver from this wastewater, the wastewater can be discharged to the sewer without additional pretreatment by the company. This unit paid for itself in less than two years with the value of silver recovered. The company also collects used film and sells it to the same recycler. The recycler burns the film and collects the silver from the residual ash. By removing the silver from the ash, the ash becomes nonhazardous (EPA, 1988).

7. RELATED TECHNIQUES

Although the term waste minimization refers to source reduction and recycling, there are many other activities associated with industrial operations that result in a reduced potential effect on human health and the environment. These operations include treatment techniques to reduce the volume and toxicity of waste streams. When the volume of a waste stream is reduced, the immediate effect is conservation of the secure storage specifically required and designated for RCRA hazardous wastes. A potential long-term effect of volume reduction is the reduced number of hazardous waste disposal sites requiring long-term monitoring or, in the worst case, cleanup. On the other hand, toxicity reduction addresses the conversion, destruction, or removal of the specific chemical constituents that caused the waste to be hazardous; the volume may or may not change. If successfully performed, toxicity reduction has the same immediate and long-term effect as volume reduction, namely, conservation of increasingly scarce secure storage space and a consequent long-term reduction in the number of hazardous waste disposal sites created and maintained. We address each of these techniques briefly in this section.

7.1 Volume Reduction

Volume reduction can be accomplished by various means, including the following:

Segregation of wastes
Conservation of water
Concentration of the wastes

Waste Minimization

Wastes (including wastewaters) generated in a manufacturing facility generally are of three types: process wastes, noncontact wastewaters, and sanitary wastes. Process wastes are by and large considered to be those wastes that contain some of the input materials/chemicals and products or by-products of manufacturing. Noncontact wastewaters include cooling tower wastes, boiler blowdowns, and other wastewaters that contain at most only trace concentrations of the chemicals used in process.

Segregation of Wastes

Segregation of solid wastes into process wastes (contaminated) and nonprocess wastes such as plant waste paper (noncontaminated) results in a reduction in the volume of hazardous wastes requiring treatment and disposal. Segregation of wastewaters into process wastewaters (contaminated with process chemicals) and noncontact and sanitary wastewaters (not contaminated with process chemicals) can result in a lower volume of hazardous waste sludge to be treated and disposed.

Conservation of Water

Good housekeeping practice includes being an efficient steward of all plant resources, including the plant water supply. Many plants view the water supply as an unlimited resource available at little cost, and often this is the case. However, a volume of water that costs little at the source may cost much more to treat or dispose of once contaminated with process chemicals. For example, tank cleanouts between batches often produce large volumes of contaminated wastewater. Conservation of water can result in a reduced volume of waste generated in a plant.

Concentration of the Wastes

Volume reduction also can be accomplished by concentrating the wastes, either by evaporation, crystallization, ion exchange, reverse osmosis, or any other physical change that does not destroy the chemical contaminant species present. Instead, the contaminants are concentrated into a reduced volume for further treatment or disposal. In some cases involving wastewater, the remaining volume may be lost to the atmosphere (evaporation without vapor condensation); in others the remaining volume is a much less contaminated water stream (evaporation with vapor condensation, ion exchange, reverse osmosis). For nonaqueous wastes such as contaminated solvents, the solvent may be distilled and recycled (true waste minimization), while the still bottoms become the concentrated stream.

7.2 Toxicity Reduction

The ultimate goal by and large of wastewater treatment is toxicity reduction. The contents of this volume deal almost exclusively with treatment to reduce toxicity, so little needs to be said in this chapter. It should be noted that toxicity reduction in wastewater streams occurs by treatment, either physical, chemical, or biological or by some combination of these. Toxicity reduction in nonaqueous streams can occur by treatment, incineration (a form of treatment that results in the destruction of organic contaminants and conversion of many inorganic species to immobile forms), and immobilization through solidification/stabilization/fixation technologies. In the case of immobilization, the toxic constituents are not removed or destroyed; they are blended into a matrix in which the tendency to migrate by leaching is minimized. This volume covers aqueous wastes in detail; nonaqueous industrial wastes are covered elsewhere (Lindgren, 1983; Hunt and Schecter, 1989).

8. WASTE MINIMIZATION PRACTICES IN INDUSTRY

Activities in waste minimization occur in industry at two levels: the corporate level and the plant level. Corporations with successful waste minimization programs have well-defined programs at both levels and emphasize interaction between the two groups.

8.1 Activities at the Corporate Level

The foundation of any well-designed waste minimization program is the corporate or management-level commitment to the program. A clear directive from senior management provides momentum, removes obstacles, and prevents misunderstandings and duplicate efforts. Conversely, the lack of a corporate commitment generally means that only sketchy ineffective policies and programs will be implemented, and then not at all locations.

The management of a company will support a waste minimization program if it is convinced that the benefits of such a program will outweigh the costs. The potential benefits include economic advantages, compliance with regulations, reduction in liabilities associated with the generation of wastes, improved public image, and reduced environmental impact. The objectives of a program are best conveyed to a company's employees through a formal policy statement or management directive. A company's upper management is responsible for establishing a formal commitment throughout all divisions of the organization. The person in charge of the company's environmental affairs

Waste Minimization 15

is responsible to advise management of the importance of waste minimization and the need for this formal commitment (EPA, 1988).

Although a growing number of companies and corporations are conducting waste minimization programs and are therefore deciding on some level of corporate management commitment, several companies have made long-term commitments at the highest level. These companies include 3M, Dow Chemical, and DuPont. In the case of all three, corporate policies have been developed and publicized, both within the corporation and outside of it.

8.2 Activities at the Plant Level

At the plant level, the policies and directives of management are carried out. Various activities are required at the plant level to achieve the goal of waste reduction. EPA has recommended that a task force be formed at each plant; the size and duties of the task force depend on plant size and company organization. Typical responsibilities of a task force at the plant level include (EPA, 1988) the following:

Get commitment and a statement of policy from management.
Establish overall program goals.
Establish a waste tracking system.
Prioritize the waste streams or facility areas for assessment.
Select assessment teams.
Conduct (or supervise) assessments.
Conduct economic feasibility analyses of favorable options.
Select and justify feasible options for implementation.
Obtain funding and establish schedule for implementation.
Monitor (and/or direct) implementation progress.
Monitor performance of the option, once it is operating.

Several of these responsibilities will be discussed in succeeding sections: These include the performance of facility assessments and an evaluation of economic feasibility. Other activities at the plant level are either obvious in scope or require company-specific guidance.

8.3 Barriers to Waste Minimization

Although waste minimization projects can reduce operating costs and improve environmental compliance, they can lead to conflicts between different groups within a company. Table 1 lists examples of jurisdictional conflicts that can arise during the implementation of a waste minimization project (EPA, 1988).

Table 1 Examples of Barriers to Waste Minimization

Production
    A new operating procedure will reduce waste but may also be a bottleneck that decreases the overall production rate.
    Production will be stopped while the new process equipment is installed.
    A new piece of equipment has not been demonstrated in a similar service. It may not work here.

Facilities /maintenance
    Adequate space is not available for the installation of new equipment.
    Adequate utilities are not available for the new equipment.
    Engineering or construction manpower will not be available in time to meet the project schedule.
    Extensive maintenance may be required.

Quality control
    More intensive QC may be needed.
    More rework may be required.

Client relations/marketing
    Changes in product characteristics may affect customer acceptance.

Inventory
    A program to reduce inventory (to avoid material deterioration and reprocessing) may lead to stockouts during high product demand.

Finance
    There is not enough money to fund the project.

Purchasing
    Existing stocks (or binding contracts) will delay the replacement of a hazardous material with a nonhazardous substitute.

Environmental
    Accepting another plant's waste as feed stock may require a lengthy resolution of regulatory issues.

Waste treatment
    Use of a new nonhazardous raw material will adversely impact the existing wastewater treatment facility.

Source: EPA (1988).

Waste Minimization 17

In addition, there are attitude-related barriers that can disrupt a program. A commonly held attitude is, "If it ain't broke, don't fix it!" This attitude stems from the desire to maintain the status quo and avoid the unknown. It is also based on the fear that a new option may not work as advertised. Without the commitment to carefully conceive and implement the option, this attitude can become a self-fulfilling prophecy. Management must declare that "it is broke."

Another attitude-related barrier is the feeling that "it just won't work." This response is often given when a person does not fully understand the nature of the proposed option and its impact on operations. The danger here is that promising options may be dropped before they can be evaluated. One way to avoid this is to use idea-generating brainstorming sessions. This encourages participants to propose a large number of options, which are then individually evaluated.

An often-encountered barrier is the fear that the option will diminish product quality. This is particularly common in situations where unused feed materials are recovered from the waste and then recycled back to the process. The deterioration of product quality can be a valid concern if unacceptable concentrations of waste materials build up in the system. One way to allay this concern is to set up a small-scale demonstration in the facility, or to observe the particular option in operation at another facility (EPA, 1988). Problems with maintaining product quality may become key factors in a company's evaluation of waste minimization options. For example, one large U.S. airline has reported to this author a dissatisfaction with the use of aqueous-based paints on ground-service equipment; lack of durability was cited as the most severe problem. To minimize waste production while still using solvent-based paints, this airline has installed special nozzles to increase the deposition rate of paints and decrease solvent releases.

9. FACILITY ASSESSMENT

The goal of the facility assessment phase is to develop a comprehensive set of waste minimization options and to identify the attractive options that deserve additional, more detailed analysis. In order to develop these options, a detailed understanding of the plant's wastes and operations is required. The assessment should begin by examining information about the processes, operations, and current waste management practices at the facility. The next step in the facility assessment is the assignment of a priority to each waste stream or operation to be studied; this step is followed by the development of waste minimization options and the screening of these options to arrive at a "short list" for further study and possible implementation.

9.1 Current Waste Management Practices

The questions that this information gathering effort will attempt to answer include the following (EPA, 1988):

What are the waste streams generated from the plant? And how much?
Which processes or operations do these waste streams come from?
Which wastes are classified as hazardous and which are not? What makes them hazardous?
What are the input materials used that generate the waste streams of a particular process or plant area?
How much of a particular input material enters each waste stream?
How much of a raw material can be accounted for through fugitive losses?
How efficient is the process?
Are unnecessary wastes generated by mixing otherwise recyclable hazardous wastes generated?
What types of housekeeping practices are used to limit the quantity of wastes generated?
What types of process controls are used to improve process efficiency?

Table 2 lists information useful in conducting the assessment. Reviewing this information will provide important background for understanding the plant's production and maintenance processes and will allow priorities to be determined (EPA, 1988).

Waste Stream Records

One of the first tasks of a waste minimization assessment is to identify and characterize the facility waste streams. Information about waste streams can come from a variety of sources. Some information on waste quantities is readily available from the completed hazardous waste manifests, which include the description and quantity of hazardous waste shipped to a treatment, storage, and disposal facility (TSDF). The total amount of hazardous waste shipped during a one-year period, e.g., is a convenient means of measuring waste generation and waste reduction efforts. However, manifests often lack such information as chemical analysis of the waste, specific source of the waste, and the time period during which the waste was generated. Furthermore, manifests do not cover wastewater effluents, air emissions, or nonhazardous solid wastes.

Other sources of information on waste streams include biennial reports and National Pollutant Discharge Elimination Systems (NPDES) monitoring reports. These NPDES monitoring reports include the volume and constituents of wastewaters that are discharged. Additionally, toxic substances release

Table 2 Facility Information Useful for Waste Minimization Assessments

Design information
    Process flow diagrams
    Material and heat balances (both design balances and actual balances) for
        production processes
    Pollution control processes
    Operating manuals and process descriptions
    Equipment lists
    Equipment specifications and data sheets
    Piping and instrument diagrams
    Plot and elevation plans
    Equipment layouts and work flow diagrams

Environmental information
    Hazardous waste manifests
    Emission inventories
    Biennial hazardous waste reports
    Waste analyses
    Environmental audit reports
    Permits and/or permit applications

Raw material/production information
    Product composition and batch sheets
    Material application diagrams
    Material safety data sheets (MSDS)
    Product and raw material inventory records
    Operator data lots
    Operating procedures
    Production schedules

Economic information
    Waste treatment and disposal costs
    Product, utility, and raw material costs
    Operating and maintenance costs
    Departmental cost-accounting reports

Other information
    Company environmental policy statements
    Standard procedures
    Organization charts

Source: EPA (1988).

inventories prepared under the "right to know" provisions of the Superfund Amendment and Reauthorization Act (SARA), Title III, Section 313, provide valuable information on emissions into all environmental media.

Analytical test data available from previous waste evaluations and routine sampling programs can be helpful if the focus of the assessment is a particular chemical within a waste stream.

Flow Diagrams and Material Balances (EPA, 1988)

Flow diagrams provide the basic means for identifying and organizing information that is useful for the assessment. Flow diagrams should be prepared to identify important process steps and sources where wastes are generated. Flow diagrams are also the foundation on which material balances are built.

Material balances are important for many projects, since they allow for quantifying losses or emissions that were previously unaccounted for. Also material balances assist in developing the following information:

baseline for tracking progress of the efforts
data to estimate the size and cost of additional equipment and other modifications
data to evaluate economic performance

In its simplest form, the material balance is represented by the mass conservation principle:

Mass in = mass out + mass accumulated

The material balance should be made individually for all components that enter and leave the process. When chemical reactions take place in a system, there is an advantage to doing "elemental balances" for specific chemical elements in a system.

Material balances can assist in determining concentrations of waste constituents when analytical test data are limited. They are practically useful when there are points in the production process where it is difficult (due to inaccessibility) or uneconomical to collect analytical data. A material balance can help determine if fugitive losses are occurring. For example, the evaporation of a solvent from a parts cleaning tank can be estimated as the difference between the solvent put into the tank and the solvent removed from the tank.

Sources of Material Balance Information (EPA, 1988)

By definition, the material balance includes both materials entering and leaving a process. Table 3 lists potential sources of material balance information. Material balances are easier, more meaningful, and more accurate when they are

Table 3 Sources of Material Balance Information

Samples, analyses, and flow measurements of feedstocks, products, and waste streams
Raw material purchase records
Material inventories
Emission inventories
Equipment cleaning and validation procedures
Batch make-up records
Product specifications
Design material balances
Production records
Operating logs
Standard operating procedures and operating manuals
Waste manifests

Source: EPA (1988).

done for individual units, operations, or processes. For this reason, it is important to define the material balance envelope properly. The envelope should be drawn around the specific area of concern, rather than a larger group of areas or the entire facility. An overall material balance for a facility can be constructed from individual unit material balances. This effort will highlight interrelationships between units and will help to point out areas for waste minimization by way of cooperation between different operating units or departments.

9.2 Determining Waste Streams and Operations to Assess

Ideally, all waste streams and plant operations should be assessed. However, prioritizing the waste streams and operations to evaluate is necessary when funds or personnel are limited. The facility assessments should concentrate on the most important waste problems first and then move on to the lower-priority problems as time, personnel, and budget permit.

Setting the priorities of waste streams or facility areas to assess requires a great deal of care and attention, since this step focuses the remainder of the assessment activity. Important criteria to consider when setting these priorities include the following (EPA, 1988):

Compliance with current and future regulations
Costs of waste management (treatment and disposal)

Potential environmental and safety liability
Quantity of waste
Hazardous properties of the waste (including toxicity, flammability, corrosivity, and reactivity)
Other safety hazards to employees
Potential for (or ease of) minimization
Potential for removing bottlenecks in production or waste treatment
Potential recovery of valuable by-products
Available budget for the waste minimization assessment program and projects

Small businesses or large businesses with only a few waste-generating operations should assess their entire facility. It is also beneficial to look at an entire facility when a large number of similar operations exist. Similarly, the implementation of good operating practices that involve procedural or organizational measures, such as soliciting employee suggestions, awareness-building programs, better inventory and maintenance procedures, and internal cost accounting changes, should be implemented on a facility-wide basis. Since many of these options do not require large capital expenditures, they can be implemented in short order.

A key aspect of evaluating which waste streams or processes to assess is the site visit. Although collected information is critical to gaining an understanding of the processes involved, seeing the site is important in order to witness the actual operation. For example, in many instances, a process unit is operated differently from the method originally described in the operating manual. Modifications may have been made to the equipment that were not recorded in the flow diagrams, equipment lists, or even the "as built" drawings.

When people from outside the plant participate in the assessment, it is recommended that a formal site inspection take place. Even when the team is made up entirely of plant employees, a site inspection by all team members is helpful after the site information has been collected and reviewed. The inspection helps to resolve questions or conflicting data uncovered during the review. The site inspection also provides additional information to supplement that obtained earlier Table 4 presents useful guidelines for the site inspection.

9.3 Development of Waste Minimization Options (EPA, 1988)

Once the origins and causes of waste generation are understood, the assessment process enters a creative phase. The objective of this step is to generate a

Waste Minimization 23

Table 4 Guidelines for Site Inspection

Prepare an agenda in advance that covers all points still requiring clarification. Proved staff contacts in the area being assessed with the agenda several days before the inspection.

Schedule the inspection to coincide with the particular operation that is of interest (e.g., make-up chemical addition, bath sampling, bath dumping, start-up, shutdown, etc.).

Monitor the operation at different times during the shift, and if needed, during all three shifts, especially when waste generation is highly dependent on human involvement (e.g., in painting or parts-cleaning operations).

Interview the operators, shift supervisors, and foremen in the assessed area. Do not hesitate to question more than one person if an answer is not forthcoming. Assess the operators' and their supervisors' awareness of the waste generation aspects of the operation. Note their familiarity (or lack of it) with the impacts their operation may have on others.

Photograph the area of interest, if warranted. Photographs are valuable in the absence of plant layout drawings. Many details can be captured in photographs that otherwise could be forgotten or inaccurately recalled at a later date.

Observe the "housekeeping" aspects of the operation. Check for signs of spills or leaks. Visit the maintenance shop and ask about any problems in keeping the equipment leak-free. Assess the overall cleanliness of the site. Pay attention to odors and fumes.

Assess the organizational structure and level of coordination of environmental activities between various departments.

Assess administrative controls, such as cost-accounting procedures, material-purchasing procedures, and waste collection procedures.

Source: EPA, 1988.

comprehensive set of options for further consideration. The process for identifying options should follow a hierarchy in which source reduction options are explored first, followed by recycling options. This hierarchy of effort stems from the environmental desirability of source reduction as the preferred means of minimizing waste. Treatment options should be considered only after acceptable waste minimization techniques have been identified.

The major waste minimization technical options are presented in Sec. 5 and 6 of this chapter.

A recent draft document and software package developed under U.S. EPA sponsorship provides guidance and methodologies for identifying and prioritizing waste minimization options at a facility (PEER, 1989). By using the program, one is able to first arrive at a process flow sheet for each process train in the facility if one does not exist already. Process units and waste streams are identified; the fate of each waste stream is determined. Then, by an interactive series of questions and answers, the program arrives at a series of options for further exploration and a priority ranking of the options.

When the list of options has been developed, feasibility evaluations for technical and economic soundness are performed. We address economic evaluations in another section of this chapter; here we present general guidance for the technical evaluation.

The evaluation of the technical feasibility of a waste minimization option is to a large degree a subjective exercise influenced by the plant operating philosophy, the outlook and prejudices of the evaluators, and other factors. Beyond the simple question, "Will it work?", are a variety of questions and issues, each of which carries some weight in the evaluation. These questions include the following (EPA, 1988):

Is the system safe for workers?
Will the product quality be maintained?
Is space available?
Is the new equipment, materials, or procedures compatible with production operating procedures, work flow, and production rates?
Is additional labor required?
Are utilities available? Or must they be installed, thereby raising capital costs?
How long will production be stopped in order to install the system?
Is special expertise required to operate or maintain the new system?
Does the vendor provide acceptable service?
Does the system create other environmental problems?

As a practical matter, by this stage a short list of options usually has "fallen out" from the efforts made. The purposes of conducting a formalized technical evaluation are to ensure that no hidden issues remain and that all options receive as objective a ranking as possible. Often by this step in an assessment there is clearly only one remaining option.

Waste Minimization

10. ECONOMICS OF WASTE MINIMIZATION

Effective waste minimization occurs only when there is a substantial driving force; the most common driving force is economic, i.e., waste minimization can save money.

Economic benefits to waste minimization can be either direct or indirect. Direct economic benefits include reduction in costs for treatment and disposal because of reduced waste volumes, reduced raw material costs due to improved inventory control and "just-in-time" manufacturing, and the elimination of some material costs and disposal costs by recycling. Direct economic benefits are generally easy to quantify.

Indirect economic benefits are more difficult to predict or quantify. These benefits originate, in part, from the avoidance of long-term liabilities for damages to human health and the environment by reducing the quantities of wastes discharged or disposed. Other indirect economic benefits include improvements in materials, labor, or energy productivity that reduce operating costs, reductions in costs associated with the presence of hazardous materials such as for worker exposures, improved use of managers' time, and benefits of waste minimization in marketing, public relations, and financial transactions (Office of Technology Assessment, OTA, 1986). Although more difficult to quantify, the indirect economic benefits are often much larger than the direct cost savings discussed above.

An economic evaluation for waste minimization options is carried out using standard measures of profitability, such as a payback period, return on investment, and net present value (EPA, 1988). Each company has its own economic criteria for selecting projects for implementation. In performing the economic evaluation, various costs and savings must be considered. As in any project, the cost elements can be broken down into capital costs and operating costs.

For smaller facilities with only a few processes, the entire assessment procedure will tend to be much less formal. In this situation, several obvious options, such as installation of flow controls and good operating practices, may be implemented with little or no economic evaluation. In these instances, no complicated analyses are necessary to demonstrate the advantages of adopting the selected options.

10.1 Capital Costs

A comprehensive list of capital cost items associated with a large plant upgrading project is presented in Table 5. These costs include not only the fixed

Table 5 Capital Investment for a Typical Large Project

Direct capital costs
  Site development
    Demolition and alteration work
    Site clearing and grading
    Walkways, roads, and fencing
  Process equipment
    All equipment listed on flow sheets
    Spare parts
    Taxes, freight, insurance, and duties
  Materials
    Piping and ducting
    Insulation and painting
    Electrical
    Instrumentation and controls
    Buildings and structures
  Connections to existing utilities and services (water, HVAC, power, steam, refrigeration, fuels, plant air and inert gas, lighting, and fire control)
  New utility and service facilities (same items above)
  Other nonprocess equipment
  Construction/installation
    Construction/installation labor salaries and burden
    Supervision, accounting, timekeeping, purchasing, safety, and expediting
    Temporary facilities
    Construction tools and equipment
    Taxes and insurance
    Building permits, field tests, licenses
Indirect capital costs
  In-house engineering, procurement, and other home office costs
  Outside engineering, design, and consulting services
  Permitting costs
  Contractors' fees
  Start-up costs
  Training costs
  Contingency
  Interest accrued during construction
    Total fixed capital costs
Working capital
  Raw materials inventory
  Finished product inventory
  Materials and supplies
    Total working capital

Source: EPA (1988).

Waste Minimization 27

capital costs for designing, purchasing, and installing equipment, but also costs for working capital, permitting, training, startup, and financing charges.

With the increasing level of environmental regulation, initial permitting costs are becoming a significant portion of capital costs for many recycling options (as well as treatment, storage, and disposal options). Many source reduction techniques have the advantage of not requiring environmental permitting in order to be implemented (EPA, 1988).

10.2 Operating Costs

The basic economic goal of any waste minimization project is to reduce (or eliminate) waste disposal costs and input material costs. However, a variety of other operating costs (and savings) should also be considered. In making the economic evaluation, it is convenient to use incremental operating costs in comparing the existing system with the new system that incorporates the waste minimization option ("incremental operating costs" represent the difference between the estimated operating costs associated with the option and the actual operating costs of the existing system without the option). Table 6 describes incremental operating costs and savings and incremental revenues typically associated with waste minimization projects (EPA, 1988).

Reducing or avoiding present and future operating costs associated with waste treatment, storage, and disposal are major elements of the economic evaluation. Companies have tended to ignore these costs in the past because land disposal was relatively inexpensive. However, recent regulatory requirements imposed on generators and waste management facilities have caused the costs of waste management to increase to the point where it is a factor in a company's overall cost structure. Table 7 presents typical external costs for off-site waste treatment and disposal. In addition to these external costs, there are significant internal costs, including the labor to store and ship wastes, liability insurance costs, and on-site treatment costs (EPA, 1988).

For the purpose of evaluating a project to reduce waste quantities, some costs are larger and more easily quantifiable. These include (EPA, 1988):

Disposal fees
Transportation costs
Predisposal treatment costs
Raw material costs
Operating and maintenance costs

Table 6 Operating Costs and Savings Associated with Waste Minimization Projects

---

Reduced waste management costs:
    Off-site treatment, storage, and disposal fees
    State fees and taxes on hazardous waste generators
    Transportation costs
    On-site treatment, storage, and handling costs
    Permitting, reporting, and recordkeeping costs
Input material cost savings:
    An option that reduces waste usually decreases the demand for input materials.
Insurance and liability savings:
    A WM option may be significant enough to reduce a company's insurance payments. It may also lower a company's potential liability associated with remedial clean-up of TSDFs and workplace safety. (The magnitude of liability savings is difficult to determine.)
Changes in utilities costs:
    Utilities costs may increase or decrease. This includes steam, electricity, process and cooling water, plant air, refrigeration, or inert gas.
Changes in operating and maintenance labor, burden, and benefits:
    An option may either increase or decrease labor requirements. This may be reflected in changes in overtime hours or the number of employees. When direct labor costs change, then the burden and benefit costs will also change. In large projects supervision costs will also change.
Changes in operation and maintenance supplies:
    An option may result in an increase or decrease in the use of O&M supplies.
Changes in overhead costs:
    Large WM projects may affect a facility's overhead costs.
Changes in revenues from increased (or decreased) production:
    An option may result in an increase in the productivity of a unit. This will result in a change in revenues. (Note that operating costs may also change accordingly.)
Increased revenues from by-products:
    A WM option may produce a by-product that can be sold to a recycler or another company as a raw material. This will increase the company's revenues.

Source: EPA (1988).

Savings in these costs should be taken into consideration first, because they have a greater effect on project economies and involve less effort to estimate reliability. The remaining elements are usually secondary in their direct impact and should be included on an as-needed basis in fine-tuning the analysis.

Waste Minimization

Table 7 Typical Costs of Off-site Industrial Waste Management[a]

Disposal of drummed hazardous waste[b]	
Solids	$75 to $110/drum
Liquids	$65 to $120/drum
Bulk waste	
Solids	$120/yd^3
Liquids	$0.60 to $2.30/drum
Lab packs	$110/drum
Analysis (at disposal site)	$200 to $300
Transportation	$65 to $85/hr @ 45 miles/hr (round trip)

[a] Does not include internal costs, such as taxes and fees, and labor for manifest preparation, storage, handling, and recordkeeping.
[b] Based on 55-gal drums. These prices are for larger quantities of drummed wastes. Disposal of a small number of drums can be up to four times higher per drum.
Source: EPA (1988).

10.3 Profitability Analysis

A project's profitability is measured using the estimated net cash flows (cash incomes minus cash outlays) for each year of the project's life. If the project has no significant capital costs, the project's profitability can be judged by whether an operating cost savings occurs or not. If such a project reduces overall operating costs, it should be implemented as soon as practical (EPA, 1988).

For projects with significant capital costs, a more detailed profitability analysis is necessary. The three standard profitability measures are:

Payback period
Internal rate of return
Net present value

The payback period for a project is the amount of time it takes to recover the initial cash outlay on the project. The formula for calculating the payback period on a pretax basis is the following:

$$\text{Payback period (in years)} = \frac{\text{capital investment}}{\text{annual operating cost savings}}$$

For example, suppose a waste generator installs a piece of equipment at a total cost of $120,000. If the piece of equipment is expected to save $48,000/year, then the payback period is 2.5 years.

Payback periods are typically measured in years. However, a particularly attractive project may have a payback period measured in months. Payback periods in the range of three to four years are usually considered acceptable for low-risk investments. This method is recommended for quick assessments of profitability. If large capital expenditures are involved, it is usually followed by more detailed analysis (EPA, 1988).

The internal rate of return (IRR) and net present value (NPV) are both discounted cash flow techniques for determining profitability. Many companies use these methods for ranking capital projects that are competing for funds. Capital funding may well hinge on the ability of the project to generate positive cash flows beyond the payback period to realize acceptable return on investment. Both the NPV and IRR recognize the time value of money by discounting the projected future net cash flows to the present. For investments with a low level of risk, an aftertax IRR of 12 to 15% is typically acceptable (EPA, 1988).

10.4 Adjustments for Risks and Liability

As mentioned earlier, waste minimization projects may reduce the magnitude of environmental and safety risks for a company, thereby providing indirect economic benefit. Although these risks can be identified, it is difficult to predict if problems will occur, the nature of the problems, and their resulting magnitude. One way of accounting for the reduction of these risks is to ease the financial performance requirements of the project. For example, the acceptable payback may be lengthened from four to five years, or the required internal rate of return may be lowered from 15 to 12%. Such adjustments reflect recognition of elements that affect the risk exposure of the company, but cannot be included directly in the analysis. These adjustments are judgmental and necessarily reflect the individual viewpoints of the people evaluating the project for capital funding. Therefore, it is important that the financial analysts and decision makers in the company be aware of the risk reduction and other benefits of the options. As a policy to encourage waste minimization, some companies have set lower hurdle rates for waste minimization projects (EPA, 1988).

Waste Minimization

Although profitability is important in deciding whether or not to implement an option, environmental regulation may be even more important. A company operating in violation of environmental regulations can face fines, lawsuits, and criminal penalties for the company's managers. Ultimately, the facility may even be forced to shut down. In this case, the total cash flow of a company can hinge on implementing an environmental project (EPA, 1988).

10.5 Economic Evaluation Software

A detailed economic evaluation of waste minimization options is best performed within the company that owns the facility being evaluated. Each firm employs company-specific procedures and rules of thumb.

However, computer software is becoming available to aid in arriving at a generalized economic evaluation. One program requires input of manufacturing process description, labor classifications and costs, material classifications and costs, facility resources and costs, and data related to the effect of the waste minimization option on labor, materials, facilities, revenues, and waste management (PEER/GBC, 1989).

11. CASE STUDIES

Many companies have already made corporate commitments to pursue waste minimization and these firms have been leaders in performing and publicizing their programs.

This section contains case studies of successful waste minimization projects completed by several firms. Common elements of these and other success stories in the waste minimization arena are summarized below; most of the elements of success have been discussed earlier in this chapter. These factors include the following:

Strong corporate commitment
Aggressive project teams at the plant level
Sound technical evaluation of options
Economic analysis showing reasonable payback
Corporate recognition of individuals involved in projects
A sound evaluation of waste practices prior to implementing the program

11.1 Waste Elimination at MEMC Electronic Materials, Inc. (Dickens, 1991)

Introduction

MEMC Electronic Materials, Inc. manufactures polished and epitaxial silicon wafers. Silicon wafers are the substrate, or base, on which microelectronic circuits (microchips) are made. MEMC is a worldwide producer of silicon with manufacturing plants in the United States, Europe, and Asia. Its customers are the manufacturers of logic and memory microchips used in everything from computers and consumer electronics to automobiles and airplanes.

Waste Elimination Strategy

MEMC has evolved to a proactive strategy for managing chemical emissions and hazardous waste produced as by-products of silicon manufacturing. The focus of the strategy is waste elimination.

Waste elimination is the elimination of chemical emissions and hazardous waste generation by changes in product design and manufacturing technology. It is an alternative to end-of-pipe waste management systems for compliance with environmental laws. Waste elimination is driven by two ideas:

1. Waste management is an unproductive drain on company resources.
2. Companies that eliminate chemical emissions and hazardous waste obtain a market advantage.

Waste elimination is a process of continuous improvement. The result over time is a large reduction in unproductive expenditures for waste management and pollution control.

Results

The MEMC manufacturing plant in Spartanburg, South Carolina has completed a number of waste elimination projects. These include process elimination, chemical substitution, substituting mechanical for chemical methods, modifying equipment and maintenance procedures, and projects for quality and yield improvement. The following highlights major results.

Chromic Acid. Prior to 1988, the only available acid etchants for evaluating crystal structure in silicon were based on chromic acid. Waste chromic acid and sludge from chromic acid treatment are regulated as hazardous waste. In 1988, MEMC developed a new etchant for evaluating silicon based on copper salts rather than chromium (Chandler, 1990). The copper-based etchant does not create a hazardous waste when treated. MEMC also determined that a process called "rod etching," which accounted for 80% of chromic acid use, was

Waste Minimization

unnecessary. Rod etching was eliminated. The copper-based acid etchant for evaluating silicon crystal structure was substituted for chromium-based etchants on all but one silicon product. An etchant with a lower chromium content was substituted for the one product still requiring chromic acid. Results are outlined in Table 8. During a period when manufacturing output increased by 10%, 96% of chromic acid used was eliminated by a manufacturing technology change.

Chromium Treatment Sludge. Table 9 outlines MEMC's reduction in chromium treatment sludge generation. Manufacturing technology changes to eliminate chromic acid use allowed changes in waste treatment technology that further reduced the generation of chromium treatment sludge. Between 1988 and 1991, the MEMC Spartanburg plant eliminated 53,800 lb/year of chromium treatment sludge. This represented an overall reduction of 81%. In addition, the chromium content of the sludge was reduced tenfold from 47,000 parts per million (ppm) by weight to 6200 ppm.

Mixed Acid and HF Air Emissions. Table 10 outlines waste elimination results for a series of quality and yield improvement projects related to preparing silicon samples for physical and chemical property tests. The most significant project involved substituting mechanical methods for a former chemical step using a mixed acid etchant containing hydrogen fluoride (HF). This alone eliminated 76,500 lb/year of mixed acid etchant use and acid waste generation. The quality improvement projects, together with elimination of the rod etching process described above, eliminated 34.5% of plant-wide, process air emissions of HF at the MEMC Spartanburg plant. The large reduction in HF air emissions was achieved solely by a manufacturing technology change. No new air pollution controls were installed.

Ozone-Depleting Chemicals. Freon 113 (trichlorotrifluoroethane) and methyl chloroform (1,1,1-trichloroethane) are ozone-depleting chemicals. They are common solvents used in the electronics industry for precision

Table 8 Chromic Acid Eliminated by Manufacturing Technology Change: MEMC Spartanburg, South Carolina Plant

Item	1988	1989
Chromic acid use, lb as CrO_3	5490	2.10
Manufacturing production[a]	1.00	1.10
Chromic acid use *eliminated*, lb as CrO_3		5830

[a]Manufacturing capacity is proprietary. 1988 assigned a value of 1.00.
Source: Dickens (1991).

Table 9 Chromic Treatment Sludge Reduction, 1988 to 1991: MEMC Spartanburg, South Carolina Plant

Item	Value
Baseline chromium sludge generation (1988), lb/year	$66.70
Chromium sludge *eliminated*	
Manufacturing technology change, drums/year[a]	73
lb/year	35,850
Waste treatment technology change, drums/year[b]	42
lb/year	17,950
Total, drums/year	115
lb/year	53,800
Annual cost *eliminated*	
Process chemicals[a]	$27,900
Personnel protective equipment[a]	16,800
Waste treatment chemicals[a,b]	4,800
Off-site waste transportation and disposal[c]	16,700
Off-site treatment to meet Third-Third Land Ban[c]	25,900
Hazardous waste disposal tax[d]	3,000
Total	$95,100

[a] Elimination of rod etching process, substitution of copper-based for chromium-based acid etchants.
[b] Modified treatment chemical recipe for copper-based acid etchants.
[c] Rates effective August 1990.
[d] State tax effective July 1990.
Source: Dickens (1991).

cleaning of semiconductor materials and packaging. At the MEMC Spartanburg plant, Freon 113 was used to clean containers called "tote pans." These tote pans were used for temporary storage of polished silicon wafers between cleaning and inspection steps. Freon 113 was also used to clean plastic cassettes. The plastic cassettes are a protective device used to carry silicon wafers between manufacturing steps. Methyl chloroform was used in one of several proprietary steps for cleaning raw materials associated with silicon crystal production.

Tote pan cleaning was eliminated by switching to a "just-in-time" product flow that eliminated the temporary storage of silicon wafers. The need for solvent cleaning of plastic cassettes was eliminated by changing the flow of cassettes through manufacturing steps. The cassettes are now cleaned using soap and water. MEMC developed special techniques for the clean

Table 10 Acid Use and Hydrogen Fluoride (HF) Air Emission Reduction, 1988 to 1991: MEMC Spartanburg, South Carolina Plant

Item	Value
Mixed acid etchant use *eliminated*, lb/year	101,210
HF air emission *eliminated*, lb/hr	0.105
Percent of plant-wide process HF emissions	34.5
Annual cost eliminated	
Process chemicals	$52,630
Waste treatment chemicals	1,430
Total	$54,050

Source: Dickens (1991).

handling of raw materials associated with silicon crystal production. These techniques eliminated the need for methyl chloroform degreasing of raw materials. An aqueous-based cleaner was substituted for the methyl chloroform cleaning step.

Table 11 outlines the results of a manufacturing technology change to reduce ozone-depleting chemical use between 1988 and 1991. The MEMC Spartanburg plant eliminated process steps involving 40.6% of Freon 113 use and 12.5% of methyl chloroform use. The changes reduced air emissions of Freon 113 by 42.0% and air emissions of methyl chloroform by 15.6%. The changes also eliminated the generation of 5400 lb/year of hazardous waste solvent.

Summary

Between 1988 and 1991, the manufacturing technology change at the MEMC Spartanburg plant:

Eliminated 96% of chromic acid use.
Reduced generation of hazardous chromium treatment sludge by 81%.
Reduced plant-wide, process air emission of hydrogen fluoride by 34.5%.
Eliminated 40.6% of Freon 113 use and 42.0% of Freon 113 air emissions.
Eliminated 12.5% of methyl chloroform use and 15.6% of methyl chloroform air emissions.
Eliminated more than $380,000/year in process chemical and waste management cost.

MEMC's waste elimination results gained positive recognition from the company's customers and peers. This recognition included the 1990 South

Table 11 Freon 113 and Methyl Chloroform Use Reduction, 1988 to 1991: MEMC Spartanburg, South Carolina Plant

Item	Value
Freon 113	
Use *eliminated*, lb/year	88,300
Percent of plant-wide use	40.6
Air emissions *eliminated*, lb/year	85,300
Percent of plant-wide emission	42.6
Hazardous waste *eliminated*, lb/year	3,800
Percent of plant-wide waste	23.6
Methyl chloroform	
Use *eliminated*, lb/year	25,300
Percent of plant-wide use	12.5
Air emissions *eliminated*, lb/year	23,700
Percent of plant-wide emissions	15.6
Hazardous waste *eliminated*, lb/year	1,600
Percent of plant-wide waste	9.0
Annual cost *eliminated*	
Process chemicals	$136,400
Ozone depleting chemical excise tax	100,500
Waste solvent disposal	600
Total	$237,500

Source: Dickens (1991).

Carolina Governor's Pollution Prevention Award. The MEMC case study is a model for other companies considering waste elimination efforts.

11.2 General Dynamics Paint Facility (California Department of Health Services, 1990)

Introduction

General Dynamics' paint production operations facility was completed in December 1988 to replace the manual mixing and hand-spraying of metal parts

Waste Minimization

in naval weapon systems. It includes computer-controlled robots (a GRI OM 5000 unit), which allow quick, automated precision painting. A proportional paint mixer was also added, which feeds preselected quantities of individual paint components directly to a paint spray nozzle, eliminating batch makeup operations. Additionally, electrostatic spray guns and automatic waste-cleaning solvent collection systems were introduced to allow for the recycle and reuse of waste paint. Spray paint booths are also available for touch-ups. Stills are used for recycling paint-cleaning solvents (see Fig. 3).

The paint facility uses both oil- and water-based paints. For oil-based paints, polyurethane thinner is used for paint thinning and equipment cleaning. A thinner containing isopropyl alcohol and xylenes is used with water-based paint. Paint waste was reduced from 42 tons in 1987 to 31 tons in 1988 and was further reduced to 17 tons in 1989. About 1000 gal of polyurethane cleaning solvent per year is now being recycled through the paint shop solvent stills, resulting in approximately 60 to 100 lb of still bottoms per week, or about 5000 lb/year. The still bottoms and waste paint are sent off-site for incineration.

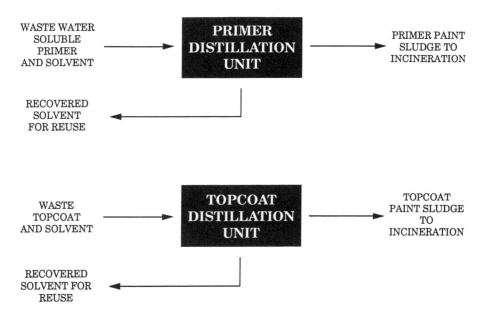

Figure 3 Production paint shop solvent recovery stills. (California Department of Health Services, 1990.)

Technical Evaluation

Paint purchases decreased from 6528 gal in 1988 to a projected 5230 gal in 1989 (based on purchases for the first six months); solvent purchases decreased from 2496 gal in 1988 to a projected 1080 gal in 1989. These decreases were mainly due to the changes in equipment and operating procedures in the paint shop, but were also partially due to changes in inventory and a decrease in production rates from 1988 to 1989.

The replacement of the conventional paint spray guns with electrostatic paint spray equipment increased paint solids transfer efficiency, thus reducing the paint usage and number of parts requiring repainting. Paint solids transfer increased from 35 to 65%, which decreased overall paint usage by approximately 40%.

Waste paint decreased from 42 tons in 1987 to 31 tons in 1988, and to 17 tons in 1989. These quantities include both waste paint and still bottoms (stills were not installed until October 1988). Currently, still bottoms are being generated at a rate of 5000 lb (2.5 tons)/year from the batch distillation of the solvent used for cleaning oil-based paint equipment and another 8000 lb (4 tons)/year from the recycling of paint solvent. If these waste tonnage rates are added to the current waste generation rate, 14 tons/year, to obtain the amount of paint waste generated with only the addition of the proportional paint mixers and increased transfer efficiencies from the electrostatic spray nozzles, then the waste generation rate would be 23.5 tons/year. However, the actual waste generation rate is 17 tons/year, with a waste reduction rate of 6.5 tons/year. Again, this calculation does not consider the decreased production rate and the fact that most of the solvents recycled in the still would have been disposed of as flammable waste instead of paint waste.

Only polyurethane solvent (for oil-based paints) is currently being recycled in the stills. When distilling the water-based paint solvent, which contains isopropyl alcohol and xylene, the recycled solvent separates into two layers: a water layer and a solvent layer. Even when the water is removed by draining, water contained within the solvent prevents the solvent from being reused for thinning and cleaning. Potentially, the water can be separated with a similar approach as that used for General Dynamics' CFC solvent skills. This method would add a water separator upstream and a molecular sieve downstream of the stills for improved water separation.

Economic Evaluation

The installation cost of the robotic painting system was $1,400,000. The system included a parts conveyor, computer-controlled robots, electrostatic

Waste Minimization

spray guns, proportional paint mixing, and cleaning solvent collection equipment. The disposal of 42 tons of waste in 1987 would have cost about $73,000 at the current disposal rates of $420/drum, plus $7000 per truckload for transportation (80 drums). The disposal costs of 21 tons in 1989 would be about $36,500. The payback period from a waste-disposal standpoint alone would be almost 40 years. This substantially overstates the payback period, however, because the savings in labor costs from painting and waste disposal and any decrease in reject parts were not included, as this information was considered proprietary.

The two solvent stills were installed for a reported cost of $14,000. These units are currently recycling 1000 gal/year of cleaning solvent (polyurethane thinner), resulting in $1500 in purchase savings and $2500 in disposal savings, for a total savings of $4000. The estimated payback period for this equipment is 3.5 years. This estimate does not consider the increased labor costs required to operate the equipment.

Conclusions

The paint facility with proportional paint mixing, robotic controlled electrostatic sprays, and cleaning solvent collection equipment appears to be operating well and is reducing worker exposure and waste paint considerably over previous levels. The main problem area is the waste solvent collection process, where water needs to be separated from the solvent (in the case of the recycling water-based primer cleaning solvent). This problem had led to a reluctance to use the solvent recycled from the water-based paint cleaning for wipe-down and cleanup because of its water content. This problem can be eliminated by adding a water separator upstream and a molecular sieve and a dryer downstream from the stills for improved water separation. Only polyurethane thinner solvent from cleaning paint equipment is currently recycled. This demonstrates that even simple waste minimization processes can create unexpected problems.

Detailed costs were not available to develop a comprehensive economic evaluation of the waste minimization methods applied to the production paint shop. The payback period of almost 40 years, calculated on the overall investment of $1.4 million, is substantially overstated because labor cost savings from paint and waste disposal and decreases in reject paints were not available, nor were saved liability and health and safety costs. The payback for the solvent stills is only about 4 years, but would be less if all cleaning solvent was being recycled.

The paint shop modifications decreased overall paint waste generated by 50% from 1987 to 1989. With additional funding to solve the water contamination problems in the recycling of water-based primer-cleaning solvent, confidence in the quality of the recycled solvent should return. This disposal of used cleaning solvents from the paint shop will eventually be eliminated, which will decrease solvent purchases and the cost of solvent waste handling and disposal.

The "transferability" of robotic technology to other industries could be economically prohibitive, given the substantial capital investment required, unless a company had sufficient paint-cleaning solvent disposal problems and could afford the costs. Parts of the system, such as the electrostatic sprays and solvent stills, are fairly inexpensive and could be easily transferred to other industries.

11.3 Petroleum Refinery Processes: Removal of Metals (California Department of Health Services, 1989)

Conversion Processes

Process Description. Many primary distillation products undergo some form of conversion operation. Conversion processes are used to purify or change the chemical and/or physical properties of the distillates, usually by breaking large molecules into smaller ones. Typical conversion processes are hydrotreating, catalytic and thermal cracking, hydrocracking, and coking.

In most refineries, usually only the naphtha and middle distillate cuts or low metals content residuum are hydrotreated. The hydrotreating process converts the heavier components of these cuts to lower boiling products by mild cracking and by adding hydrogen to the molecules using a fixed-bed catalytic reactor. The reactor products are culled, and the hydrogen with impurities and high-grade products are separated. Among the catalysts most commonly used in hydrotreating is cobalt molybdate with various promoters on silica-alumina supports.

Hydrocracking is a catalytic method of converting refractory middle-boiling or heavy sour feedstocks into high-octane gasoline, reformer charge stock, jet fuel, and/or high-grade fuel oil. This process has a high degree of flexibility in adjusting production to meet changing product demands. It is one of the most rapidly growing refinery processes. Catalysts commonly used are nickel-silica alumina, tungsten sulfide-silica alumina, and iron-HF clay.

Conversion Process Metal Wastes. Conversion process wastes originate from the operation of the fluid catalytic cracker, hydrotreating operations, hydrocracking operations, and coking operations. A fluid catalytic cracker (FCC) catalyst is continuously regenerated by burning off the coke formed on the catalyst during the cracking process. The fuel gas from the regenerator

Waste Minimization

passes through a series of cyclones that recover most of the catalyst. This recovered catalyst is then returned to the reactor vessel. Because of air pollution regulations, refineries have installed electrostatic precipitators or equivalent tertiary separation devices to significantly limit catalyst fines in the regenerator flue gas. These catalyst fines are either landfilled or in some cases sold. They are regenerated on a continuous basis, but are generally disposed of on an intermittent basis.

A number of refinery processes require the use of a fixed-bed catalyst. These processes include catalytic reforming, hydrodesulfurization, hydrotreating, hydrocracking, and others. These catalysts become inactive in six months to three years and are eventually replaced in the reactors with a fresh catalyst during a unit shutdown. Many of these catalysts contain valuable metals that can be recovered economically. Some of these metals, such as platinum and palladium, represent the active catalytic component; other metals, such as nickel and vanadium, are contaminants in the feed that are deposited on the catalyst during use. After valuable metals are recovered (a service usually performed by outside companies), the residuals are expected to be disposed of as solid waste.

The major contaminating metals found on a catalytic cracking catalyst are vanadium, nickel, copper, chromium, and iron. Small amounts of these metals are present in the crude petroleum and, except for some of the iron, all are in the form of metal-organic compounds. Some of these compounds are volatile, and when the vacuum gas oil feed to the catalytic cracking units is prepared, they appear in the gas oil. A fraction of the iron, and probably chromium, found on the catalyst is the result of erosion and corrosion either in the lines or the equipment.

In the catalytic cracking unit regenerator, where coke is burned off the spent catalyst, the organic portion of the catalyst molecules is burned off and the metals are oxidized to an inorganic oxide that remains on the catalyst. Corrosion and erosion products may be mixed with the catalyst as fine particles or may also be deposited on the catalyst surface. The heavy metals, vanadium and nickel, and to a lesser extent, iron and copper act as the dehydrogenation catalyst and produce excessive quantities of undesirable coke and light gases, especially hydrogen. In many cases, these metal contaminants are the primary reason for discarding part of the equilibrium catalyst. Fresh catalyst is then added to maintain an acceptably low level of contamination.

Source Reduction. Operators try to control conversion processes such that the fines production rate about equals the poisoning rate in order to avoid a net removal of the equilibrium catalyst. Source reduction strategies for reducing FCC catalyst losses include the following.

Demetalize Gas Oil Charged to Cat Cracker. The catalyst withdrawal rate can be reduced by decreasing the rate of catalyst deactivation by removal of metals before gas oil is charged to the reactor. A mild hydrogenation or hydrotreating process has been used on the catalytic cracking feed in some units. This feed treatment removes some of the metal compounds and provides other benefits to the catalytic cracking operation, such as removing sulfur, which reduces the amounts of sulfur emissions from the regenerator, and increasing the yields of desirable products such as gasoline.

Minimize Use of Aeration and Purge Steam. Excess fines will be generated by the overuse of aeration and steam purges during cracker operation.

Tank Cleaning

Solids from leaded as well as nonleaded products settle to the bottom of storage tanks, where they remain until they are removed. This accumulated sludge is removed whenever the tank service is changed, the sediment content of the stored produce exceeds specifications, or the tank itself needs inspection or repair. The characteristics of the deposited sludge varies with the type of product stored in the tank. Leaded tank bottoms are considered RCRA hazardous and special handling methods are used when the sludge is removed. It is generally removed at intervals varying between once a year and once every five to seven years and spread on special concrete pads for weathering. Weathered sludge is disposed of in a Class I landfill.

Source Reduction. Suggestions for reducing tank cleaning wastes include the following source reduction strategies.

Eliminate Lead in Gasoline. The current trend of eliminating lead in gasoline, coupled with the possibility of a ban on leaded gasoline, will lower the toxicity of leaded gasoline tank sludges.

Install Storage Tank Agitators on Crude Oil Tanks. Agitation of the vessel contents prevents the deposit of settleable solids and hence reduces the need for cleaning. It must be noted that agitation does not by itself reduce the amount of waste generated; it simply transfers the solids downstream to the crude unit. They eventually end up either in asphalt or coke.

Use Corrosion-Resistant Materials. Some of the sludge generated is the result of corrosion or deterioration of the storage tank intervals. Installation of a liner or using materials of construction that are more resistant to the corrosive elements of crude oil reduces sludge production.

Prevent Oxidation of Crude Oil. Gums and resins are the result of air-oxidizing unsaturated compounds in the crude oil. Air oxidation can be minimized in crude storage tanks by providing a nitrogen blanket over the surface

Waste Minimization

of the oil, or more commonly, by the use of floating roofs. The floating roof should preferably be of the double cover type with liquid seals. Many storage tanks are already equipped with floating roofs as a result of air emission regulations.

Recycle Sludge for Organic Content. Conventional cleaning of crude tanks often relies on the mechanical removal of sludge and its subsequent land disposal. The use of warm oil and circulating cleaning techniques reduces the ultimate sludge volume and allows for recovery of considerable quantities of valuable crude tied up in the sludge. Here, a light gas oil, clean crude stock, or other available low-viscosity straight run distillate is warmed, mixed with dispersant additive, and circulated through the tank. This process resuspends the sludge and dissolves the crude that is entrapped in the sludge, which amounts to 60 to 90% of oil by volume. Again, as in the case of agitation, the solids are transferred to the part of the process where they can be more efficiently separated from valuable entrained liquid.

Reduce Sludge Volume. At large refineries, the sludge volume can be reduced by concentrating it using vacuum, gravity belt, or automated plate and frame filters (centrifugal filters were not found to be very effective). Also, the sludge volume can be reduced significantly by using the patented Chevron recovered oil process. This process is said to be more effective than conventional filtration methods. Other methods involve thermal/chemical emulsion breakup using steam or indirect heating. One such method was used at a Vickers Petroleum refinery where the deemulsifier was added to the sludge, which was then pumped through a steam heater into a conical bottom decanter. Other techniques include ultrasonic deemulsification, solvent extraction (e.g., using the BEST process available from Resources Conservation Co.) and electroacoustic dewatering.

Sludge volume can be reduced at small refineries by using the Victor extraction process. Here, the sludge is transferred into a mixing tank and agitated for an extended period of time with steam and air. This process separates the residual oil trapped in the sludge, which can then be taken off from the top of the tank. The solids, significantly smaller in volume by comparison to the original sludge, settle to the bottom. This process works well for granular-type sludges but is not very effective for clay-bearing sludges.

Sulfur Recovery from Refinery Process Emissions

Hydrogen sulfide (H_2S) occurs as a contaminant in many different refinery gas streams, including those from petroleum cracking and hydrodesulfurization, as well as natural gas and coal gasification streams. Sulfur recovery processes absorb H_2S from these streams and can produce elemental sulfur for sale or disposal. The removal of sulfur from waste gas streams must be efficient enough to allow the remaining "tail gas" to be vented to the atmosphere, following catalytic incineration. There are many different technologies currently in use, including amine or caustic scrubbing, and the Claus, Pritchard, Stretford, Unisulf, Sulfolin, LO-CAT, Sulferox, Hiperion, SCOT, Merichem, and Beavon processes (Dalrymple, 1989). Most of these processes do not generate hazardous metal wastes. Heavy metal waste streams have, however, been a chronic problem in the Stretford process, which employs pentavalent vanadium to oxidize sulfite to sulfur.

Stretford plants have been in operation for 30 years. There are hundreds of such plants worldwide, used in a variety of sulfur removal operations. In a Stretford process, the hydrogen sulfide in the gas stream is absorbed and oxidized to elemental sulfur in aqueous phase, using pentavalent vanadium that is subsequently reduced from a pentavalent to tetravalent form. Later in the process, the vanadium is reoxidized back again, using anthraquinone disulfonic acid (ADA) as a catalyst, and the elemental sulfur is floated to the surface of the solution and removed.

Waste Generation. In addition to the reactions that produce elemental sulfur, competing reactions also occur that produce undesirable by-products such as sodium thiosulfate. This is detrimental, because the thiosulfate remains in solution, and its concentration can generally be reduced only by bleeding off a portion of the solution inventory. This solution purge waste stream is hazardous, largely because it also contains vanadium compounds. The key to reducing the metal content of the waste stream is to reduce the rate of thiosulfate formation.

The undesirable formation of thiosulfate is aggravated by several factors, including:

Insufficient residence time of the solution in the absorber/reaction tank, where elemental sulfur is produced.
Insufficient vanadium in the Stretford solution to convert the sulfur compounds present to elemental sulfur.
Insufficient ADA in the solution to convert vanadium back into its pentavalent state. This, in turn, limits the rate of elemental sulfur production and results in increased thiosulfate formation.

The rate of thiosulfate formation also increased with increasing temperature and pH. Many Stretford plants employ a sulfur melter to separate the elemental

Waste Minimization

sulfur from Stretford solution. This subjects the solution to 150°C temperatures, which enhance the generation of thiosulfate.

Source Reduction. Source reduction options useful for reducing the metal content of the waste stream, as well as increasing the yield of elemental commercially salable sulfur, including the following methodologies.

Lengthening Elemental Sulfur Production Time. Elemental sulfur is produced from hydrogen sulfide through several reactions in the process absorber and reaction tank. If some of the sulfur in the solution is not fully converted to its elemental form by the time it enters the oxidation tank, then the oxygen added to the tank through air sparging will result in thiosulfate formation. This undesirable occurrence can be limited by lengthening the residence time in the absorber/reactor tank, or if that is not possible, then by adding a settling tank after the absorber/reactor in which elemental sulfur can continue to be formed.

Increasing Vanadium and ADA Content. Raising the concentration of vanadium in the Stretford solution can increase the rate of elemental sulfur formation in the absorber/reactor and thus lower the concentration of sulfur compounds that are converted to thiosulfate in the oxidizer. It may also be necessary to increase the ADA concentration as well, in order to convert the additional tetravalent vanadium that will be formed back into its pentavalent state.

ADA Isomer Selection. Limited attention is often given in refineries to the isomer of ADA used. 2,6-ADA is a commonly used isomer, although it has been found inferior to 2,7-ADA in converting vanadium to its pentavalent form. If this conversion is not performed efficiently, the elemental sulfur production rate will fall, and thiosulfate formation will increase. More attention to procuring only 2,7-ADA could augment the efficiency of the Stretford process.

Sulfur Separation. It is important that elemental sulfur separate from the rest of the solution in the oxidizer tanks, rising to the surface and being removed in the froth. It has been reported that this separation is sometimes poor, leaving more sulfur in the oxidizer solution available for conversion to thiosulfate.

A continuous drip of a small amount of diesel oil (typically less than 1 mL/min) into the oxidation tank solution aids in the separation of sulfur into the froth, as well as producing a thick, easy-to-remove froth. Sometimes diesel oil addition leads to overfrothing, however, and silicone or ocenal must be added as antifrothing agents.

Use of Filter Instead of Melter. Significant thiosulfate formation can occur due to the high temperatures in a sulfur melter. Replacing the melter with a filter can eliminate this problem and also provides a way of efficiently removing commercially salable sulfur from the solution.

Reduce Feed Gas SO_2 Concentration. Some feed gas streams to Stretford plants are rich in sulfur dioxide, resulting in troublesome thiosulfate formation. This is frequently the case if a Claus process precedes the Stretford process. Reducing the sulfur dioxide concentration in the feed gas to a Stretford plant leads to less formation of thiosulfate and other undesirable salts. This can be accomplished through reducing SO_2 to H_2S by hydrogenation. One approach is to install a reducing gas generator and a Beavon or Shell Clause offgas treating (SCOT) process reactor upstream of the Stretford unit.

Process Substitution. The SCOT process can replace the Stretford unit altogether, eliminating the vanadium and thiosulfate-bearing waste stream problems and producing elemental sulfur at a faster rate, and generally at a higher level of quality than Stretford process sulfur. There are also several processes derived from Stretford vanadium chemistry, including the Unisulf and Sulfolin processes. The Unisulf process, commercialized by Unocal in 1985, employs carboxylic acid and an aromatic sulfonate complexing agent and has virtually eliminated the by-product salt formation problems of the Stretford processes, according to Unocal reports. Unisulf plants use no sulfur melter, eliminating that source of thiocyanate formation.

The Sulfolin process solution contains organic nitrogen compounds in addition to vanadium. The process was developed by SASOL South Africa and Linde AG. It is a new process with limited data available in the literature, although reduced by-product salt making rates have been claimed.

Another process option is the LO-CAT process, which employs chelated iron liquid redox chemistry and has been popular for smaller operations. Solution compositions include iron, proprietary chelates, a biocide, and a surfactant that facilitates sulfur sinking to the bottom of the oxidizer, where it is removed as a slurry. Other chelated iron processes include Sulferox and Hiperion.

Waste management methodologies for refining processes are summarized in Table 12.

Waste Minimization

Table 12 Source Reduction Waste Management in Petroleum-Refining Processes

Waste stream	Management alternatives
Conversion wastes	Demetallize oil to cat cracker
	Minimize aeration and purge steam
Tank cleaning	Eliminate lead in gasoline
	Install storage tank agitators on crude oil tanks
	Use corrosion-resistant materials
	Prevent crude oil oxidation
	Refine sludge
	Reduce sludge volume
Tail gas sulfur recovery wastes	Lengthen elemental sulfur production time
	Increase vanadium and ADA content
	ADA isomer selection
	Sulfur separation using oil drip
	Use of filter instead of melter
	Reduce feed gas SO_2 concentration
	Process substitution

Source: California Department of Health Services (1989).

11.4 Waste Reduction in a Small Business: Kenmonth Engine Shop (California Department of Health Services, 1988)

This shop specializes in crankshaft repair and is larger than the average automotive repair shop. Because the work is specialized, requires expensive equipment, and operates most efficiently at a bulk rate, the facility is unlike most other neighborhood repair shops. However, all repair shops must use some method to clean parts. This case study shows a successful alternative to using jet spray washers for parts cleaning. Smaller businesses can minimize hazardous waste generation by taking this example and scaling it down to meet their business needs.

When crankshafts come into the Kenmonth shop in bulk for repair, they are coated with oil, dirt, and grease. A thorough cleaning is required to remove all surface contaminants. The repaired crankshaft must be shipped out clean or its service life will be greatly reduced. Prior to May of 1985, this company used two caustic jet spray washers to clean parts. Although a universal tool for cleaning metal parts, the washer was not the most efficient and cost-effective choice, as it left residual contaminants in hidden surfaces of the crankshafts. A large amount of labor was required to load the parts into the jet spray washers and make sure all inner part surfaces were thoroughly cleaned after removal. In addition, costs to heat the jet spray washers on a round-the-clock basis and to dispose of spent caustic have gradually increased. Disposal costs alone have climbed from $0.14/gal in 1984 to $2.50/gal in 1988.

As a result of increasing industry regulation under RCRA and a better awareness of the environmental hazard posed by hazardous waste generation, the shop owner committed himself to reduce his contribution of hazardous wastes. The primary means for achieving waste reduction was the purchase and installation of a "burn out oven," which eliminated the need for one of the company's two jet spray washers. The oven can be used on about 50% of the parts coming into the shop. Smaller parts and specialized works are sometimes cleaned in the remaining jet spray washer. The high temperature in the oven first bakes the oil to a combustible temperature. After combustion, the temperature is increased again to secondarily burn airborne particulates before release into the atmosphere. The oven meets all state air emission standards.

Parts removed from the oven are coated with a fine ash residue, which is easily removed by machines that bombard the parts with small glass beads or metal shot. The glass beads and metal shot do not adhere to the parts as they would if a jet spray washer was used. Any remaining ash is blown off the part using compressed air, instead of wiping off deeply embedded oil and grease with a rag. Ash, spent beads, and spent metal shot have all been approved for transport and disposal to a local sanitary landfill by the transporter and landfill operator.

Although not the primary motivating factor in deciding to install the oven, a cost savings was realized in less than one year. The owner felt an obligation to take the lead in this area since smaller shops look to him for advice on successful shop management. Table 13 shows the cost involved.

The natural gas required to operate the oven costs $300 to $500/month less than it cost to heat the jet spray washer it replaced, which required 24-hr heating to eliminate cold morning start-ups. The amount of caustic waste requiring disposal has been reduced from approximately 200 to 22 gal/month for the entire shop.

Smaller ovens are currently available for use in smaller parts shops. This example demonstrates that a small capital investment can pay for itself

Waste Minimization

Table 13 Kenmonth Engine Shop Costs

Purchase of oven	$8000
Installation cost	$1000
Total costs	$9000
Natural gas (varies seasonally)	$ 400/month
Elimination of employee	$1000/month
Reduction in caustic waste volume	$ 495/month
Total savings	$1895/month
Payback period	$9000 ÷ $1895/mo = 5 months

Source: California Department of Health Services (1988).

in less than one year. However, the oven does require a glass bead cabinet or shot peen machine to remove ash residue. These machines range in price from $3000 to $40,000. This cost would need to be considered if one of the machines was not currently installed in the shop. In the case study above, only the oven was necessary. It reduced waste, saved energy, and helped to generate a higher-quality final product through an improved cleaning process.

11.5 Waste Reduction in an Automotive Assembly Plant (California Department of Health Services, 1988)

Introduction

This facility is a large auto assembly plant owned and operated by one of the major auto manufacturers. The plant manufactures approximately 60 cars each operating hour and employs 4500 people. The waste reduction successes described below reflect more than this company's commitment to avoiding waste. The U.S. auto industry has become fiercely competitive and this plant, like many others, has been in a fight for survival. For the employees at this plant, waste reduction and its attendant economic benefits are part of the struggle to preserve their livelihood.

Auto Painting

This plant's single most significant waste reduction measure has been in the way it paints cars. The plant employs a two-stage painting process. The base coat provides the color; the second coat, or clear coat, provides protection and finish. When this process was installed in 1985, it was designed so that the base coat was applied with a traditional spray-painting technique and the clear coat was applied with an electrostatic spray-painting technique. Although it was recognized that the electrostatic technique is significantly more efficient than the traditional approach, plant engineers did not believe they could achieve the desired qualities if the base coat was applied electrostatically.

The electrostatic process outwardly resembles the conventional approach; the paint is atomized as it leaves the nozzle and is blown toward the car. However, in the electrostatic process, the paint is given a positive charge as it leaves the nozzle, while the cars are held at a highly negative potential. Consequently, each drop of paint is attracted toward the car. Painting processes are rated according to their transfer efficiency (the percentage of the paint used that actually coats the object). Whereas the conventional painting process had only a 30% transfer efficiency, the electrostatic process has over 90% transfer efficiency. The electrostatic process uses one-third as much paint as the conventional process to put the same amount of paint on the car.

In both processes, the cars are painted from above; air, also blown from above, carries the overspray paint through metal grates below the cars (see Fig. 4). The pigments in the air stream are absorbed in flowing water beneath the grate and carried off. After leaving the painting operation, the water circulates through the "sludge farm." There, pigments and other paint

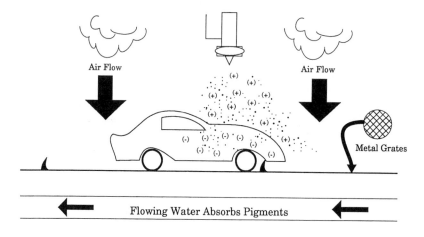

Figure 4 Electrostatic painting process. (California Department of Health Services, 1988.)

Waste Minimization

components are removed from the water by dissolved air flotation. The material skimmed from the top is managed as a hazardous waste; the cleaned water is pumped back into the plant.

The immediate motivation for the process modification was an air pollution regulation. The plant uses a solvent-based paint containing, in particular, toluene and methyl ethyl ketone (MEK). In both painting processes, the solvent is volatilized when the paint is atomized for application. The solvent vapor is carried by the airflow through the plant's smoke stacks into the surrounding atmosphere. The local air pollution regulatory agency informed the plant's environmental staff that they must achieve a large reduction in the plant's organic vapor emissions or shut down. The environmental staff identified two options to avoid closing down: They could either purify the air coming out of the paint booth with a scrubbing technique, such as activated carbon, or they could develop a more efficient painting technique, such as electrostatic painting. The latter would reduce emissions because the volume of solvent released is approximately proportional to the amount of paint used to paint cars.

The environmental staff was presented with a classic waste reduction dilemma. The treatment alternative, scrubbing the air, was very expensive but certain to work. The source reduction alternative, applying the base coat electrostatically, appeared to be less expensive but there was no guarantee that the plant could achieve the desired surface quality with the technique. Because cars are purchased largely on the basis of appearance, this was no small issue. Recognizing that the additional cost of the treatment alternative would have a real impact on the plant's profitability, the environmental staff and paint department agreed to take the chance.

Conversion to Electrostatic Paint Technique

In 1987, at a cost of $2 million and after much design work, the plant converted the base coat application to the electrostatic technique. Plant staff worked around the clock using test "funny cars" to optimize the application and to ensure that it would give the desired quality. They were successful and eventually the conventional spray jets were replaced.

Because the plant is using less paint, there has been a commensurate reduction in the amount of paint waste generated and, perhaps most important, in the amount of solvent released to the atmosphere. To calculate the savings due to this improvement, the plant staff must consider "avoided costs." For example, the air pollution agency charges the plant according to the number of tons reactive organic compounds released to the atmosphere. At approximately the same time that the plant was installing the new painting technique, the agency raised its fees from $71/ton to $241/ton and eliminated a 10-ton annual exemption. Thus, the plant's annual bill from the agency probably went up despite enormously reduced emissions. To fairly assess the savings, the plant's accountants must calculate the "avoided cost" of generating emissions at the old rate and paying them at the new fee schedule.

Similarly, virtually every cost associated with hazardous waste generation has skyrocketed, including generator fees, disposal costs, and transportation costs. The avoided cost calculation for reduced hazardous waste generation must take into account each of these increases.

Using these techniques, the plant estimates that this process modification will pay for itself in two years. Even though the air emissions motivated these changes, the greatest savings were in paint purchase costs, followed by waste disposal costs, and finally air emission costs. Had the plant conservatively chosen to take the treatment option and scrub the air, only the air pollution agency fees would have been reduced and those savings would have been more than replaced by the costs of managing the scrubber waste. The plant would have suffered a short- and long-term loss.

In large part, the success of this source reduction technique is due to the determination and innovation of the plant staff. Certain specialized finishes desired on some cars cannot be achieved by the electrostatically applied base coat alone. For these cars, the plant applies a thin overcoat by the conventional method (before the application of the clear coat). Because this layer is so thin, the amount of waste and emissions generated are small. It may be flexibility, above all else, that determines the success of waste reduction.

Dewatering Paint Sludge

The second waste reduction measure from this plant is a traditional treatment option. As discussed earlier, the paint sludge is removed from the recirculating water by dissolved air flotation. For several years, this very wet sludge was sent to a hazardous waste landfill at great expense. The environmental staff realized that there was no reason to pay a premium price for the disposal of water. They installed a plat and frame filter press. Now, the wet sludge is pumped into the filter press, where the water is squeezed out, leaving a moist, if not dry, filter cake.

Because of the importance of maintaining low water content in hazardous waste landfills, the environmental benefit is clear. Moreover, in this case, the water squeezed out in the press is returned to the recirculating system and presents no discharge burden.

Waste Minimization

General Waste Reduction (Good Housekeeping)

In their drive to reduce costs, the plant's environmental staff is reviewing the entire plant operation for other waste reduction opportunities. They have found several that prove the potential of simple housekeeping changes. For example, one of the final stages of manufacturing a car is to add gas, oil, antifreeze and various other automotive fluids. Like any process involving fluids, there is spillage. At considerable expense, the various fluids were sent off-site for disposal or recycling. When workers in this operation were informed of the costs due to this waste, they were able to reduce the rate of waste antifreeze generation in half, from 5000 gal/bimonthly to 5000 gal/quarter.

The production staff places sealant on the various holes and edges of the car for water-proofing. The environmental staff found that barrels of sealant were being replaced when they were perhaps only 90% consumed. Although this saved the production staff valuable time, it led to the generation of large amounts of waste sealant. Now all the sealant is used.

An automobile assembly plant involves many different operations and generates a wide variety of waste streams. This facility's experiences demonstrate that in every area of plant operations, there are likely to be significant opportunities to reduce waste generation and thereby improve the economic efficiency of the plant.

11.6 Solid Waste Volume Reduction at MEMC Electronic Materials, Inc. (Dickens, 1991)

Introduction

MEMC Electronic Materials, Inc. manufactures polished and epitaxial silicon wafers. Silicon wafers are the substrate, or base, on which microelectronic circuits (microchips) are made. MEMC is a worldwide producer of silicon with manufacturing plants in the United States, Europe, and Asia. Its customers are the manufacturers of logic and memory microchips used in everything from computers and consumer electronics to automobiles and airplanes.

Nonhazardous Solid Waste Management

Silicon wafer manufacturing generates a large volume of nonhazardous solid waste. This waste includes scrap and off-specialization silicon, used packaging from raw materials and supplies, spent manufacturing supplies, trash, and wastewater treatment sludge. Office, maintenance, and support activities generate office waste, food waste, and various scrap metals, scrap plastics, and used machine parts. Construction activities generate construction debris. MEMC's manufacturing plants cannot operate without access to local sanitary landfills for solid waste disposal. However, many communities face a crisis with respect to landfill capacity. Reliable and affordable disposal of nonhazardous solid waste is a growing problem.

In 1990, the MEMC manufacturing plant in Spartanburg, South Carolina experienced a large cost increase for nonhazardous solid waste disposal due to increased sanitary landfill fees. In response, the company undertook a solid waste study. The purpose of the study was twofold:

1. Establish a baseline generation rate and composition for nonhazardous solid waste.
2. Identify opportunities to reduce the volume of nonhazardous solid waste landfilled.

The solid waste study included a cardboard and office paper recycling trial. The baseline period for solid waste generation was the first quarter (January–March) of 1990.

Results

Waste hauling invoices were used to quantify the generation rate and disposal cost of wastewater treatment sludge, plant trash, and construction debris. Trash sampling and a review of accounting and manufacturing records were used to identify major plant trash components and to quantify their generation rates. Accounting and utility records were used to quantify the volume and value of waste materials already being recycled. Figures 5 and 6 illustrate the baseline solid waste generation and composition results.

Baseline Solid Waste Generation. From Fig. 5's data, we see that the annualized, baseline (first quarter 1990) generation of nonhazardous solids at the MEMC Spartanburg plant was 13,300 yd^3/year. Of the total waste generation, 12,160 yd^3/year were landfilled. The remaining waste consisting of scrap silicon and empty plastic and steel chemical drums was collected and sold for recycling. The annual cost of solid waste disposal was $113,070. This disposal cost was offset by $55,570/year in revenue from the sale of scrap silicon and empty chemical drums. The net, baseline cost of solid waste disposal including revenue from a recycled material sale was $57,500/year. Plant trash accounted for 66% of the landfilled solid waste volume. Wastewater treatment sludge represented 25% of the landfilled solid waste volume. However, wastewater treatment sludge disposal was 57% of the baseline cost of solid waste disposal. The unit price for landfill disposal of wastewater treatment sludge is four times the unit price for trash and construction debris.

Waste Minimization

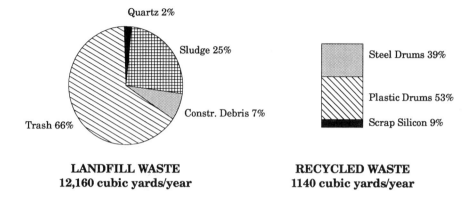

Figure 5 Baseline solid waste generation: MEMC Spartanburg, South Carolina plant. 13,300 yd^3/year, first quarter 1990 annualized, $57,500 net annual disposal cost including revenue from recycling. (From Dickens, 1991.)

From Fig. 6, we note that the annualized, baseline generation of plant trash was 7980 yd^3/year. Waste packaging materials, waste office paper, and spent manufacturing supplies were the largest plant trash components. Waste cardboard and office paper accounted for 30% of the plant trash volume landfilled, wood pallets and skids 8%, various high-density polyethylene (HDPE) plastics 4%, and waste polystyrene packaging 3%. Various low-density polyethylene (LDPE) and vinyl plastic materials represented 8% of the plant trash volume landfilled. Used plastic bags and "bubble pack" packaging materials were the primary components of this waste category. Included in "other materials" were 51,000 lb/year of spent graphite machine parts that have potentially high value if sold for scrap.

Recycling Trial. The cardboard and office paper recycling trail conducted during June 1990 was a success. During the trial, MEMC employees collected and recycled 85% of waste cardboard generated and 73% of waste office paper generated. The volume of plant trash hauled to landfill disposal was reduced by 24%. The annualized weight of cardboard collected and recycled was 70 tons/year. The annualized weight of office paper collected and recycled was 38 tons/year. Cardboard and office paper recycling are now permanently established procedures at the MEMC Spartanburg plant.

Volume Reduction Analysis. MEMC classified the plant trash components identified in Fig. 6 into three categories: readily recyclable, potentially

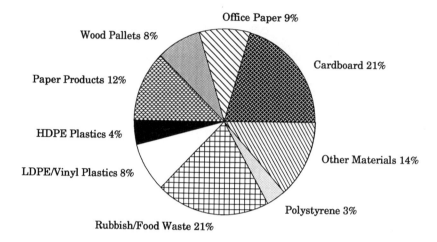

Figure 6 Baseline plant trash composition: MEMC Spartanburg South Carolina Plant. 7980 yd^3/year, first quarter 1990 annualized prior to recycling efforts, all material landfilled. (From Dickens, 1991.)

recyclable, and not recyclable. The purpose was to target recycling efforts toward solid waste components that would obtain the greatest reduction in the volume of solid waste landfilled. A recycling market exists and local brokers are available to take "readily recyclable" solid waste materials. They are collected and sold in the form generated. The value of readily recyclable materials is sufficiently high to pay for their handling and off-site transportation. Readily recyclable waste materials generated at the MEMC Spartanburg plant include cardboard, office paper, wood pallets and skids, and spent graphite machine parts. Brokers were found to buy these materials and provide equipment for handling, material storage, and off-site transportation.

"Potentially recyclable" solid waste materials are recyclable, but the market for materials in the form generated is small or does not exist. The materials must be processed by MEMC or an outside contractor to make them suitable for recycling. The value of potentially recyclable materials is generally insufficient to pay for their handling and off-site transportation. The cost of processing and transporting potentially recyclable materials must be measured against the material's cost of landfill disposal. The landfill cost savings, if sufficiently high, may subsidize the cost to process and transport these materials. Potentially recyclable waste materials generated at the MEMC Spartanburg plant

Waste Minimization

include polystyrene packaging, various HDPE chemical and product packages, clean plastic bags, and certain used paper products.

Volume Reduction Results. Figure 7 illustrates first quarter 1991 solid waste generation at the MEMC Spartanburg plant. The data represent the result of efforts to divert readily recyclable solid waste materials away from landfilled disposal. Although the total generation of nonhazardous solid waste was essentially unchanged (13,320 yd^3/year), MEMC recycling efforts reduced the volume of plant trash landfilled by 31.3% compared to the first quarter 1990 baseline period. The total volume of nonhazardous solid waste landfilled (plant trash, wastewater treatment sludge, and construction debris) was reduced by 21.2%. Manufacturing production and capital construction expenditures during the first quarter of 1991 were similar to those from the baseline period.

Compared to the baseline period, the annualized revenue from recycled material sales increased from $55,570 to $57,690/year. The annualized cost of solid waste landfill disposal decreased from $113,070 to $99,390/year. The net annualized cost of solid waste disposal decreased from $57,500 to $31,700/year, representing a savings of $25,800/year. Revenue from the sale of spent graphite machine parts was 29% of these savings. The annual revenue from the sale of waste cardboard and office paper is small, less than $500/year.

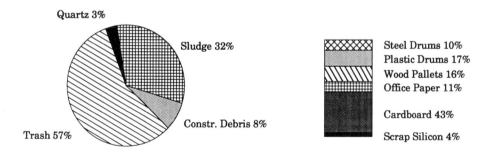

Figure 7 1991 Solid waste generation: MEMC Spartanburg South Carolina plant. 13,320 yd^3/year, first quarter 1991 annualized, $31,700 net annual disposal cost including revenue from recycling. (From Dickens, 1991.)

Transportation and handling required to recycle waste wood pallets and skids cost $2600/year. However, the landfill cost savings associated with waste cardboard, waste office paper, and waste wood pallet recycling are more than $14,000/year. Landfill cost avoidance is the biggest economic incentive for solid waste material recycling.

Waste Source Reduction. In addition to recycling, MEMC is pursuing waste source reduction to reduce its generation of nonhazardous solid waste. During the fourth quarter of 1990, MEMC completed the conversion of its manufacturing production and material purchasing systems from off-line batch computer processing to on-line, real-time computer systems. This conversion greatly reduced the volume of paper required to track manufacturing production, purchase orders, and purchased material distribution. The generation and collection of waste office paper at the MEMC Spartanburg plant decreased 33% between the June 1990 recycling trial and first quarter of 1991. This reduction eliminates the use of more than 12 tons/year of paper forms.

Summary and Conclusions

The MEMC Electronic Materials, Inc. manufacturing plant in Spartanburg, South Carolina reduced its volume of plant trash landfilled by 31.3% between the first quarters of 1990 and 1991. The corresponding reduction in the total volume of nonhazardous solid waste landfilled (plant trash, wastewater treatment sludge, and construction debris) was 21.2%. The net annualized cost of solid waste disposal was reduced by $25,800/year. Waste source reduction eliminated the use of more than 12 tons/year of paper forms.

Several conclusions from the MEMC solid waste study and recycling effort were unexpected:

1. Landfill cost avoidance is the greatest economic benefit of solid waste recycling. Revenue from the sale of waste cardboard and office paper is less than $500/year. However, cardboard and office paper recycling avoids more than $11,000/year in landfill disposal costs.
2. Landfill cost savings can subsidize the cost of recycling low-value waste materials. The MEMC Spartanburg plant spends $2600/year to collect and ship waste wood pallets and skids to recycling. The recycling of waste wood pallet and skids avoids more than $3000/year in landfill disposal costs.
3. Solid waste recycling is a large materials-handling problem and represents disruption to established waste-handling routines. Substantial management effort and the cooperation of employees are required to make recycling programs work at large industrial plants.

4. The majority of nonhazardous solid waste landfilled by MEMC is related to wastewater treatment and to the packaging of raw materials and manufacturing supplies. The elimination and reuse of packaging materials are MEMC's greatest short-term opportunity for solid waste volume reduction. Long-term solid waste volume reduction will require new technology to eliminate or reduce MEMC's generation of wastewater treatment sludge.

The MEMC case study is a model for other companies considering solid waste volume reduction.

12. SUMMARY

The benefits of waste minimization to waste generators are substantial. The benefits to overall human health and the environment are significant. The cost and disruption at the plant level, although not insignificant, are generally outweighed by a decrease in long-term liabilities and cleanup costs. At present, the technologies and methodologies for accomplishing waste minimization are in place. Regulation and economic considerations are destined to make waste minimization an integral part of all industrial processes.

In this chapter, we have laid the groundwork for waste generators to examine their options with a goal of first preventing pollution at the source by source reduction and secondarily recycling and reusing the wastes generated. In the remaining chapters of this volume, waste treatment unit operations will be discussed and the treatment of wastes generated by specific industries considered.

REFERENCES

California Dept. of Health Services (CDHS) (1988). Economic Implications of Waste Reduction, Recycling, Treatment and Disposal of Hazardous Wastes. Toxic Substances Control Div., Alternative Technology Sect., pp. 9–29.

CDHS (1989). Reducing California's Metal-Bearing Waste Streams. Toxic Substances Control Div., Alternative Technology Sect., Pasadena, CA, pp. 11-10 through 11-16.

CDHS (1990). Alternative Technologies for the Minimization of Hazardous Waste. Toxic Substances Control Div., Alternative Technology Sect., Pasadena, CA, pp. 47–50.

Chandler, T. C. (1990). MEMC etch—a chromium trioxide-free etchant for delineating dislocations and slip in silicon. *J. Electrochem. Soc., 137,* 944 (March).

Dalrymple, D. A. and Trofe, T. W. (1969). An Overview of Liquid Redox Sulfur Recovery. *Chem. Eng. Prog., 85,* 3 (March).

Dickens, P. S. (1991). Personal Communication from MEMC Electronic Materials, Inc., Spartanburg, S.C.

Doyle, P. (1987). Hazardous waste management: An update. In *A Legislator's Guide* (S. Schwoch, ed.). Nat. Conf. of State Legislators, Washington, D.C., p. 1.

Environmental Protection Agency (EPA) (1988). Waste Minimization Opportunity Assessment Manual. HWERL, Office of Res. and Devel., Washington, D.C., EPA/600-2-88-025, pp. 1–26.

EPA (1987). Waste Minimization—Environmental Quality with Economic Benefits. HWERL, Office of Res. and Devel., Washington, D.C., EPA/530-SW-87-026, pp. 1–22.

Federal Register (1989). Pollution Prevention Policy Statement. Washington, D.C., Vol. 54, pp. 3845–3847.

Hunt, G. E. and Schecter, R. N. (1989). Minimization of hazardous waste generation. In *Standard Handbook of Hazardous Waste Treatment and Disposal* (H. M. Freeman, ed.). McGraw-Hill, New York, pp. 5.3–5.27.

Lindgren, G. F. (1983). *Guide to Managing Industrial Hazardous Waste.* Butterworth, Boston, Mass.

Office of Technology Assessment (OTA) (1986). Serious Reduction of Hazardous Waste. Summary, Washington, D.C., p. 48.

PEER Consultants and George Beetle Comp. (1989). User's Guide for Economic Assessments Software. Prepared for U.S. EPA-CERI, Cincinnati, Ohio, pp. 1–29.

PEER Consultants and Univ. of Dayton Research Institute (1989). User's Guide for *SWAMI*—Strategic Waste Minimization Incentive. Prepared for U.S. EPA-CERI, Cincinnati, Ohio, Version 1.1, pp. 1–3.

Schoenberger, R. J. and Corbin, M. H. (1985). Technologies for hazardous waste reduction—A state of the art review. *Proceedings of 3rd Annual Hazardous Materials Management Conference*, Philadelphia, Pa., p. 371.

2

Stormwater Management and Treatment

Robert Leo Trotta

O'Brien & Gere Engineers, Inc., New York, New York

Constantine Yapijakis

The Cooper Union, New York, New York

1. CONSIDERATIONS FOR STORMWATER MANAGEMENT AND TREATMENT

1.1 Pollution Aspects and Considerations

The pollution aspects of stormwater are related to the substances that become entrained in it from its point of origin to its point of discharge into a water body. Stormwater originates from the clouds and its first contamination is from pollution sources contained within the air we breathe. Most notable and well known is the pollution related to acid rain. Acid rain is generally stormwater that has absorbed airborne contaminants propagated by the burning of sulfur-bearing fuels used for heating and power generation. The oxidation of the sulfur and subsequent reaction with atmospheric water vapor produces sulfuric acid. This is but one example of a mechanism that contributes to the contamination of stormwater. Further details with respect to acid rain and other pathways involving the entrapment of pollutants in stormwater are discussed in Sec. 3 of this chapter.

In industry there are many compounds that in the presence of water and other substances could lead to the development of acidic, caustic, or poisonous characteristics in stormwater. Of particular interest in this regard is the possible entrainment of nutrients, organics, inorganics, heavy metals, pesticides, volatile organics, oils, greases, and other pollutants. The contaminants can enter into stormwater in the form of liquids, floatables, grit, settleable solids, suspended solids, soluble substances, and dissolved gases. These substances in significant concentrations can have an adverse impact on fishlife and plantlife contained within the water body receiving the stormwater discharge, as well as wildlife that utilizes the water resources. Furthermore, when such water bodies are either tributary to or directly used as drinking water supplies, the contaminated stormwater could contribute to the destruction of the surface water supply.

Similarly, groundwater drinking water supplies can be polluted by contaminated stormwater. The stormwater enters groundwater supplies through points of recharge from surface waters and through percolation into the soil.

1.2 Regulatory Considerations

In the United States, federal laws have dictated the course of measures implemented on the federal, state, and local levels to control discharges into the nation's surface waters. These laws primarily center on the implementation of control strategies for discharges that would otherwise cause the degradation of the quality of these waters. In the past, the laws focused on wastewater discharges; however, more recent considerations have been with respect to combined sewer overflows and industrial stormwater discharges.

Combined sewer overflow is that discharge to water bodies from combined sewers which occurs as a result of a storm event. Combined sewers are those sewers that normally convey sanitary flows during dry weather conditions but handle stormwater as well during wet periods (precipitation events). The overflow from combined sewers occurs as a result of the exeedance of the capacity of the sewer system to convey both sanitary and storm flows. Currently industries that are connected to such systems are regulated through pretreatment regulations administered on a local level in accordance with federal and state program requirements.

The 1987 amendments to the Clean Water Act mandated the establishment of a permit system for point sources of stormwater discharges into waters of the United States. The permit requirements developed by the U.S. Environmental Protection Agency (EPA) currently mandate the issuance of State National Pollutant Discharge Elimination System permits (NPDES permits) for the following categories of stormwater discharges (40 CFR 126.26):

Stormwater Management and Treatment

1. A discharge with respect to which a permit has been issued prior to February 4, 1987
2. A discharge associated with industrial activity
3. A discharge from a municipal separate storm sewer system serving a population of 250,000 or more
4. A discharge from a municipal separate storm sewer system serving a population of 100,000 or more but less than 250,000
5. A discharge that the administrator or state, as the case may be, determines has contributed to a violation of a water quality standard or is a significant contributor of pollutants to U.S. waters

Although not stipulated at this time, other categories of discharges may require permitting after October 1, 1992, as implied by the current regulations. Categories 1, 2, and 5, as mentioned, will have a primary impact on the industrial and business sector, whereas categories 3 and 4 are likely to have a secondary impact as described here.

Naturally, industrial facilities and businesses currently permitted under category 1 will continue to meet permit requirements as mandated by the regulatory agency. Industries falling under category 2 and businesses classified under category 5 need to comply with NPDES permit requirements.

These requirements include permit filing, monitoring, reporting, and compliance stipulations. In cases where a stormwater discharge causes or significantly contributes to the contravention of a water quality standard, mitigative measures would be required to meet the stipulations of the NPDES permit. Although no discharge permits are required of industries and businesses discharging into separate municipal storm sewer systems, any requirements that may be placed on such systems can impact industries and businesses. Municipalities falling under categories 3 through 5 may be subject to discharge requirements according to the NPDES permit limitations imposed. In such cases, through the enactment of ordinances governing discharges into separate municipal stormwater systems, industries and businesses discharging stormwater into such systems need to be aware of federal, state, and local stormwater discharge requirements.

After October 1, 1992, industries discharging into separate municipal storm sewer systems with populations less than 100,000 may be subject to secondary requirements due to the possible regulation of these municipal discharges. Similarly, local ordinances and regulations would need to be investigated as well as the federal and state regulations.

2. QUANTITY AND QUALITY

2.1 Hydrologic Considerations

Meteorologists collect data on, report on, and work with the total depths of rainfall events of various durations. Engineers, on the other hand, use the average rainfall intensity (ratio of total depth and duration of an event) as the primary parameter for their work, implicitly assuming that the intensity of a rainfall event is constant during its occurrence. Extensive presentations of the following concepts may be found in any book on hydrology for engineers (see, e.g., Viessman et al., 1977, and Linsley et al., 1982).

Rainfall Depth, Duration, and Frequency

Many different empirical formulas have been proposed by researchers to describe the presumed relationships between rainfall intensity and the duration frequency of an event or between rainfall depth and duration frequency. Such relationships are derived from statistical analysis either of point rainfall data, i.e., precipitation events as measured by a single rain-gage station, or of data from networks of rain gages. The point data and their evaluation results are statistically adequate to define the main temporal variations of the characteristics of storm events. One observation is that as the duration of a storm event decreases, the average rainfall intensity increases given a specific frequency of return. Another observation useful in design is that, as the frequency of the return increases, the average rainfall intensity decreases given a specific duration. Data from networks of rain gages and their evaluation results are statistically sufficient to define the main spatial variation characteristic of storm events. The observation is that the more limited the area over which a storm event is occurring, the higher the value of the average rainfall intensity as compared to the maximum observed point rainfall intensity within the event area. For design purposes, the ratio of the spatial average to the point temporal average rainfall intensity (corresponding to identical frequencies of return) is required in order to adjust a design storm event point depth to account for spatial variation.

Probable Maximum Rainfall

Certain critical storm events are used in estimating flood flow peak design values by U.S. water resources agencies such as the Corps of Engineers. As reported by Riedel et al. (1967), one such critical storm event is the probable maximum precipitation. This is defined as the critical depth-duration-area rainfall relationship for a specific area during the seasons of the year, resulting

Stormwater Management and Treatment 65

from a storm event of the most critical meteorological conditions. The probable maximum rainfall is based on the most effective combination of factors that control rainfall intensity. Annual probable maximums may be less important than seasonal maximums, in flooding situations that may occur in combination with snowmelt runoff.

Evapotranspiration and Interception of Rainfall

Evaporation is the process by which precipitated water is lost to the runoff process by transference from land and water masses of the earth to the atmosphere, in the form of vapor. Transpiration is water loss to the atmosphere through the action of plants that absorb it with their roots and let it escape through pores in their leaves. From the practical viewpoint of water resources engineers, only total evapotranspiration (i.e., combined evaporation and transpiration) is of interest. Various investigators have proposed theoretical, analytical, or empirical methods for estimating evapotranspiration losses, but no system has been found acceptable under all encountered conditions. An additional part of the precipitation volume from a storm event is intercepted by the vegetation cover of a drainage area until it evaporates and, thereby, it is lost to the runoff process. The volume of intercepted water depends on the storm event character, the species and density of plants and trees, and the season.

Depression Storage and Infiltration Losses

Precipitation that is also lost to the surface runoff process may infiltrate into the ground or become trapped in the many ground depressions from where it can only escape through evaporation or infiltration. Due to the fact that there is extreme variability in the characteristics of land depressions and insufficient measurements, no generalized relationships with enough specified parameters for all situations are possible. Nevertheless, a few rational models and values of the range of depression storage losses have been reported in the literature. On the other hand, infiltration losses are a very significant parameter in the distribution of the water volume from a storm event. As accurate as possible estimates of infiltrating volumes must, therefore, be made since they affect the timing, distribution, and magnitude of precipitation surface runoff. The type and extent of the vegetal cover, the condition and properties of the surface crust and the soil, and the rainfall intensity are among the factors that may influence the rate of infiltration f. No satisfactory general relationship exists. Instead, hydrograph analyses and infiltrometer studies are methods used for infiltration capacity estimates. For small urban areas that respond rapidly to storm inputs, more precise values of infiltration rates are sometimes needed, whereas on large watersheds where long-duration storm events generate the peak flow conditions, average or representative values of f may suffice.

2.2 Surface Runoff

Runoff Flows and Hydrographs

When considering stormwater management, surface runoff is the main concern. However, the relationship between precipitation and runoff is most complex and influenced by such storm event characteristics as pattern, antecedent events, and watershed parameters. Many approximate formulas, therefore, have been developed and empirical methods such as the rational formula or site-specific equations can estimate the peak runoff rate, in cases where it is sufficient for the analysis and design of simple stormwater systems. Calculations of runoff volumes using sound rational equations based on physical principles and hydrographs are necessary in cases where a more detailed analysis of the system hydrology and hydraulics is needed. A hydrograph is a continuous graph showing the magnitude and time distribution of the main parameters, stage and discharge, of surface runoff or stream flow. It can be, therefore, a stage hydrograph or a discharge hydrograph (more common) and it is influenced by the physical and hydrological characteristics of the drainage basin. The discharge shown by a hydrograph at any time is the additive result of the direct surface runoff, interflow, groundwater or base flow, and channel precipitation. A typical hydrograph is shown in Fig. 1.

Drainage Basin Characteristics

The shape of the flood hydrograph from a catchment area is a function of the hydrologic input to that region and of the catchment characteristics, such as area, shape, channel, and overland slopes, soil types and their distribution, type and extent of vegetative cover, and other geological and geomorphological watershed features. One of the primary measures of the relative timing of hydrologic events is basin lag t_1. Basin lag is defined as the time between the center of mass of the rainfall excess producing surface runoff and the peak of the hydrograph produced. The lag time is influenced by such parameters as the shape and average slopes of the drainage area, the slope of the main channel, channel geometry, and the storm event pattern. Various investigators have proposed relationships predictive of basin lag, but Snyder's equation (1938), based on the data from large natural watersheds, is the most widely used and adapted by others

$$t_1 = Ct \, (LcaL)^{0.3} \qquad (1)$$

where

 t_1 = basin lag (hr)
 Ct = coefficient depending on basin properties
 Lca = distance (miles) along the main stream from the base gage to a point opposite the basin centroid
 L = maximum travel distance (miles) along the main stream (1 mile = 1609 m)

The Soil Conservation Service (SCS, 1972) defines t_1 as

$$t_1 = 0.60 tc \tag{2}$$

where tc (hr) is the time of concentration, another primary measure of the relative timing of hydrologic events. The time of concentration is usually defined as the sum of the overland travel time from the furthest basin point and the channel travel time to the outlet of concern.

Runoff and Snowmelt Runoff Determination

Water resource engineers are involved in estimating stream flows using one of two approaches. The first, an indirect approach in which runoff is estimated

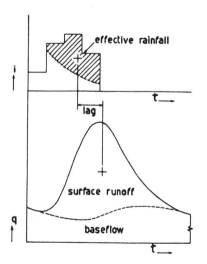

Figure 1 Rainfall/runoff relationship.

based on observed or expected precipitation, will be discussed in Sec. 2.3. The second method is based on the direct analyses of recorded runoff data without consideration of corresponding rainfall data. These types of analyses are usually frequency studies to evaluate the probability of occurrence of a specific runoff event, to determine the risk associated with a design or operation alternative. Such frequency analyses are usually determining maximums or floods and minimums or droughts. However, when existing runoff records are short-term or incomplete, the frequency analyses cannot be very reliable. In certain cases, sequential generating techniques or time-series analyses are used to develop synthetic records of runoff for any desired length of time. In many areas, such as mountainous watersheds, snowmelt runoff is the dominant source of stream flows. For instance, Goodell (1966) has reported that as much as 90% of the annual water supply volume in the high-elevation watersheds of the Colorado Rockies may be originating in snowfall accumulations. Some of the greatest flood flows may be caused by a combination of very large rainstorms and simultaneous snowmelt. Adequate knowledge of the extent and other characteristics of snow packs within a watershed, therefore, is very important in streamflow forecasting. Investigators have followed various approaches to runoff determination from snowmelt, which range from simple correlation analyses that ignore the physical snowmelt process to sophisticated methods using physical equations. The U.S. Army Corps of Engineers (1956) conducted extensive studies that produced several general equations for snowmelt (in./day) during rain-free periods and periods of rain, both for open or partly covered areas and for heavily forested watersheds. (Note 1 in./day = 2.54 cm/day).

Overland Flow Routing

Watershed overland flow simulation, as well as flood forecasting and reservoir design, generally use some type of flow-routing methodology. Routing may be employed to predict the temporal and spatial variations of the outflow hydrograph from a watershed receiving a known volume of precipitation. There are two types of routing: hydrologic, which employs the continuity equation with a relationship between storage and discharge within the system, and hydraulic, which uses both the continuity and momentum equation. The latter better describes the flow dynamics through use of the partial differential equations for unsteady flow in open channels. In hydrologic routing, watershed runoff is considered modified by two kinds of storage, channel and reservoir, and the watershed can be considered (Nash, 1957) as reservoirs in series with an individual relationship between storage and outflow. The assumption is that each reservoir is instantaneously full and discharges into the one following, and so on. The Muskingum method or the concept of routing a time-area histogram can also be used to derive an outflow hydrograph from a watershed (Viessman et al., 1977). In hydraulic routing, the two routed flow components (the overland and channel flow) are considered and the watershed is described mathematically by defining the various phases of flow of the effective rainfall through its boundaries. The resulting computer programs are very complex and, therefore, most applications use simplifications in overland flow routing. Empirical equations are usually used to estimate the lag or overland flow travel time t_0. For instance, the Federal Aviation Agency (1970) uses the following equation for airfield drainage problems, but it has also been used frequently for overland flow in urban basins:

$$t_0 = 1.8(1.1-C)L^{0.50}/(S^{0.35}) \qquad (3)$$

where

t_0 = overland travel time (min)
C = rational formula runoff coefficient
L = length of overland flow path (ft)
S = average surface slope (%)

Another equation applied to surface runoff from developed areas and proposed by Morgali and Linsley (1965) is

$$t_o = 0.94(L \cdot n)^{0.6}/(i^{0.4})(S^{0.03}) \qquad (4)$$

where

n = Manning's roughness coefficient (Table 1)
i = effective precipitation intensity (in./hr)
S = average overland slope (ft/ft)

Table 1 Manning Roughness Coefficients (From Kibler et al., 1982)

Type of surface	Manning's n
Dense grass or forest	0.40
Pasture or average grass cover	0.20
Poor grass, moderately bare surface	0.10
Smooth, bare, packed soil, free of stones	0.05
Smooth impervious surface	0.035

The above equation needs to be solved by iteration since both i and t_0 are unknown. Table 1 presents the values of Manning's roughness coefficient recommended by Kibler et al. (1982) for small urban or developing watersheds. An equation recommended by the Soil Conservation Service (SCS, 1975) for agricultural and rural watersheds is shown below:

$$t_0 = 60(L^{0.8})[(1000/CN) - 9]^{0.7}/1900\, S^{0.5} \tag{5}$$

where

L = hydraulic length of longest flow path (ft)
CN = SCS runoff curve number (Table 2)
S = average watershed slope (%)

In mixed areas the formula overestimates t_0, and the SCS recommends the use of factors to correct for channel improvement (Fig. 2) and impervious areas (Fig. 3). McCuen et al. (1981) presented revised lag factors, found to yield more accurate estimates, in place of the SCS ones (Fig. 4). Finally, for channel flow in a catchment area, the well-known Manning's formula may be used to estimate velocities and channel travel time.

Land Use Effects

Drainage basin characteristics, such as slope, size of impervious portion, soil and rock type, and vegetal cover, affect the magnitude and distribution variation of runoff. Therefore, any modifications of these due to human actions and land use changes will have varying impacts on both the quantity and quality of runoff. Land use changes, in particular, that alter both the form of the drainage network and the watershed surface characteristics (Hall, 1984) may increase or decrease the runoff volume from a given site, as well as the peak and overland travel time of a flood. Activities that impact on the infiltration rate and surface storage of a catchment area are most important considering their effect on flow volume, peak rate, and overland lag. Industrial operations that may cause such impacts on stormwater management can include wildscape clearing and grading for buildings and parking lots, felling of forests and drainage of swamps to open up land, and stormwater drainage infrastructure built where there was once a rural area. In such cases, the natural drainage systems are altered and supplemented by manmade stormwater drainage and flood alleviation schemes such as channels, storm drains, flood embankments, and flood storage or infiltration ponds. In general, land use practices that decrease flow volume also decrease the peak rate of flow, and vice versa (Viessman et al., 1977). On the

Table 2 SCS Runoff Curve Numbers (From SCS, 1975)

Land use	Cover Treatment or practice	Hydrologic condition	Hydrologic soil group A	B	C	D
Fallow	Straight row	—	77	86	91	94
Row crops	Straight row	Poor	72	81	88	91
	Straight row	Good	67	78	85	89
	Contoured	Poor	70	79	84	88
	Contoured	Good	65	75	82	86
	Contoured	Poor	66	74	80	82
	Contoured and terraced	Good	62	71	78	81
	Contoured and terraced					
Small grain	Straight row	Poor	65	76	84	88
	Contoured	Good	63	75	83	87
	Contoured and terraced	Poor	63	74	82	85
		Good	61	73	81	84
		Poor	61	72	79	82
		Good	59	70	78	81
Close-seeded legumes[a] or rotation meadow	Straight row	Poor	66	77	85	88
	Straight row	Good	58	72	81	87
	Contoured	Poor	64	75	83	85
	Contoured	Good	55	69	78	84
	Contoured and terraced	Poor	63	73	80	82
	Contoured and terraced	Good	51	67	76	81
Pasteur or range	Contoured	Poor	68	79	86	89
	Contoured	Fair	49	69	79	84
	Contoured	Good	39	61	74	80
		Poor	47	67	81	88
		Fair	25	59	75	83
		Good	6	35	70	79
Meadow		Good	30	58	71	78
Woods		Poor	45	66	77	83
		Fair	36	60	73	79
		Good	25	55	70	77
Farmsteads		—	59	74	82	86
Roads Dirt[b]		—	72	82	87	89
Hard surface[b]		—	74	84	90	92

[a] Close-drilled or broadcast.
[b] Including right of way.

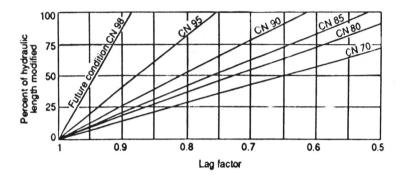

Figure 2 Lag adjustment factors for Eq. (5) when the main channel has been hydraulically improved. (From SCS, 1975.)

other hand, reductions in the time lag or concentration time of a drainage basin affect the frequency or reduce the return period of a certain flow.

2.3 Design Considerations

Industrial parks and individual industrial sites (including agricultural industry activities) comprise either urbanized drainage areas or small rural watersheds. Methods that have been found appropriate for stormwater management in these cases include peak flow formulas, urban runoff models, and small

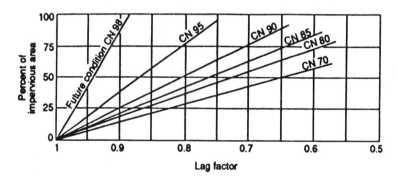

Figure 3 Lag adjustment factors for Eq. (5) when impervious areas occur in the watershed. (From SCS, 1975.)

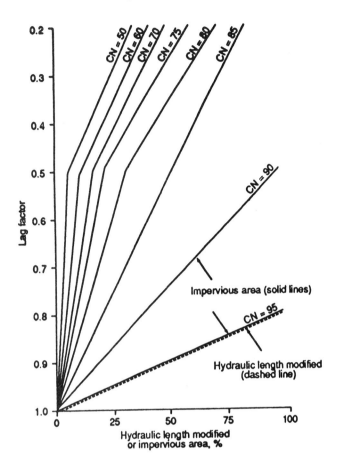

Figure 4 Proposed lag factor vs. hydraulic length modified or impervious area. (From McCuen et al., 1981.)

watershed simulation procedures. Some of these are described in the following subsections.

Rational Formula

The most common empirical procedure used for designing small drainage systems is the rational formula

$$Q = CiA \tag{6}$$

where

Q = peak runoff flow (cfs)
C = runoff coefficient (Table 3), ratio of runoff/rainfall
i = average effective rainfall intensity (in./hr) with a duration equal to the time of concentration
A = drainage area (acres)

Assumptions made for the application of the rational formula include: (1) return periods for rainfall and runoff are considered to be equal; (2) runoff coefficient selected is considered constant for the entire design storm and also

Table 3 Runoff Coefficients, C

Description of area of surface	C factor
Business	
Downtown	0.70–0.95
Neighborhood	0.50–0.70
Residential	
Single-family	0.30–0.50
Multiunits, detached	0.40–0.60
Multiunits, attached	0.60–0.75
Residential (suburban)	0.50–0.70
Apartment	0.50–0.70
Industrial	
Light	0.50–0.80
Heavy	0.60–0.90
Parks, cemeteries	0.10–0.25
Playgrounds	0.20–0.35
Railroad yard	0.20–0.35
Unimproved	0.10–0.30
Pavement	
Asphaltic and concrete	0.70–0.95
Brick	0.70–0.85
Roofs	0.75–0.95
Lawns, sandy soil	
Flat, 2%	0.05–0.10
Average, 2–7%	0.10–0.15
Steep, 7%	0.15–0.20
Lawns, heavy soil	
Flat, 2%	0.13–0.17
Average, 2–7%	0.18–0.22
Steep, 7%	0.25–0.35

from storm to storm; (3) design rainfall intensity is read from a locally derived intensity/duration/frequency curve; (4) rainfall intensity is considered constant over the entire watershed and design storm event; and (5) in practice, a composite weighted average C is estimated for the various surface types of the study area.

SCS TR-55 Method

As mentioned previously, the Soil Conservation Service (SCS, 1975) report on "Urban Hydrology for Small Watersheds," known as Technical Release No. 55, provides a simple rainfall/runoff method for peak flow estimates based on the 24-hr net rain depth and the time of concentration t_0. This is a graphical approach assuming homogeneous watersheds where the land use and soil type are represented by a single parameter, the runoff curve number (CN). The SCS peak discharge graph shown in Fig. 5 is applied only when the peak flow is designed for 24-hr, type II storm distributions (typical of thunderstorms experienced in all U.S. states except the Pacific Coast ones).

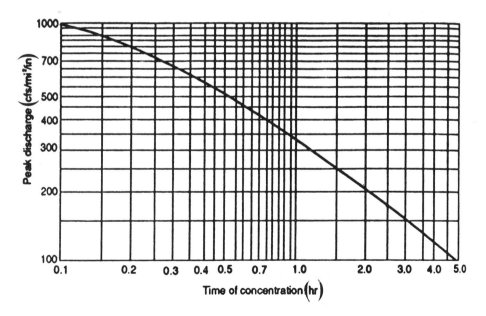

Figure 5 Peak discharge vs. time of concentration t_0 for 24-hr, Type II storm distribution (1 ft^3/mile^2/in. = 0.0043 m^3/sec/km^2/cm). (From SCS, 1975.)

Unit Hydrograph

One method that has been used extensively to predict flood peak flows and flow hydrographs from storm events is the unit hydrograph method. The unit graph for a watershed is defined as the hydrograph showing the runoff rates of a given one-day storm event producing a 1-in. depth of runoff over the watershed. Time periods other than one day are also used for the derivation of unit hydrographs. The assumption made, generally not true, in the application of this method is that the rainfall is distributed in the same spatial and temporal pattern for all storm events. The development of unit hydrographs for other than the derived duration is accomplished by the use of a method called the S hydrograph, which employs a unit hydrograph to form an S hydrograph resulting from a continuous rainfall.

Synthetic Unit Hydrograph Formulas

Peak flows draining from urban and small catchment areas can be estimated from critical or design storm information using the synthetic unit hydrograph technique. One technique employed by the U.S. Corps of Engineers (1974) uses Snyder's method of synthesizing unit hydrographs, according to which a storm with a duration given by

$$tr = t_0/5.5 \tag{7}$$

and a lag time given by

$$t_0 = Ct\,(LLca)^{0.3} \tag{8}$$

produces a peak flow for 1 in. of excess rain given by

$$Qp = 640 CpA/(t_0) \tag{9}$$

where

- tr = duration of the unit rainfall excess (hr)
- t_0 = lag time (hr) from the centroid of unit rainfall excess to the peak of the synthetic unit hydrograph
- Ct = coefficient representing variations of watershed slopes and storage
- L = length (miles) of the main stream channel from the outlet to the farthest point (1 mile = 1609 m)
- Lca = length (miles) along the main channel to the point opposite the watershed centroid
- Qp = peak flow of synthetic unit hydrograph (cfs) (1 cfs = 0.02832 m^3/sec)

Stormwater Management and Treatment

Cp = coefficient of peak flow accounting for watershed retention or storage capacity

A = watershed area (mile^2) (1 mile^2 = 2.59 km^2)

According to another, similar method developed by the Soil Conservation Service (Mockus, 1957) for constructing synthetic unit hydrographs, the peak flow produced from a storm event with a duration D is equal to

$$Qp = 484A/tp \tag{10}$$

where

tp = time (hr) from the beginning of the effective rain to the time of the peak runoff flow, which by definition is $tp = t_0 + D/2$

D = duration (hr) of rainfall, equal to $0.133tc$, where tc is the time of concentration

Both of the above empirical equations apply for a certain duration D of 1-in. net rain. The Qp from either formula can be multiplied by the actual net p (peak flow) for other storm events with equal D but different depths. Peak flows for storm events with longer or shorter durations D have to be estimated using unit hydrograph methods.

Urban Runoff Models

Urban runoff computer modeling attempts to quantify all relevant phenomena from rainfall to resulting runoff. This requires the determination of a design storm minus the losses to arrive at a net rainfall rate, the use of overland flow equations to find and route the gutter flow, the routing of flow in stormwater drains, and the determination of the outflow hydrograph. Most urban runoff models deal only with single storm events and, if the errors made are small and noncumulative, the predicted runoff is valid. More recently, the trend has been a continuous-time simulation of many storm and dry periods. The following urban watershed models are representative of many others used.

The Road Research Laboratory and Illinois Simulator is an urban runoff model that uses the time-area runoff routing method; it was developed in England. Another very widely accepted and used storm runoff model is the EPA Stormwater and Management Model (SWMM), which is designed to simulate the runoff of a catchment area for any predescribed storm event pattern. It can determine, for short-duration rainfalls, the locations and magnitudes of local flood flows and also the quantity and quality of runoff at several locations both in the system and in receiving water bodies. Finally, the University of Cincinnati Urban Runoff Model is similar to the SWMM and divides a drainage area into subcatchments with closely matched characteristics whose flows are routed overland into gutters and sewers.

Design Storms

Most designs of major structures involving hydrologic analyses utilize a flood magnitude that is considered critical. When possible, stream flow records are analyzed, but usually design flood hydrographs have to be synthetized from available storm records using rainfall/runoff procedures. Typical storm depths required for major structure design are the probable maximum precipitation or PMP (discussed in Sec. 2.1) and the standard project storm or SPS. The latter is crucial in the design of large dams and its value is usually obtained from the record of large storm events in the neighborhood of the drainage area of study. The SPS is patterned after a storm event that has caused the most critical rainfall depth/area/duration relationship and also includes the effect of snowmelt. In general, the SPS rainfall is approximately 50% of the PMP. Finally, frequency curves can be plotted and used in major and minor structure design in cases where extensive records are available. Customarily, frequency-based floods are not a part of the design criteria for major structures, but they are commonly used in minor structure design.

Stormwater Retention Basins

As previously discussed, land use changes impact on stormwater runoff. Volumes and peak flows will increase following urbanization, i.e., industrial park development, when previously natural pervious land is covered by such structures as buildings, roadways, and parking lots, and when natural storage areas or depressions have been eliminated and the vegetal cover removed. On the other hand, it may be desirable at an industrial site to collect contaminated runoff and hold it either for treatment or slower release to a water body. Detention ponds are constructed to alleviate some of the above problems and serve as holding and treatment facilities. Their primary objective is to reduce the peak flows of surface runoff and the peak loads of pollutants they carry into receiving water bodies. The design of detention basins is carried out on the basis of hydraulic and hydrologic principles, but it is usually not known which design storm would result in the largest retention storage volume. Determination of the required size must, therefore, be accomplished by designing for several critical storm events with various intensities and durations and for several antecedent soil moisture conditions and flow release rates. If the storm event runoff can be approximated by a triangular hydrograph (when the

Stormwater Management and Treatment

duration of an event is equal to or less than the time of concentration tc), the required storage (Boyd, 1982) in the retention basin is

$$S = 0.083(Ts + tc)(Qp - Q_0) \tag{11}$$

where

S = storage volume required (acre-ft)
Ts = duration (hr) of rainfall event
tc = time of concentration (hr)
Qp = peak inflow rate (cfs) (1 cfs = 0.2832 m³/sec)
Q_0 = peak outflow rate (cfs)

If the storm event runoff can be approximated by a trapezoidal hydrograph (when duration is longer than the time of concentration), the required storage (Boyd, 1982) in the retention basin is

$$S = 0.083[Qp\,Ts - Q_0\,(Ts + tc) + Q_0^2\,tc/Qp] \tag{12}$$

All parameters and units are the same as above. Of course, more elaborate routing techniques and other methods could also be used, but the above approximate, simple method has been found to give very good results.

2.4 Quality Considerations

The traditional target of water pollution control, in general, and industrial pollution control, in particular, has been discharges from point sources because they were relatively easy to monitor and treat. But, for the past decade or so, more and more attention is being paid to contributions of pollutants from nonpoint sources, including industrial activities, due to the stormwater runoff process. Runoff erodes, washes off, and carries all sorts of pollutants from large surface areas, pervious and impervious, and eventually discharges them into receiving water bodies.

Point and Nonpoint Source Pollution

Conventional pollution control is geared toward point sources that provide relatively concentrated pollutants and steady flows which do not fluctuate excessively. Nonpoint sources, on the other hand, are much more diffuse and governed by randomly occurring and intermittent storm events. Point sources of pollution discharge into surface water bodies at specific points, and they can be measured and their pollutional load directly estimated. Such typical point sources include industrial and sewage treatment plant discharges, combined sewer overflow, and collected and piped leachate from sanitary landfills.

Diffuse or nonpoint sources typically include urban runoff, construction activities, mining operations, agriculture and animal farming, atmospheric deposition, erosion of virgin lands and forests, and transportation; of these, the major problems stem from urban runoff and agricultural activities (Council on Environmental Quality, CEQ, 1980). Pollution from urban runoff contains such quality parameters as organic pollutants, heavy toxic metals, coliforms and pathogens, suspended solids, oil and grease, nitrogen and phosphorus, and toxic priority pollutants. Obviously, industrial zones and highly dense urban areas produce higher pollution loads and contribute a greater variety of pollutants. Agricultural activities primarily contribute nitrogen and phosphorus, pesticides and insecticides, organic substances, pathogens, and soil erosion products. Their concentrations depend on the application of fertilizers and pesticides and various tillage activities. Finally, atmospheric deposition has also been designated (CEQ, 1980) as a major nonpoint source of lead, phosphorus, PCBs, and acidity (in the form of acid precipitation).

Estimates of Stormwater Pollutant Loadings

Pollutant loadings resulting from nonpoint sources are commonly expressed in terms of mass/area/time (i.e., tons/mile^2/day) as compared to mass/time (i.e., lb/day) for point sources. Following the identification and listing of all the contributing diffuse sources in the study area, their loads have to be characterized and quantified. Wu and Ahlert (1978) defined three levels of detail at which estimates of pollutant loads contributed by stormwater runoff might be required: mean yearly loads, assumed to be spread uniformly over both wet and dry periods; interevent loads, which take into account the variations that occur from storm to storm; and intraevent loads, which consider the transient water quality state during an individual storm event. Concentration value from nonpoint source sampling and runoff flows are usually used to estimate a pollutant load per individual storm. An average or expected mean concentration for the storm event is multiplied by the volume of runoff, and these load per storm values from many representative storm events are used for the calculation of a yearly pollutant discharge from the study area (Whipple et al., 1983).

Another technique for estimating yearly pollutant loads has been presented by Smith and Stewart (1977) among others, and it employs simple modeling by regression analysis. Plots of log-load vs. log-runoff volume from a group of storm events are used for regression analysis that reflects the log-normal distribution of the data and may be employed to predict a pollutant load based on a runoff volume. As mentioned at the beginning the this section, yearly loads are commonly normalized on the basis of the catchment areas, but other

Stormwater Management and Treatment

methods are also used to express loads as a function of curb length, population density, drainage density, and a specific land use (Whipple et al., 1983). The latter is quite adequate as a first approximation, but it may often lead to results that will deviate significantly from measured values (Krenkel and Novotny, 1980). Typical pollutant loads from various nonpoint sources and land uses are presented in a later section.

Water Quality Models

Wu and Ahlert (1978) categorized the models used for estimating stormwater pollutant loadings into the four types: zero-order or empirical methods, direct methods, statistical methods, and descriptive methods. A variety of diffuse source-estimating models are available in each of the four categories, and a few representative ones are mentioned here. The choice and use of a specific model is a function of the resources available to a project and the particular characteristics of the nonpoint sources involved. The empirical methods constitute the simplest approach and involve the application of unit load rates for the particular water quality parameter to the catchment areas with different land uses within a watershed. These rates are taken, if possible, from local water quality monitoring reports or from the literature on similar (hydrologically and based on land use) watersheds. The direct methods calculate the average pollutant loads as the product of the average flow and concentration measured from the study area, if we assume that the flow and concentration are independent variables. The mean concentration can be found from the literature or derived on the basis of flow-weighting from grab samples or estimated with the help of a statistical method.

Classical statistical methodologies, such as multiple linear regression analysis or discriminant analysis, can be applied to available water quality data for the prediction of pollutant loads form nonpoint sources. In studies where multiple linear regression analysis has been used, the dependent variables were total storm event loads or mean concentrations, and the independent variables were such parameters as climate, land use, topography, and season (Hall, 1984). Finally, the descriptive methods fall into two types: first, loading functions based on the universal soil loss equation (which will be discussed in a later section) that predict sediment loads on which potency factors can be applied to estimate other pollutant loadings; and second, simulation methods allowing for transient effects during storm events, on the basis of the modeling of dust and dirt accumulation and their removal rates by both rainfall and street-sweeping practices. Regarding the prediction of stormwater runoff loadings in an urbanized drainage area, among the more widely known and applied water quality models are the Stormwater Management Model (SWMM) of Lager et al. (1971) and the Storage, Treatment and Overflow Runoff Model (STORM) of the U.S. Army Corps of Engineers (1974). Although the water quality components of these two models are practically the same, STORM provides continuous simulations of storm events over an extended time frame, as opposed to SWMM that handles isolated events. However, STORM has by far the simpler hydrologic component.

Major Quality Parameters of Concern

A detailed overview of pollution parameters stemming from various diffuse sources discharging stormwater runoff contaminated by industrial activities is presented in Sec. 3. The type of pollution and the extent and diversity of contamination, as well as the measures to be undertaken for prevention and treatment, depend of course, on the particular type of activity and the chemicals utilized, manufactured, stored, or transported. In each case, care should be taken to identify and subtract background mineral and organic natural pollution concentrations in order to account for the actual pollutant loading contribution by the industrial activities themselves. Major water quality parameters of concern will undoubtedly, include BOD, suspended solids, pathogens, mineral oil and grease, heavy metals, nutrients such as N and P, insecticides and pesticides, but also all sorts of trace manmade substances from the EPA's priority pollutants list. Of greater concern should be the pollutants that may bioaccumulate and concentrate in deposits, soils, or sludges, to eventually impart long-term and difficult to clean up toxicity in surface and ground fresh water bodies.

2.5 Erosion, Scouring, Sedimentation

Erosion Process and Controlling Factors

The land surface loses soil particles continually and they are transported downstream with overland runoff and in stream flow until deposition occurs in lakes, estuaries, or coastal areas. As reported by Novotny and Chesters (1981), soil particles by themselves comprise a major pollutant, in the form of suspended solids, turbidity, and sedimentation of waterways. Additionally, eroded soil (and especially its fines) is a primary carrier of many other pollutants from contaminated industrial sites: heavy metals, nutrients, insecticides and pesticides, PCBs, and other organic and inorganic toxic substances. Also, large amounts of particulate sediment result in the runoff from pervious and impervious urban areas, and it contains pollutants emanating from traffic, combustion processes, other air pollution sources, and all kinds of urban litter and spills. The rate of erosion and, therefore, its capacity as a carrier of other pollutants are affected and controlled by many factors. Some of the more important include the rainfall regime, vegetal cover of the watershed, soil type and its infiltration capacity, and land slope. On the other hand, industrial activities that may cause excessive erosion will include various agricultural practices, intensive cultivation close to a stream, residential or commercial construction, unstable road banks, surface mining, and animal feed lots close to a stream.

Suspended Sediment Transport

Eroded soil particles and the pollutants they carry move as suspended sediment or washload in the flowing water in overland flow or streams and as bed load that slides and rolls along the channel bottom. The processes are not independent because suspended material at a river stretch may turn into bed load at another. Measurements of sediment movement in lowland, largely agricultural areas indicate that washload may account for 90 to 95% of the total sediment load (Novotny and Chesters, 1981). The transport of washload (the generally accepted limit of particles is 0.06 mm, i.e., clay and silt fractions) depends more on the availability of such sediments from upstream sources and not flow characteristics. However, suspended sediments (including silts and clays) will settle in low-velocity zones in slow-moving streams or lakes and reservoirs, where the transport and settling rates of fine sediments are controlled by flow conditions. Williams and Berndt (1972) proposed an empirical formula for channel sediment delivery based on studies of Texas watersheds and assuming that sediment deposition depends on particle settling velocity

$$DR = exp[BTi(Di)^{0.5}] \qquad (13)$$

where

- DR = delivery; close to unity where transport is controlled by sediment availability; and DR < 1 where sediment transport capacity of stream is exceeded due to flow conditions
- B = routing coefficient determined from field data; range is 4.9 to 6.3, with an average value of 5.3
- Ti = travel time (h) from source of sediment i to watershed outlet
- Di = the median particle diameter of the sediment (mm) for the source (subwatershed) i

The DR factor is applied to the soil loss estimate that can be obtained from the Universal Soil Loss Equation (discussed in a later section). On the other hand, bed-load transport estimates have been based on the equation proposed by duBoys (Linsley et al., 1982)

$$Gi = Y T_0/w (T_0 - Tc) \qquad (14)$$

where

Gi = rate of bed-load transport per unit width of stream (lb/ft - sec) (1 lb/ft-sec = 1.488 kg/m-sec)
Y = empirical coefficient depending on the size and shape of sediment particles (Table 4)
w = specific weight of water (64 lb/ft^3)
T_0 = shear at stream bed (lb/ft^2), equal to WHS, where H(ft) = stream depth and S(ft/ft) = energy slope (1 lb/ft^2 = 4.8824 kg/m^2)
Tc = shear value at which transport begins (Table 4)

A DR-watershed size relationship has been proposed by Roehl (1962) and used as a first-step guide (Fig. 6). Finally, a correlation of DR with channel density and soil texture, proposed by McElroy et al. (1976), is shown in Fig. 7.

Drainage Basin Sediment Production

Average annual sediment production values from a catchment area depend on factors such as soil type, land uses, topography, and the existence of lakes and reservoirs. Stream flow sampling can yield relationships of suspended sediment discharge and water flow, such as the typical sediment-rating curve shown in Fig. 8. With the long-term sediment/flow relationship established, it can be combined with a long-term flow-frequency curve to obtain average annual production values. Data from over 250 drainage areas from

Table 4 γ, Empirical Coefficient and τ_c Shear Values

Particle diameter (mm)	Y		Tc	
	ft^6/lb^2-sec	m^6/kg^2-sec	lb/ft^2	kg/m^2
1/8	0.81	0.0032	0.016	0.078
1/4	0.48	0.0019	0.017	0.083
1/2	0.29	0.0011	0.022	0.107
1	0.17	0.0007	0.032	0.156
2	0.10	0.0004	0.051	0.249
4	0.06	0.0002	0.090	0.439

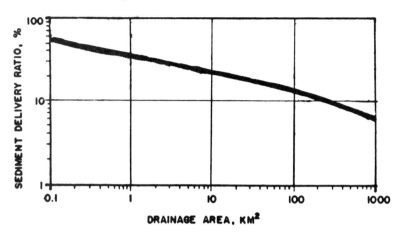

Figure 6 Sediment delivery ratio factor vs. the watershed area. (From Roehl, 1962.)

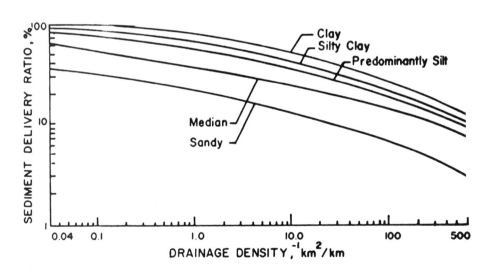

Figure 7 Sediment delivery factor vs. drainage density and soil texture. (From McElroy et al., 1976.)

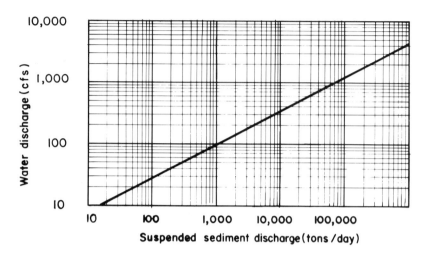

Figure 8 Typical sediment-rating curve.

around the world were analyzed by Fleming (1969) who proposed the following relationship:

$$Qs = aQ^n \tag{15}$$

where

Qs = mean annual suspended sediment load (tons) (1 ton = 907.2 kg = 2000 lb)
a,n = coefficients based on various vegetal covers (Table 5)
Q = mean annual stream flow (cfs) (1 cfs = 28.32 L/sec)

Errors of up to plus or minus 50% may be expected from such relationships and, although they offer an estimate of the order of magnitude of sediment yields, their results should be compared with sediment data on similar watersheds in the same region, if possible (Linsley et al., 1982).

Universal Soil Loss Equation

The Universal Soil Loss Equation (USLE) (Weschmeier and Smith, 1965) has been traditionally used in erosion modeling to predict sediment loads resulting from upland (sheet) soil erosion for a certain time period

$$A = RK\,(LS)\,CP \tag{16}$$

Table 5 Coefficients Based on Vegetal Cover

		a	
Vegetal cover	n	For Qs (tons)	For Qs (metric tons)
Mixed broadleaf and coniferous	1.02	117	106
Coniferous forest and tall grassland	0.82	3,523	3,196
Short grassland and scrub	0.65	19,260	17,472
Desert and scrub	0.72	37,730	34,228

where

- A = soil loss (ton/ha) for a storm event
- R = rainfall erosion factor
- K = soil erodibility factor
- LS = slope-length factor
- C = vegetal cover factor
- P = erosion control practices factor

As discussed previously, this soil loss estimate has to be adjusted by the sediment delivery ratio (DR). An extended discussion of the use of the USLE and estimates of the various factors may be found in Novotny and Chesters (1981) or Wanielista (1978).

3. PATHWAYS FOR CONTAMINATION

Contaminated stormwater runoff or nonpoint pollution accounts for more than 50% of the total water quality problem (Novotny and Chesters, 1981). In many areas, diffuse pollution sources such as runoff from agricultural activities, strip mining, urban stormwater, and runoff from construction sites are becoming major water quality problems. Nonpoint pollution involves not only the usual pollution parameters, but also serious problem contaminants such as PCBs, acid rain, and pesticides that do not have parallel in the traditional point source environmental pollution control. The following discussion presents most of the major nonpoint contaminant sources, with an emphasis on industrial activities, the kinds of problems caused, and some considerations for their control. Figure 9 illustrates generally classified pathways for stormwater contaminants at an industrial site.

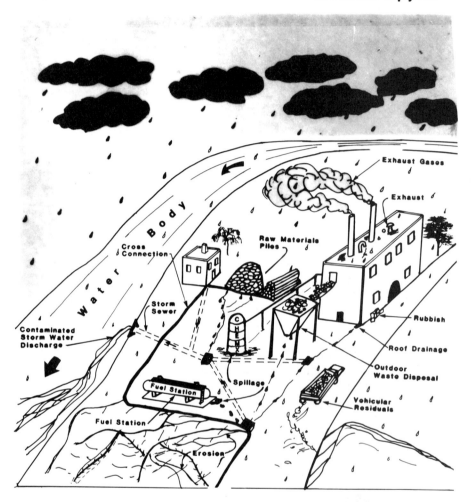

Figure 9 Typical pathways for stormwater contamination. (Courtesy of O'Brien & Gere Engineers, Inc., 1990.)

3.1 Atmospheric Impacts

Atmospheric contaminants, in dissolved, gaseous, or particulate form, enter stormwater runoff through either the process of precipitation or as dustfall; gases also enter by direct absorption at the earth's surface. The deposition rates of particulate atmospheric contaminants in U.S. urban areas vary from 3.5 to

Stormwater Management and Treatment

over 35 tons/km^2/month (Krenkel and Novotny, 1980), and the higher rates are found in congested industrial areas and business districts. In addition to particulate matter, many other contaminants are contained in, transported by, or deposited from atmospheric fall out, such as nitrogen and phosphorus, sulfur dioxide, toxic heavy and trace metals, pesticides and insecticides, fungi and pollen, methane and mercaptans, fly ash, and soil particles. Dustfall rates vary significantly from area to area and are largest in the central United States, with the geometric means of dustfall values ranging from 2.8 to 144 tons/mi^2/month (Wanielista, 1978). Although contaminant concentrations in rural dustfall are related closely to soil conditions, urban dustfall is related more to local air pollution problems. Fly ash from industrial coal-burning activities and disintegration of urban litter is one more important source of atmospheric contaminant contribution, especially in the vicinity of industrial and urban centers (Novotny and Chesters, 1981).

3.2 Acid Precipitation Impacts

One particularly important contamination due to atmospheric precipitation results from the combustion or organic fuels (especially coal) containing sulfur and nitrogen, which appear in gaseous end-products (SO_x and NO_x). These gases react with H^+ in atmospheric moisture to produce acid rain (pH of less than about 5.6, which is the pH of "normal" rainwater in equilibrium with dissolved CO_2). Although the major part of acidity in precipitation is attributed to energy production (power plants), nitrates and therefore NO_x are attributed mostly to agricultural and traffic sources. For instance, while in New York State and parts of New England, the data showed that 60 to 70% of acidity is due to sulfuric acid and 30 to 40% to nitric acid, these proportions are reversed in areas of heavy traffic, such as southern California (Glass et al., 1979). The lowest pH values of precipitation are usually measured near large coal-burning power plants or smelting operations and in heavy-traffic corridors. Since the 1950s, the trend to meet local air pollution standards by building taller stacks has worsened the acidity of rain (longer travel times in the atmosphere, longer reaction times for pollutants) and turned the local problems into regional ones. Acidification of lakes in watersheds with low-buffer capacity soils (lakes in watersheds with low-buffer capacity soils, those lacking $CaCO_3$) has been occurring in the northeastern United States, southeastern Canada, and Sweden. The resulting low pH has caused severe decreases in or elimination of fish populations and has adversely affected the biota of streams and lakes and the terrestrial ecosystems of watersheds.

3.3 Housecleaning and Site Drainage

During normal housecleaning operations inside an industrial plant or at the site, there are intentional or accidental releases of pollutants that may find their way into either surface runoff or the stormwater drainage system of the industrial site. Regular operations in and around a plant involve cleaning up spills, washing vessels and all sorts of containers, washing floors in the production buildings and warehouses. In many cases, the drains may be connected to the stormwater drainage system of the site, thereby causing direct contamination. On the other hand, accidental spills at the parking lots, unloading areas, driveways, and roads within the site and intentional discharges of waste storage and disposal areas provide a variety of pollutants that contaminate the surface runoff originating at the site.

3.4 Raw Material Stockpiles

Bulk materials are often stockpiled outdoors at industrial sites, mining locations, or transportation facilities, i.e., coal stockpiled in coal terminals or power plants. In such cases, several impacts occur from discharges of untreated leachate generated by precipitation and/or contaminated storm water runoff into surface or subsurface water bodies. Contaminants found in these discharges would depend on the nature, purity, and time of exposure of the stockpiled bulky raw materials. There are two pathways, producing two types of contaminated wastewaters. First, when precipitation runoff occurs, it causes particulates to wash off and be carried away from the stockpile surface. Second, rainwater or snowmelt slowly percolates through the stockpile, dissolving some of the chemicals in addition to concentrating pile-bound particulates and appearing as leachate. There is a complex relationship between runoff and leachate, i.e., in the outer crust of the stockpile, it is possible that the two terms become synonymous as runoff percolates below the surface and reemerges to join the main surface streams. The quality of contaminated runoff from a stockpiling area or a bulk material transport terminal, in general, would not meet the federal or state criteria for discharges into a surface water body without treatment. The first approach to the problem should be prevention, collection, reduction of runoff and leachate volume, and the second, treatment of the resulting contaminated stormwater prior to leaving the industrial facility. Sheets of plastic and other material or permanent cover structures should be used for reduction of the volume and degree of contaminant concentration in the stormwater drainage or leachate. Lining of the stockpile areas and installation of collection and containment piping, ditches, berms, and other structures

Stormwater Management and Treatment

could alleviate problems and aid in the subsequent treatment of the stormwater, which will depend on the nature of contamination.

3.5 Spent Material Stockpiles

Many industries have designated locations outdoors where they store, stockpile, or dispose of their wastes. These may include an area adjacent to the buildings where drums with waste are stored or an open pit for liquid wastes, an area within the industrial site where spent solid wastes or powder are stockpiled, and/or an infiltration pond or aerated lagoon where liquid wastes are discharged. In such cases, several pathways of contamination can exist due to precipitation carrying away pollutants through surface runoff or leachate reaching surface or subsurface water bodies. Contaminants in the stormwater runoff would depend on the nature, mixture, and time of exposure of the spent materials and waste disposal systems. The mechanisms of contamination are the same as discussed in Sec. 3.4. Similarly, the first approach to the problem should be prevention, proper storage and disposal of the industrial wastes, collection and reduction of runoff and leachate volume, and the second, treatment of the resulting polluted stormwater.

3.6 Roadway Drainage

There is significant awareness of highway and roadway stormwater runoff as a nonpoint source threatening the quality of water resources. The impact from this source of contamination can generally be separated into changes in the quantity of runoff due to the creation of large impervious areas and changes in the quality of runoff due to a change in the character of the catchment surface involving depositions from vehicular traffic and accidental spills of chemicals. The accumulation on highways and roadways, railroad tracks and yards, and urban streets of materials that can be removed by stormwater runoff, such as high amounts of heavy metals attributed to emissions and to the breakdown of road surface materials and vehicle parts, asbestos from clutch plates and brake linings, motor oil and fuel and gasoline spills, tire and wheel abrasion particles, spills of various chemicals due to traffic accidents, and deicing salts are all contributors of pollution. In general, contaminants found in highway and roadway runoff are similar in type and quantity to those found in stormwater runoff from urban areas and, in particular, industrial zones (Sylvester and DeWalle, 1972).

3.7 Land Use Impacts

The land use effects on stormwater runoff quality depend on the prevailing activities taking place within the area of concern, their intensity, and the resulting nonpoint sources of contamination. The problem of land use and its effect on water quality is primarily associated with urban, industrial, and agricultural developments. In rural areas, animal barnyards and feedlots may cause severe contamination of water bodies, and so can overfertilization and intensive pesticide application in farmland without adequate erosion controls. Land uses usually requiring more intensive control measures typically include industrial areas, mining operations, animal farms, and construction areas. Nevertheless, areas classified into a single land use category can have diverse characteristics, such as topography, soil types, and slopes, and they therefore can generate wide-ranging volumes of flow and quantities of contaminants (Novotny and Chesters, 1981). Several of these land use categories and their impact with regard to stormwater runoff pollution are discussed in the following sections.

3.8 Mining Drainage

Stormwater runoff generated from some mining operations may pose certain serious problems to the quality of water resources, but mining cannot be viewed as a homogenous source of nonpoint contamination (EPA, 1984). Many different minerals are mined, coal and metal ores among them, each causing its own type of pollution problems. Mining nonpoint sources include runoff and leachate from abandoned mines, as well as from inactive access roadways and old tailings and spoil piles, resulting in sediment, salts, metals, and acid drainage discharges. Even though active mining operations also cause similar pollution problems, they are considered to be point source problems and are regulated under state and federal National Pollutant Discharge Elimination System (NPDES) permits (EPA, 1984). In addition, the Surface Mining Control and Reclamation Act (SMCRA) of 1977 includes requirements for the collection and treatment of active coal mine runoff to meet point source discharge criteria.

In strip-mining operations, enormous, bare surface land areas are exposed and the consequence is huge soil erosion yields. On the other hand, acid mine drainage has lethal and sublethal effects on the biota within surface water bodies. Methods for controlling pollution from active mines are available and required by SMCRA for all new mines, the key to prevention being proper site planning. Also, methods are available for solving many contamination

Stormwater Management and Treatment 93

problems related to surface mining, such as regarding land areas and adding topsoil for revegetation in abandoned mines to control the excessive erosion and runoff of pollutants from the area. Additional best management practices would include (EPA, 1984, and Krenkel and Novotny, 1980) the sealing of abandoned mines and/or diverting the surface runoff to reduce drainage contamination; mixing of fine and coarse materials to stabilize mill tailings; equalizing the flow of and treating acid mine drainage by neutralization; compounding of hazardous materials using asphalt or concrete or capping with clay to assure permanent storage and leachate reduction; and containment of leached materials by use of ditches, dikes, and impoundments.

3.9 Construction Site Runoff

Nonpoint source pollution resulting from construction activities has very high localized impacts on water quality. Sediment is the main construction site contaminant, but the stormwater runoff may contain other pollutants such as fertilizers and nutrients, pesticides and insecticides (used at construction sites), petroleum products and construction chemicals (cleaning solvents, paints, asphalt, acids, etc.), and debris. Erosion rates from construction sites may be 10 to 20 times and runoff flow rates can be up to 100 times those from agricultural lands (EPA, 1984). Some of the pollution control methods that could be used are protection of disturbed areas from rainfall and flowing runoff water, dissipation of the energy of runoff, trapping of transported sediment, and good housekeeping practices to prevent the other pollutants mentioned above from being transported by stormwater runoff. Finally, each construction project should be planned and managed by considering drainage problems and contamination, avoiding critical areas on and adjacent to the construction area, and attempting to minimize impacts on natural drainage systems.

3.10 Agricultural Industry

The nature and extent of agricultural nonpoint source pollution are directly related to the way and intensity with which the land is used. For instance, raw cropping usually involves not only a great deal of land disruption, but also the application of fertilizers and pesticides. According to the EPA (1984), therefore, agricultural activities constitute the most pervasive cause of water pollution from nonpoint sources. Actually, pollution from agriculture has various sources, each with different associated impacts, which may be categorized as follows: nonirrigated croplands, both row (i.e., corn and soybeans) and field (i.e., wheat); irrigated croplands; animal production on rangeland and pastureland; and livestock facilities. The latter two activities will be discussed in Sec. 3.11. The discharged contaminants from agricultural croplands include eroded sediments and washed out fertilizers, nutrients and organics from manure applications, traces of pesticides and herbicides, and leached plant residues. Pesticides and other organic chemicals become airborne, especially when sprayed by plane or helicopter, and travel long distances before they precipitate on the earth or directly on surface water bodies. Excessive applications of fertilizers or manure to cropland result in nitrogen, phosphorus, and potassium contributions to runoff that accelerate the eutrophication of lakes and reservoirs and cause high nitrate levels in ground water. On the other hand, irrigation return flows would carry salts and other minerals to surface water bodies or aquifers by percolation, and manure application would contribute bacterial contamination. Since the majority of pollutants from agricultural activities is carried by stormwater runoff, the usual soil conservation practices such as contour farming, strip cropping, terracing, and crop rotation are the most effective ways of controlling pollution. Also, the use of grassed waterways, runoff diversions and retention basins, crop management practices, timing of pesticide and fertilizer applications, and management of quantity and timing of irrigation water are additional pollution control techniques. Finally, manure should be incorporated immediately into the soil and not spread on frozen ground to reduce runoff losses.

3.11 Animal Farming

Rangeland and pastureland used in livestock farming can contribute significant amounts of sediment and nutrients to stormwater runoff, especially when overgrazing is allowed. Rangeland and pastureland erosion is a fundamental problem and, therefore, management practices that maintain or improve the condition of such lands can significantly reduce their erodibility. Manure from livestock contributes nitrogen, phosphorus, potassium, ammonia, fecal bacteria, and other contaminants that pollute water resources. Runoff from more contained livestock areas, such as feedlots and barnyards, contributes much higher concentrations of pollutants, but the NPDES permit program regulates only the large operations (EPA, 1984). Wherever livestock has access to highly erodible areas and water courses, pastures may become a local pollution hazard, unless the animals are kept away by fencing. Finally, good feedlot waste management would include runoff diversions, ponds, and scraping that control liquid and solid animal waste.

3.12 Silvicultural Nonpoint Sources

Virgin or very lightly developed forests and woodlands are the best protection from the sediments and contaminants carried by stormwater runoff. Such areas often generate very small amounts of runoff because their soil can absorb large quantities of water due to high permeability, they can retard and retain surface runoff, and due to tree canopy and groundcover, they can also significantly reduce soil erosion losses. Forest and rangelands contribute contamination that includes sediments, wildlife and vegetation decay end-products, as well as pesticides and nutrients depending on forest management practices. Silvicultural activities are actually comprised of a number of different operations (i.e., road building, pesticide and herbicide application, harvesting and logging, removal of trees from the harvesting site, and preparing the site for revegetation), each of which has a different potential for nonpoint source pollution (EPA, 1984). Clearcutting or uncontrolled logging operations may greatly reduce the resistance of a forest to soil erosion. Observations and records show that almost all sediment contributed to surface water bodies from forests originates in clearcut areas and logging roads (Beschta, 1978), especially roads that disrupt or infringe on natural drainage channels.

3.13 Industrial Urban Areas

Industrial activities in urban areas range from workshops and light manufacturing contributing relatively small amounts of contaminants to heavy and wet industries such as steel mills, cement manufacturing, meat packing, textiles, and beer production. Their discharges of pollutants to water bodies result in minor to major problems and originate from point and nonpoint sources. The latter have been presented in detail in previous sections. The main source of nonpoint industrial pollution is air pollutant release and subsequent deposition within an industrial area, but in general, assessing the contribution of nonpoint pollution from manufacturing activities is quite complex. The extent of soil contamination in certain industrial sites is of such magnitude that it necessitates the collection and treatment of the entire volume of stormwater runoff. Pollutants emanating from industrial areas include heavy toxic metals (such as copper, cadmium, lead, zinc), particulates, inorganic and organic priority pollutants, nutrients, petroleum products and other liquid contaminants from intentional dumping and accidental spills and leaks, either at transfer and manufacturing areas or at storage and waste disposal sites.

Observation and sampling data indicate that particulate and other pollutant loads in industrial areas are, in general, higher than those for residential/

commercial areas. For instance, the sampling program of New York City's Industrial Pretreatment Program (Metcalf & Eddy, 1984) showed that during the one to two months of summer vacation (shutdown of most operations), there was a dramatic reduction (anywhere from 70–90%) in the concentrations of all heavy metals in samples of stormwater runoff from industrial areas. This result signifies that a considerable amount of contaminants attributed to stormwater runoff is still industrial in origin, indirectly contributed through spills and intentional discharges. Similar results were also reported for heavy metals (Cr, Cu, Ni, Pb, and Zn) in urban runoff by Wilber and Hunter (1975), who found significant increases in concentration when land use percentages increased for industrial use from 5 to 12% and for commercial use from 42 to 58% (residential land use decreased from 53 to 30%).

3.14 Landfill Runoff and Leachate

The usual method of industrial solid waste disposal involves landfilling either in sanitary or secure landfills. Nevertheless, prior to the 1970s when landfill leachate was recognized as a potential source of groundwater and surface water pollution, only about 25% of the landfills in the United States were classified as sanitary, and none as secure landfills (Novotny and Chesters, 1981). As previously discussed in Sec. 3.4 and 3.5 with regard to stockpiles of raw and spent materials, the generation of leachate results from the infiltration and percolation of rainfall, stormwater runoff, and groundwater through an operational or closed solid waste landfill. In general, it is not possible to exactly predict leachate characteristics because of variable landfill constituents; however, leachate contamination may severely pollute neighboring ground and surface waters. Solid waste disposal sites produce a highly mineralized leachate containing pollutants such as heavy toxic metals, chlorides, nutrients, volatile organics, cyanides, chlorinated hydrocarbons, and other priority organic pollutants. Techniques of leachate pollution control would include minimization of leachate formation by reducing water infiltration; collection and treatment; detoxification or immobilization of hazardous industrial wastes prior to landfilling; development of landfill sites on uplands, not on floodplains, and in soils of low permeability.

4. MANAGEMENT

No program for management would be adequate without first considering the strategy by which to avoid the initial problem of handling contaminated

Stormwater Management and Treatment

stormwater. That is, first, plan for the construction and operation of facilities that seek to avoid the contamination of stormwater through inadvertent contact with (1) raw materials, (2) manufacturing processes, (3) finished products, and (4) waste materials present at the industrial site. Second, when stormwater is likely to contact contaminants at a site, measures for minimizing such contact to the best practical means are recommended to reduce the quantity of contaminated stormwater that may be generated. The following sections describe the elements of planning and implementing a strategy for contaminated stormwater management. When treatment becomes a necessity, Sec. 5 and 6 of this chapter should be reviewed, as well as the many chapters on treatment processes that comprise this handbook.

4.1 Site Planning and Practices

The goal of the planning process is to reduce the incidence of contamination of stormwater. In instances where it is not practical to eliminate the contact of stormwater with contaminating substances, the secondary goal of minimizing the exposure and containment of contaminated stormwater should be reached. In this case, the quantity of contaminated stormwater would be minimized.

Volume/Rate Reduction

The volume of stormwater to be exposed to contaminants can be reduced through the minimization of impervious areas. Open areas where rainfall does not contact with potential sources for contamination could be graded to minimize erosion. In addition to grading, ground cover should be utilized, which would impede the velocities of sheet flow as well as aid in the retention and subsequent evaporation process. Through increased retention by ground cover, flow reduction would be furthered through the increase in the quantity of percolation. Nonpoint drainage should be channeled to discharge into natural water courses, wetlands, or areas for seepage.

Through the use of porous pavements in yard areas requiring hard surfaces, it may be possible to maximize the natural drainage of uncontaminated stormwaters. In areas where the underlying soils might be impermeable, an underdrain system may be necessary or conventional collection systems could be utilized. More important for paved or otherwise impervious areas is to identify areas that are not exposed to sources of contamination and to route the stormwater collected from such areas separately from areas that may have contaminated stormwater. In so doing, the quantity of contaminated stormwater to be later handled will be reduced.

Another means for reducing the rates of flows with subsequent volume reduction is through the use of storage or retention areas. This method is elaborated in the following subsection.

Storage/Retention

The storage/retention of stormwater can be utilized for (1) flow reduction, (2) flow equalization, (3) source control, and (4) increasing the effectiveness of natural drainage. Storage/retention can be provided economically when flat rooftops are available, where drainage can be routed to natural or manmade depressions (basins) or through the use of dikes or earthen berms. Limited storage can be provided by taking advantage of the volume within stormwater sewers/pipelines when available (ASCE and WPCF, 1969).

Flow equalization and flow reduction can be utilized to reduce the possibility of erosion, which would otherwise occur through the discharge of stormwater during peak rainfall intensities. Basically, the storage capacity provided would act as a shock absorber to decrease the peaking effects through dampening. The discharge of stormwater from storage/retention areas could be controlled hydraulically through the use of weirs, orifices, or tipping devices.

The discharge of stored flows at controlled rates on open lands could facilitate the percolation process, as well as deliver stormwater to the points of discharge at a lesser rate. Either way, the impact on the discharge water course would be reduced, as opposed to receiving the stormwater at high rates. In the event that stormwater treatment is to be provided, storage/retention would help decrease the size of facilities by decreasing the peak hydraulic requirements.

Finally, storage/retention of stormwater is useful in the segregation of flows from contaminated and uncontaminated areas. Stormwater could be captured and conveyed separately from areas that may contain sources of contamination. Many site-specific issues have to be examined prior to the use of storage/retention. The actual case-by-case intent and goals would govern the exact methodology to be implemented.

Housekeeping Practices

Good housekeeping practices can have a positive effect on the minimization or elimination of sources of stormwater contamination. In general, a program to keep yard areas and stormwater collection and conveyance items clean and free of refuse, raw materials, products, solids, or other materials that may contaminate stormwater should be maintained.

The routine cleaning of parking areas, storage yards, stockyards, streets, and sidewalks to remove debris and potentially harmful materials should be

Stormwater Management and Treatment

performed. Catch basin and storm sewer cleaning should be performed as needed to remove deposits that can contaminate stormwater and/or decrease the capacity of the conveyance system.

When contaminating materials come into contact with areas subjected to stormwater, these materials should be cleaned up prior to a storm. Equipment and cleaning materials should be kept on site where hazardous materials may be utilized. Workers should be trained in the procedures necessary to provide a proper clean-up or decontamination of areas where contaminating materials have been inadvertently placed.

4.2 Cross Connections

Cross connections include the connection of sanitary drains, process/production area drains, or floor drains from material/product area drains, or floor drains from material/product areas to separate storm sewers. These types of connections can be found in existing older facilities, as well as those that have been renovated or expanded. Such connections are generally illegal and cause the contamination of stormwater. These connections may also result in the discharge of contaminants during dry weather.

When planning a stormwater management program, a check for cross connections can be made by examining storm sewer lines during dry weather. If any flow is present in a storm sewer during dry weather, it very likely may be due to cross connection(s) or infiltration sources. Infiltration is the flow of groundwater into a storm sewer through cracks in manholes or piping, when the groundwater table is above the storm sewer/structures. In order to ascertain whether dry weather storm sewer discharge is likely to be from a cross connection or groundwater source, an analytical testing program would need to be implemented.

Upon the determination that cross connection(s) are present, the identification of points connected would need to be made. Dye or smoke testing could be used to determine the points of connection. Actual protocol for locating cross connections would vary case by case.

4.3 Spill Prevention Program

Spill prevention programs are most utilized in the hazardous waste field, where specific measures have been implemented for the prevention/containment of hazardous material spillage. Spills can occur due to (1) the rupturing of a storage vessel, (2) overfilling a vessel, (3) breaks in conveyance pipelines or equipment, and (4) inadvertent discharge via an open drain connection. The majority of spill prevention protocols developed to date have been in the area of liquid spill prevention; however, many of the same concepts apply to solid material spills, where such solids are stored in vessels as well (hoppers, silos, etc.).

When potentially contaminating liquids or materials are stored either in vessels (liquids/solids) or stockyard piles (solids) subjected to stormwater, spill prevention and containment should be implemented. The measures to be implemented should include both the storage area and areas designated for the delivery of liquids/materials (transfer or unloading areas).

Areas where delivery vehicles will be parked during the transfer of liquids/materials should be within a diked area or otherwise enclosed area where drainage can be isolated in the event of a spill. Drains from these areas should be normally closed and routinely opened to allow uncontaminated stormwater to be released and spillage or contaminated stormwater to be collected. In the event this area becomes contaminated, the extent of contamination will be limited to the delivery area from where stormwater can be pumped out and decontaminated.

The storage vessels subjected to stormwater should be contained within a diked area or containment wall/tank suitable to contain the contents of the vessel should a rupture of the vessel or interconnecting piping occur. Similarly, this would limit the quantity of stormwater that may be contaminated and facilitate the collection and disposal of contaminated water. As an example, aboveground oil storage facilities have been regulated by state legislature, local regulations, and fire codes. These provide an excellent example of preventive measures and generally offer protection for stormwater.

5. ON-SITE VS. OFF-SITE TREATMENT

Depending on the regulatory requirements governing the need for treatment of contaminated stormwater, along with the possible availability of off-site treatment, a decision as to whether to provide on-site or off-site treatment will need to be made. Therefore, prior to proceeding, an industry should (1) investigate whether or not off-site treatment is available, (2) identify if off-site facilities provide the level of treatment required, (3) ascertain tipping fees or costs to be incurred, and (4) determine the reliability of the off-site facility. Once options have been identified, the direction to be taken will be a function of reliability, environmental soundness, regulatory acceptability, economics, and company policy.

Stormwater Management and Treatment 101

Off-site treatment facilities include (1) municipal wastewater treatment plants, (2) private residential wastewater treatment plants, and (3) private commercial wastewater treatment facilities. The ability of any of these facilities to provide industrial stormwater treatment would be a function of its available treatment capacity and regulatory permit requirements.

Small industrial sites may be able to truck collected contaminated stormwater to a treatment facility, whereas larger sites would need to construct conveyance pipelines. Depending upon the proximity of the off-site facilities, conveyance requirements and associated costs would need to be considered. In certain instances where several industries requiring treatment are in close proximity of one another, a central off-site treatment facility should be considered. In such cases, privatization of such a facility may offer an advantage to the utilizing industries.

The above discussion briefly presented a few of the various considerations to be weighed prior to making a decision on off-site vs. on-site treatment. In fact, many other factors need to be considered on a case-by-case basis. In particular, the quantity of stormwater to be treated and the needs for bypassing flows are to be considered as well. In order to properly evaluate the requirements and measures to be implemented, a detailed study of the needs for treatment may be a necessity.

6. CONTAMINATED STORMWATER TREATMENT

Treatment of contaminated stormwater varies with the type and degree of contamination found to be present. As a general guide to contaminated stormwater treatment, the numerous books and reports describing treatment processes and specific contaminant handling should be referenced. Books describing unit operations and processes that may be applicable in the handling of specific types of pollutants in industrial wastewater might be especially helpful. Hence, this section will not attempt to reiterate the wealth of information contained in the volumes of this handbook, but rather to describe basic processes and concepts applicable to contaminated stormwater treatment.

6.1 Batch vs. On-Line Treatment Capability

The variability of stormwater flow due to varying storm intensity (in./hr) and magnitude (total cumulative inches of precipitation) makes the sizing of treatment units a more complicated matter as compared to sizing such units for sanitary wastewater flows. One consideration in designing a system is with respect to handling the contaminated stormwater (CSW) on a batch or on-line basis. Batch treatment is a method of treating contaminated water in treatment units on a cyclic operational basis. A specific volume and CSW would enter the units at a designated rate and reside in the treatment units a defined length of time while undergoing treatment. Following the treatment of the first volume (batch) of CSW, subsequent volumes would be processed until the total volume has been treated. Storage is inherent to this type of treatment scheme since the entire volume of CSW generated by a storm may not likely be treated by a single batch run. Hence, storage may be necessary to contain CSW until it can be processed.

On-line treatment capacity refers to a treatment system that is designed on a rate per unit time basis. Unit processes would then need to be designed to operate at a rate equal to the rate at which CSW is generated or through the use of flow equalization some lesser throughput rate. Flow equalization in this case would be provided by a storage basin designed to contain the CSW sent to the plant that would exceed the plant's capacity. In most cases, a simple overflow weir arrangement could be used to route flows in excess of the design capacity of the on-line treatment system to the flow equalization storage basin.

Whether batch treatment, complete on-line capacity, or on-line capacity plus flow equalization is to be utilized is a function of (1) the quantity of CSW to be handled over time, (2) space requirements for each system, (3) reliability, (4) ease of operation, and (5) overall capital and operating costs. Since requirements will be different for each application based on design storm criteria, treatment stipulations, and contaminated stormwater characteristics, judgment as to which methodology should be utilized will vary on a case-by-case basis.

6.2 Treatment Processes

Although depending on the characteristics of contaminated stormwater a variety of treatment methods may be applicable, for the purposes of this section the eight most promising processes are discussed broadly. These include (1) skimming, (2) screening, (3) sedimentation, (4) concentration, (5) filtration/straining, (6) dissolved air flotation, (7) biological, and (8) chemical treatment. The following subsections provide a brief description of these treatment methods.

Skimming

Skimming can remove a wide variety of floatable materials including wood, paper, plastics, styrofoam, grease, oil, and other buoyant materials. In order to accomplish skimming, a basin of adequate size (10–60-min retention time) is

Stormwater Management and Treatment 103

necessary to allow for floatable material to surface. Normally a barrier wall or partition is helpful at the surface and extending into the water columns to help trap the floating material. The stormwater entering the tank would flow beneath the wall/partition while floatables would collect upstream. Through the use of an adjustable weir or a slotted pipe (skim trough) located on the upstream side, the floatables can be removed from the surface of the basin. Where sedimentation is being utilized for solids removal, skimming can be incorporated into the same tankage.

Skimmings can be diverted to an overflow structure where grease and oils can be removed, as well as solid floatable materials. Following the removal of floatable solids through mechanical separation utilizing a rake, basket, or conveyor or other means to remove the solids from the liquid, the remaining floating greases and oils can be drained by gravity or pumped from the surface of the stormwater. The remaining stormwater can be returned to the head end of the skimming tank. When floatable solids are not as present, a commercially available oil/water separator can be utilized. Approximately 80 to 90% of the grease and oils should be removed as well as 80 to 95% of the floatable solids. In certain cases, capture may be enhanced by injecting air at a low rate to the bottom of the skimming basin.

Screening

Various types of screening equipment are available to remove debris that may be too large for skimming or material that cannot be skimmed. In their simplest form, round and rectangular bars arranged in a parallel 30- to-60-deg incline framework within a channel (bar screens) can provide for screening. Debris will deposit on these bar screens and can be removed manually with a rake. Mechanically cleaned screens are available that can either be run for continuous cleaning or put on a timer for discontinuous screening. Bar spacings vary between 10 to 75 mm, depending on application. In practice, several sets of screens in series have been used successfully. Normally, coarse screens, with 50- to 75-mm spacing, preceding medium screens, 10 to 25 mm, can provide a significant degree of screening.

Sedimentation

Grit, settleable and suspended solids, can be removed from contaminated stormwater using the sedimentation treatment process. This process relies on gravity to remove particles and materials. Depending on treatment and solids-handling requirements, it may be desirable to remove grit separately from suspended and settleable solids. Normally, grit may require mechanical means of conveyance, while suspended and settleable solids can be pumped to a dewatering device or truck.

A typical sedimentation tank for suspended/settleable solids removal may be rectangular or circular in configuration, with a water depth of 10 to 12 ft and surface areas varying on the basis of particle settling characteristics to achieve from 300 to 1200 gpd/sf surface loading rates. In certain cases, chemical flocculants can be added to enhance the settling of suspended solids. Removal efficiencies expected for settleable solids range from 30 to 90%. Suspended solids removal efficiencies range from 20 to 60% without chemical addition and 60 to 70% with chemical addition. Alum, ferrous sulfate, lime, and polymers are examples of chemical additions that may enhance settling.

In addition to the removal of settleable and suspended solids, sedimentation will remove a portion of the biological oxygen demand (BOD) associated with organic solids removal. Hence, 25 to 40% BOD removal may be achieved without chemical addition, whereas a slightly higher amount may result with chemical addition. Other compounds and contaminants such as metals and pesticides will be removed in proportion to their presence in the solids removed.

Sedimentation basins/tanks can double as storage tanks. This would be possible through the dewatering of the tank once flows to it have ceased. During the start of a storm, the tank would begin to fill up. Overflows from the tank would therefore be treated through sedimentation. As previously mentioned, the removal of floatables could be incorporated into the design. Figure 10 illustrates a conventional system for sedimentation.

Concentration

Various types of pollutant concentrators are known as methods for treating contaminated water. These devices through circular, spiral, or helical motion impart secondary fluid motion through long-path geometric flow patterns, thereby enabling their separating solids from the contaminated water. The secondary flow stream containing the solid particles entrained is collected and conveyed separately from the main water throughput. Hence, the solids can be conveyed for further treatment or degritting/dewatering. Three such devices currently applicable to stormwater treatment include the vortex regulator, swirl concentrator, and spiral flow (helical bend) regulator. Each of these three devices is illustrated in Fig. 11.

The vortex regulator was developed in England and basically consists of a circular channel. The flow enters at the periphery and exits at the center of the device. The circular flow pattern causes the solids to move toward the center where they are collected and removed.

Stormwater Management and Treatment

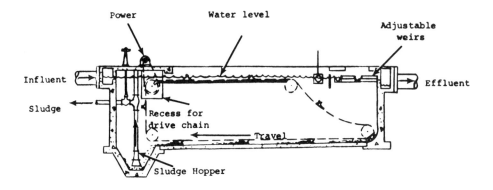

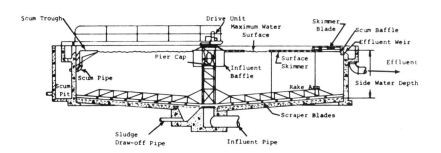

Figure 10 Rectangular and circular sedimentation basins. (From EPA, 1980.)

Similarly, with the swirl concentrator the flow enters at the periphery and exits toward the center. Solids following the secondary flow stream are collected separately. A scum ring of baffle at the surface of the swirl concentrator can be used to remove floatable materials.

The spiral flow regulator, otherwise known as a helical bend device, is also illustrated in Fig. 11. Through the use of an arc, this device causes the

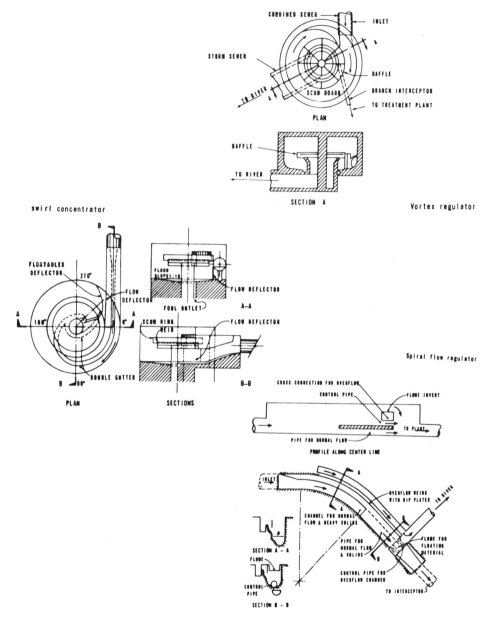

Figure 11 Vortex regulators, swirl concentrator, and spiral flow (helical bend) regulator. (From EPA, 1974b.)

development of secondary fluid motion, thereby separating the stream of solids from the main flow stream. The bend angle normally is between 60 to 90 deg.

Of these three devices, the vortex regulator and swirl concentrator appear to have better application for the removal of solids from contaminated stormwater flow found at industrial facilities. Although all three devices provide treatment, the helical bend device is more suited to combined sewer overflow (CSO) handling rather than solely contaminated stormwater. Hence, it should be considered when combined sewer systems are utilized.

The vortex regulator and swirl concentrator are applicable outside of a combined sewer system. However, when they are used solely for contaminated stormwater treatment, consideration should be given to the discharge of the solids flow stream to a sanitary or combined sewer if allowable. In other cases, the secondary solids stream will have to receive additional treatment or degritting/dewatering. Reportedly (EPA, 1974b), up to 98% suspended solids removal for vortex regulators may be possible. Solids removal efficiencies for swirl concentrators (Wanielista, 1978) range from 40 to 60% for suspended solids and 50 to 90% for settleable solids. Generally, these devices tend to offer an economic advantage over sedimentation or straining.

Filtration/Straining

Various types of filtration/straining equipment are available. Some of these include gravity filters, pressure filters, microstrainers, drum screens, and disk screens. Filtration generally uses a filter media or fine fabric to separate solids from contaminated water. Straining generally utilizes a wire mesh or coarse fabric to provide a lesser degree of removal efficiency. Strainers provide a removal efficiency ranging from 10 to 55% for suspended solids and 60 to 95% for settleable solids depending on screen sizing. Filtration equipment including microstrainers (depicted in Fig. 12) can provide 50 to 95% suspended solids removal. Surface loading rates for filtration and fine media microstrainer units range from 2 to 10 gal/min/ft^2. Rotary drum strainers and high-rate coarse screen microstrainers have surface loading rates ranging from 20 to 100 gal/min/ft^2. Specific performance details and sizing requirements can be obtained from manufacturers.

Dissolved Air Flotation

The flotation process is useful for removing solids, oils, and grease from contaminated water. Dissolved air is released into the influent contaminated water that enters the flotation tank. The dissolved air bubbles attach to solid particles, oil droplets, or grease, thereby decreasing their specific gravity.

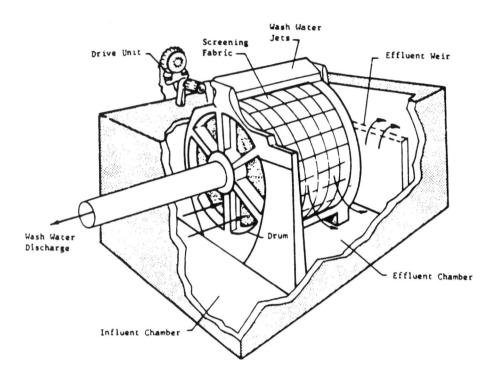

Figure 12 Microstrainer. (From EPA, 1980.)

Through buoyancy, these materials float to the surface of the tank where they can be skimmed. Figure 13 illustrates a typical dissolved air flotation unit.

Removal of suspended solids using the dissolved air flotation process ranges from 40 to 65% without the use of chemicals and 80 to 93% with chemicals such as alum, ferric chloride, or polymers. Oil and grease removal ranges from 60 to 80% without chemical addition and 85 to 95% with chemical addition. Common surface loading rates range from 500 to 8000 gal/d/ft^2. Imperial Oil Company, Inc. presented a case history for its treatment of storm runoff by dissolved air flotation (Wang and Mahoney, 1989; Wang et al., 1989).

Biological Treatment

Generally, the colloidal and soluble substances found in contaminated stormwater can contain biological oxygen demand (BOD) that cannot be

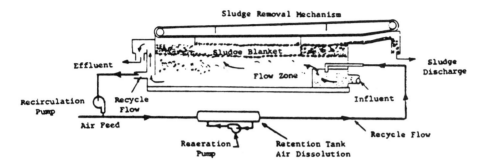

Figure 13 Air flotation thickener. (From EPA, 1974b).

removed through sedimentation or filtration/straining. In order to remove these substances biologically, the stormwater would need to contact with microorganisms that through metabolic processes remove the BOD. Unlike wastewater treatment, in which a continuous supply of waste is available to be treated, the discontinuous supply of contaminated stormwater presents a unique problem that needs to be addressed in order to make biological treatment applicable. The biological treatment process relies on the supply of contaminants within the wastewater as a food source to keep the microorganisms alive and thriving. Hence, the lack thereof during dry weather could cause the failure of such a treatment system.

Several methods of maintaining a biological treatment system during dry weather include (1) during dry weather using the system to treat sanitary flows, (2) using a combination of storage and treatment to maintain the system, and (3) if another biological treatment system is on-site, using the waste biological solids to maintain the level of microorganisms. In the first case, the substitution of sanitary flow for storm flow, if adequate sanitary flow is present, would maintain the biology of the system. During dry weather, the treated sanitary discharge could be made to the existing sanitary municipal system if available.

In the second case, storage is used along with a biological treatment system sized to work in a manner that keeps it running during dry periods. This treatment system would operate during dry weather using the contaminated stormwater captured during wet weather within the storage tank. Naturally, the system would operate at average to peak capacity during a storm.

In the third case in which a sanitary treatment system is on-site or nearby, the level of active organisms would be maintained by importing organisms from the other facility. This method would likely be subject to the operation of the facility supply, the microorganisms, and may not be as reliable as the other two strategies.

Biological systems that may have better application for contaminated stormwater treatment include trickling filters, rotating biological contactors, and aerobic/facultative lagoons. These are illustrated in Figs. 14 through 16. In certain cases, land/wetland application may be applicable as well. All of these systems have the common element of long residence time for solids and ability to maintain biomass. Expected treatment efficiencies range from 75 to 95% BOD_5 reduction. Details for the performance of these systems are widely available in EPA publications and other chapters in this handbook.

Chemical Treatment

Chemical treatment includes (1) the use of flocculants, and coagulants to augment the sedimentation or flotation process; (2) carbon adsorption; (3) ion exchange; and (4) chlorination for disinfection. The following is a brief description of each of these processes.

Flocculants are chemicals that facilitate the massing together of particles into a clump known as a floc. Flocs tend to settle faster than the individual particles, thereby increasing sedimentation efficiency. Flocculants can be either organic or inorganic in origin, and they include activated silica, certain clays, fine sands, and trade products such as magnafloc, purifloc, and superfloc.

Coagulation is the process by which colloidal particles are destabilized. Normally, colloidal particles are charged and tend to repel each other, thereby maintaining themselves in suspension. Through the use of a coagulant, the electrical charge of the particles can be neutralized. Once neutralized, the particles will no longer repel one another and begin to coagulate/flocculate and settle (precipitate). Coagulates include such compounds as aluminum sulfate ferric chloride, ferric sulfate, and copper sulfate.

Activated carbon has the ability to adsorb contaminants from the stormwater. This adsorptive property lends itself to the removal of soluble BOD, organics, and toxic substances. Carbon can be used in powdered or granular form. In the powdered or granular form, it may be added to the contaminated stormwater directly and in a mixed contacting environment remove impurities. The granular form can also be used in packed bed or as a filtration media, whereby the contaminated stormwater may be passed through and impurities adsorbed. Activated carbon is also somewhat effective in removing various inorganic contaminants, e.g., copper, lead, cyanide, and chromium.

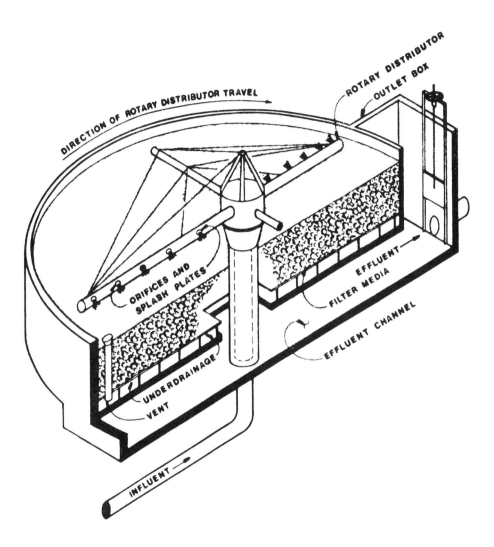

Figure 14 Trickling filter. (From EPA, 1977.)

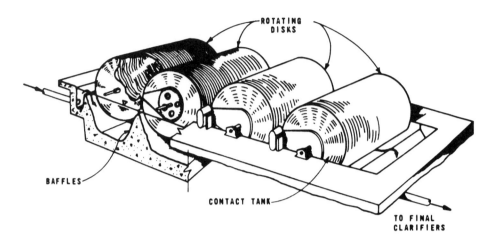

Figure 15 Rotating biological contactors. (From EPA, 1974b.)

Ion exchange is only suitable for use after all particulate matter has been removed from the contaminated stormwater. This is due to the subsequent plugging of the ion exchange media that would result if this matter is not removed. Ion exchange media come in two types: namely, cationic and anionic. Cationic media are typically used to remove positively charged ions, whereas anionic media remove negatively charged ionic substances from the water. Generally, contaminants from electroplating and metal-finishing industries may be effectively removed using the ion exchange process (e.g., cadmium, chromium, and nickel).

Chlorination is but one means for disinfecting contaminated stormwater. The object of the disinfecting process is basically to kill pathogenic bacteria. This disinfecting method has been known to be the most common and economical method of choice in the wastewater treatment field. However, methods should be investigated that may, on a case-by-case basis, offer some advantage in certain instances. Conventionally, a 15-min contact time has been used to facilitate the required kill; however, the contact time can be decreased by increasing the intensity of mixing as well as the dosage of chlorine.

Common varieties of chlorine used for chlorination are chlorine gas, chlorine dioxide, and sodium or calcium hypochlorite. Chlorine gas, though a better disinfectant, is hazardous and, therefore, not the first choice. Both chlorine dioxide and sodium hypochlorite have found wider use in chlorination.

Stormwater Management and Treatment

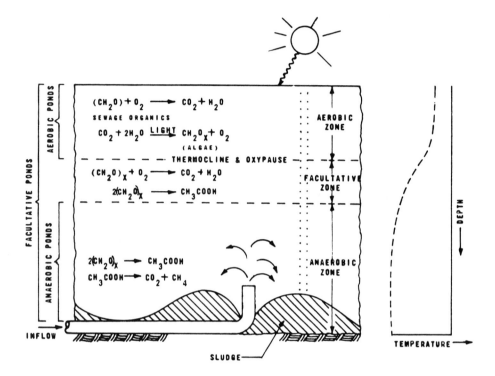

Figure 16 Facultative lagoon. (From EPA, 1974b.)

7. CASE STUDIES

In the following, the information and conclusions from several papers relevant to stormwater pollution from industrial activities and facilities are summarized as typical case studies of what has already been discussed in this chapter.

7.1 Hazardous Material Spill Control

Mandel (1987) reported that no convenient method exists to neutralize and solidify spills of acids and bases and to render these materials as noncorrosive solids. Also, there is no convenient method for adsorbing common fuel and organic solvent spills and for elevating their flashpoint to above 60°C (140°F.) Furthermore, to date there is still room for improvement of the methods for application of clean-up chemicals at a safe distance from a spill in order to minimize the risk to emergency personnel. Present techniques of spill clean-up utilize mostly manual application methods that, in most cases, only absorb the spilled materials or employ time-consuming applications of neutralizing non-solidifying materials. In most instances, therefore, the end result of the clean-up effort yields a waste material that is still considered hazardous and has to be handled and disposed of as such.

In a market survey of persons employed in U.S. industries utilizing or producing hazardous materials, with responsibilities in environmental quality, environmental engineering or management, and industrial hygiene and safety, the following information was collected from the 1800 respondents. Some of the major industries surveyed were fertilizers and agricultural chemicals, chemicals and petrochemicals, explosives, drugs and cosmetics, food and beverages, fats and oils, plastics, paints, petroleum refineries, fibers and textiles, soaps and detergents, and semiconductors. The respondents listed the major areas in which spills occurred at their facilities and the percentage of the number of spills in each area as follows: shipping 9%, receiving 8%, warehouse 11%, manufacturing 30%, process piping 14%, tank storage 17%, labs 10%, and off-site transport 4%. The average spill size in these typical industrial facilities was determined to be 11.2 gal, with a spill frequency per plant being eight spills per year. Interestingly, the number of spills reported above 55 gal (a standard drum) was statistically insignificant.

The respondents indicated that, in their view, the spill incidents were unpredictable and posed a threat to the environment and safety and caused business interruption and profit loss. They further reported that in the surveyed plants the hazardous materials mainly fell into three broad chemical classes: acids, bases, and organics. Therefore, the bulk belonged to the corrosive and ignitable classes of hazardous materials. According to the information provided in the questionnaires, current methods of spill clean-up included containment, soak-up or absorbing, neutralizing, increasing viscosity, or diverting to a sump. Most of these methods were considered time-consuming and dangerous.

7.2 Overland-Flow Treatment of Feedlot Runoff

The results of a six-month pilot-scale study in Oklahoma (EPA, 1974a), which was followed by a six-month field test, indicated that the overland-flow method of wastewater treatment has potential for treatment of runoff from beef cattle feedlots in warm, subtropical, humid regions. Confinement feeding of beef cattle is used extensively in the plains states of the United States, and rainfall runoff from feedlots has been identified as a major cause of fish kills in this

Stormwater Management and Treatment

region. Therefore, this runoff needs to be retained and treated before it can be discharged to surface water bodies without serious impacts, because it contains high concentrations of suspended solids, organics that create BOD, and nutrients that accelerate eutrophication. However, retention and treatment of the runoff at cattle feedlots located in this subhumid area of the plains states are major problems.

The application of wastewater to the land for reuse or treatment can be an efficient and economical management approach when land is readily available, as it is in the vicinity of many feedlot operations. One method of land-based wastewater management that has been used successfully is the overland-flow treatment under intermittent or continuous use. The system can be managed to achieve the efficient removal of suspended solids, BOD, and nutrients from concentrated wastewaters.

The pilot-scale study was conducted on plots 9-m (30-ft) long with a 4.5% slope, where the native vegetation of mixed grasses was left undisturbed. The study showed that liquified loadings of 5 to 7.5 cm (2–3 in.)/week applied at instantaneous loading rates of less than 2.5 mm (0.10 in.)/hr appeared suitable for field testing with dosing frequencies in the range of daily to three times per week. The field study was conducted at a 12,000-head capacity feedlot and utilized a four-component train for runoff collection and treatment. The treatment train included, as can be seen in Fig. 17, collection lagoons, a storage reservoir, the overland-flow areas, and a final polishing pond. Data from the six-month operation are summarized in Tables 6 and 7, corroborated the results of the pilot-scale study, and indicated that inclusion of the final polishing pond substantially improved the overall performance of the treatment system.

7.3 Petroleum Refinery Runoff Management

Stalzer and McArdle (1983) reported on a SOHIO Toledo refinery, 2 km^2 (500-acre) area, stormwater runoff management study in which nonsegregated oily sewer system collected both process wastewater and stormwater runoff. The handling of runoff at such a large petroleum refinery site requires adequate diversion and impounding capacity to store excessive storm flows for treatment after the runoff subsides. The paper presented the case history of upgrading the refinery's wastewater treatment system to efficiently manage the runoff while maximizing overall treatment efficiency. One of the primary objectives of the study was to estimate the peak flow rate and total impounding volume required for design storm events using the Illinois Urban Drainage Area Stimulator (ILLUDAS) for stormwater runoff computer modeling.

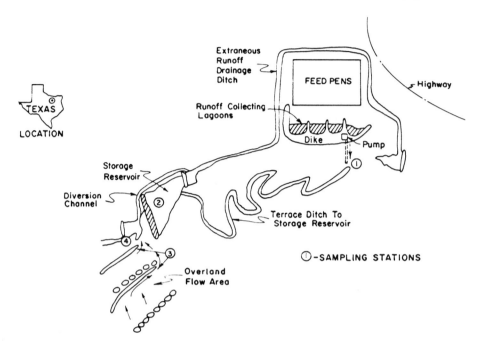

Figure 17 Feedlot and treatment train. (From EPA, 1974a.)

Table 6 Water Quality Data for the Runoff-Collecting Lagoons

Date	Parameter concentration (mg/L)					
	TDS	TSS	COD	BOD	T-P	T-N
10-1-70	1316	174	620	165	21.3	39.9
11-4-70	1122	107	314	15	17.0	14.7
12-2-70	1059	216	386	23	11.1	18.9
1-5-71	959	267	396	—	11.3	25.4
2-4-71	1128	174	403	65	13.6	28.1
2-24-71	1169	136	298	31	12.6	35.6
3-25-71	1046	292	569	80	7.9	31.1
Mean	1114	195	426	63	13.5	27.7

Source: EPA (1974a).

Table 7 Water Quality Data for the Farm Pond Discharge

Date	Days of operation	Parameter concentration (mg/L)					
		TDS	TSS	COD	BOD	T-P	T-N
10-1-70	0	342	10	78	2	0.4	4.4
11-4-70	35	437	19	71	2	0.2	3.5
12-2-70	63	477	11	92	2	0.2	3.6
1-5-71	97	848	24	149	5	1.0	6.7
2-4-71	107	832	8	166	5	0.5	6.3
2-24-71	127	780	6	134	6	0.4	5.1
3-25-71	158	874	6	183	12	1.0	8.0
Mean	—	656	12	125	5	0.5	5.4

Source: EPA (1974).

By establishing a hydraulic profile of the oily sewer at the wastewater treatment system, the diversion and impounding system, and primary wastewater treatment units, it was determined that the influent weir to the oil/water separator controlled the water level in the sewer for a considerable distance upstream. A water level sensor at a lift station downstream of the oil/water separator was used to monitor changes in flows and determine the sequence for opening the diversion valves. This was based on the lift pump having a maximum capacity of 0.5 m^3/sec (8000 gpm). For use as a design basis, the 25-year, 1-hr and 10-year, 24-hr storm events were chosen to establish the design peak flow rate (4.3 m^3/sec or 152 cfs) and total impounding volume (based on a peak rate of 1.93 m^3/sec or 68 cfs), respectively. To determine the required impounding capacity from the 10-year storm hydrograph, the amount of the runoff that can be treated during the storm was determined based on the maximum allowable wastewater treatment system flow rate and the dry weather flow rate. This flow rate was then subtracted from the hydrograph of the storm, with the area under the resultant hydrograph equal to the required impounding volume.

Because of the variability in dry flow rates, the base flow used to calculate the required impounding volume was not the average, but a 90% maximum dry weather flow based on four years of data to ensure some conservatism in the design. Even though the chance of a design storm actually occurring at the same time as the 90-percentile dry flow rate may have low probability, the additional cost for extra impounding volume must be weighed against the liabilities of not having sufficient volume. On the other hand, to enhance the conservative nature of the overall design, peak flows used in designing for diversion into the impounding basin and overflow from the basin under flood conditions were based on the 100-year storm. As flows increase in the oily sewer due to stormwater runoff, a dam in the sewer with a flow meter and valve combination allows a preset maximum flow to the wastewater treatment system. As a result, the water level upstream of the dam increases until it begins spilling over a diversion weir into the impounding basin, thereby maintaining a constant optimum flow rate through the treatment system.

7.4 Pesticide Loads in Surface Runoff

Li et al. (1990) developed two simple models or loading functions for estimating mean annual pesticide mass loads in surface runoff from agricultural areas. The loading functions were regression equations derived from 100-year simulation runs of a daily pesticide runoff model. Simulation runs were made for 1920 different cases, including 12 locations in the eastern United States and 80 combinations of pesticide half-lives (3–150 days) and adsorption partition coefficients (0.1–200 cm^3/g). Published research data indicated that surface runoff, soil erosion, chemical persistence, and strength of adsorption to soil particles are major determinants of pesticide runoff, whereas runoff and soil erosion are, in turn, determined by soil, ground cover, and weather patterns. The hydrology component was based on the U.S. Soil Conservation Service (SCS) curve number (CN) runoff equation and a daily version of the Universal Soil Loss Equation (USLE).

Two different regression equations explained the highest fractions of observed variation in pesticide runoff. Regression Model A, $P = a_0 + a_1 X$, explains 71 to 94% of pesticide runoff variation (P = mean annual pesticide runoff, percent of application) and is a particularly simple function, requiring only mean annual soil erosion (X in mg/ha) to obtain an estimate of pesticide runoff. Model B, $P = b_0 + b_1 X + b_2 Qm$, explains 85 to 96% of pesticide runoff variations and requires both mean annual soil erosion (X) and surface runoff volume (Qm = mean surface runoff during month of pesticide application, mm). The regression parameters a_0, a_1, b_0, b_1, and b_2 are provided in tables calculated by Li et al. (1990). Except for the more persistent and strongly adsorbed pesticides, standard errors of estimate are substantially smaller for Model B, indicating that it will generally provide a more accurate estimate of pesticide runoff. However, Model A is easier to use as soil erosion is routinely calculated by the Universal Soil Loss Equation.

7.5 Treatment of Base Metal Mining Drainage

Huck et al. (1974) reported on operational experience with a base metal mine drainage treatment pilot plant, established at the site of a northeastern New Brunswick, Canada mining and smelting mill. The objective was to develop and demonstrate new and existing technology for the removal of heavy metals and the neutralization of such effluents. Three minewaters, characterized as strong, weak, and moderately strong, were evaluated (see Table 8). The use of a polymeric flocculant improved settling, yielded greater reliability of operation, and appeared to reduce metal concentrations in all cases. The optimum treatment configuration for the three minewaters is shown in Fig. 18 and consists of a once-through operation using polymer and two-stage neutralization.

Table 8 Minewater Characterization[a]

Drainage	Strong minewater		Weak minewater		Moderately strong minewater	
	Mean	Range	Mean	Range	Mean	Range
Constituent						
pH	—	2.4–3.2	—	2.8–3.3	—	2.3–2.9
Sulphate	7,090	1,860–14,300	1,020	810–1,790	4,530	2,350–7,290
Acidity	6,030	4,550–9,650	800	510–1,530	5,290	2,600–15,000
Lead	4.3	0.9–9	1.3	0.1–5	1.2	0.03–3
Zinc	1,160	735–1,590	108	22–175	540	390–723
Copper	10	15–17	20	12–52	50	24–76
Iron	1,580	815–3,210	68	24–230	720	350–1,380
Suspended solids	—	75–260[c]	—	5–90	—	5–190
Ionic ratios[d]						
Zn:Fe	0.7		1.6		0.8	
Cu:Fe	0.006		0.3		0.07	
Fe:SO$_4$	0.2		0.07		0.2	

[a]Based on daily means. All constituents except pH reported in mg/L.
[b]For the first run. The second run was similar, although the iron was somewhat lower and the range was narrower.
[c]Based on data for the month of June only.
[d]By weight.
Source: From Huck et al. (1974).

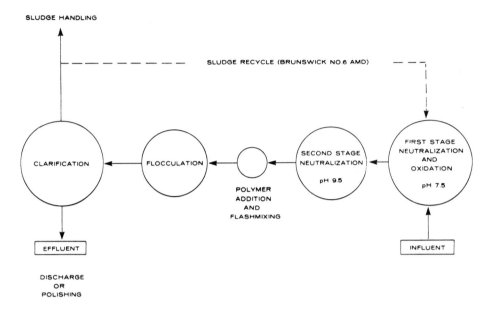

Figure 18 Optimum process configuration. (From Huck et al., 1974.)

The following conclusions were drawn from the data obtained in the study:

1. Using conventional neutralization, precipitation (optimum precipitation pH was 9.5), and sedimentation, effluent metal concentrations of less than 0.3 mg/L lead, 0.6 mg/L zinc, 0.1 mg/L copper, and 0.6 mg/L iron were achieved with all three base metal mine drainage wastewaters.

2. The initial minewater metals concentration strength had little effect on the effluent metal concentrations.

3. The improvement of effluent metal concentrations with a low ratio ($\leq 2:1$) sludge recycle appeared to be greatest for the weakest minewater. The improvement was similar to that achieved with polymer addition. High recycle reduced dissolved zinc but increased total metals.

4. Reactor contact times and the number of neutralization stages did not significantly affect results. However, as precipitation pH (8.9 to 10.5 gave approximately equal effluent qualities) control under neutralization load variation is better with staging, two-stage treatment was therefore chosen as the optimum process configuration.

Table 9 Treatment of Storm Runoff by Dissolved Air Flotation and Sand Filtration

Treatment efficiency	Test 1	Test 2	Test 3
Influent arsenic (mg/L)	0.0015	0.518	1.010
Removal by flotation (%)	33.3	79.5	90.1
Removal by flotation/filtration (%)	100.0	90.3	90.6
Influent turbidity (NTU)	3.4–3.6	3.5	3.0
Removal by flotation (%)	35.7	45.7	30.0
Removal by flotation/filtration (%)	93.0	93.7	93.3
Influent color (unit)	48–49	49.0	50.0
Removal by flotation (%)	50.5	40.8	43.0
Removal by flotation/filtration (%)	98.9	97.9	98.0
Influent O&G (mg/L)	29.2	29.3	28.5
Removal by flotation (%)	52.7	51.5	43.2
Removal by flotation/filtration (%)	60.2	61.1	74.7
Influent COD (mg/L)	85.0	86.0	83.0
Removal by flotation (%)	37.6	46.5	32.5
Removal by flotation/filtration (%)	58.8	53.5	51.8
Influent pH (unit)	7.0–7.1	7.1–7.1	7.2–7.2
After flotation (unit)	6.9–7.0	7.1–7.2	6.9–7.1
After flotation/filtration (unit)	6.9–7.0	7.0–7.1	7.0–7.1

[a]Chemical flotation treatment: Sodium aluminate, 15 mg/L as Al_2O_3, and either ferric chloride or ferric sulfate, 15 mg/L as Fe.
[b]Sand specification: ES = 0.85 mmn, UC = 1.55, depth = 11 in.
Source: Wang and Wang (1990).

5. Calcium sulphate deposition presented difficult but surmountable problems in the operation of the treatment process, and it could be controlled partly through sludge recycling. With minewater sulphate loadings of less than 2500 mg/L, calcium sulphate scaling on reactor surfaces was minimal and a daily cleaning of the pH probes was adequate. With concentrations in excess of this, probes required cleaning every 8 hr and light coatings formed on pipes and tanks within a few weeks.

6. Effluent polishing techniques such as sand filtration, coprecipitation, lagooning, or the use of an alternative reagent demonstrated that better effluent quality may be attained.

7.6 Arsenic Removal from Combined Runoff and Wastewater

The feasibility of removing hazardous soluble arsenic (+5) and other conventional pollutants from combined storm runoff and process wastewater by oil/water separation, dissolved air flotation (DAF), filtration, and granular activated carbon (GAC) adsorption was fully demonstrated for an oil-blending company in New Jersey (Wang and Wang, 1990). The oil separated from the raw combined wastewater by the American Petroleum Institute (API) oil/water separators was virgin and therefore skimmed off, dried, and reused. The oil/water separator effluent containing 1.01 mg/L of arsenic, 3 NTU of turbidity, 50 units of color, 28.5 mg/L of oil and grease (O&G), and 83 mg/L of chemical oxygen demand (COD) was fed to a dissolved air flotation (DAF) clarifier for removal of arsenic by 90.1%, turbidity by 30%, color by 43%, O&G by 43.2% and COD by 32.5%. Either ferric chloride or ferric sulfate was an effective coagulant for arsenic removal.

The same oil/water separator effluent was also successfully treated by a DAF-filtration clarifier. Reductions of arsenic, turbidity, color, O&G, and COD were 90.6, 93.3, 98, 74.7, and 51.8%, respectively. Although DAF and DAF-filtration both proved to be excellent pretreatment processes, granular activated carbon (GAC) post-treatment removed 100% of soluble arsenic. Table 9 presents some important operational data generated at the oil-blending company (Wang and Wang, 1990).

REFERENCES

American Society of Civil Engineers (ASCE) and Water Pollution Control Federation (WPCF) (1969). *Design and Construction of Storm and Sanitary Sewers.* New York.

Beschta, R. L. (1978). Long term patterns of sediment prediction following road construction and logging in Oregon coast range. *Water Resour. Res., 14,* 1011.

Boyd, M. J. (1982). In *Urban Stormwater Quality, Management and Planning* (B. C. Yen, ed.). Water Res. Publ., Littleton, Colo., p. 370.

Council on Environmental Quality (CEQ) (1980). *Environmental Quality—1979.* 10th Ann. Rep. CEQ, U.S. Government Printing Office, Washington, D.C.

Federal Aviation Agency (FAA) (1970). Department of Transportation Advisory Circular on Airport Drainage. Rept. A/C 150-5320-5B, Washington, D.C.

Fleming, G. (1969). Design curves for suspended load estimation. *Proc. Inst. Civil Eng. 43*, 1–9.

Glass, N. R., Glass, G. E., and Rennie, P. J. (1979). Effects of acid precipitation. In *Proceedings 4th Annual Energy Research and Development Conference.* ACS, Washington, D.C.

Goodell, B. C. (1966). Snowpack management for optimum water benefits. In *Proceedings of ASCE Water Resources Engineering Conference*, Denver, Colo.

Hall, M. J. (1984). *Urban Hydrology.* Elsevier, London.

Huck, P. M., Le Clair, B. P., and Shibley, P. W. (1974). *Operational Experience with a Base Metal Mine Drainage Pilot Plant.* Tech. Dev. Rep. EPS 4-WP-74-8, Water Pollution Control Directorate for Environment, Canada, Sept.

Kibler, D. F., et al. (1982). Recommended Hydrologic Procedures for Computing Urban Runoff from Small Developing Watersheds in Pennsylvania. Institute for Research on Land and Water Resources, University Park, Pa.

Krenkel, P. A. and Novotny, V. (1980). *Water Quality Management.* Academic Press, New York.

Lager, J. A., Shubinski, R. P., and Russell, L. W. (1971). Development of a simulation model for stormwater management. *J. Water Pollution Control Fed., 43*, 2424.

Li, W., Merrill, D. E., and Haith, D. A. (1990). Loading functions for pesticide runoff. *Res. J. WPCF, 62*, 16–26 (Jan./Feb.).

Linsley, R. K., Kohler, M. A., and Paulhus, J. L. (1982). *Hydrology for Engineers.* McGraw-Hill, New York.

Mandel, F. S. (1987). Novel hazardous material spill control agents and methods of application. In *Proceedings HAZMACON '87*, Santa Clara, Calif. April 21–23, p. 442.

McCuen, R. H., Rawls, W. J., and Wong, S. L. (1981). Evaluation of the SCS Urban Flood Frequency Procedures. Prelim. draft rep. prepared for U.S. SCS, Beltsville, Md.

McElroy, A. D., Chie, S. Y., Nebgen, J. W., Aleti, A., and Bennett, F. W. (1976). Loading Functions for Assessment of Water Pollution from Nonpoint Sources. EPA/600/2-76/151, Washington, D.C.

Metcalf & Eddy, Inc. (1984). *New York City Industrial Pretreatment Program Report.* Task 7—Sampling Program. New York.

Mockus, V. (1957). Use of Storm and Watershed Characteristics in Synthetic Hydrograph Analysis and Application. U.S. Dept. of Agriculture, Washington, D.C.

Morgali, J. R. and Linsley, R. K. (1965). Computer simulation of overland flow. *J. Hydraul. Div. ASCE, 91.*

Nash, J. E. (1957). The form of the instantaneous unit hydrograph. *Int. Assoc. Sci. Hydrol., 3*, 14–121, Publ. 45.

Novotny, V. and Chesters, G. (1981). *Handbook of Nonpoint Pollution—Sources and Management.* Van Nostrand Reinhold, New York.

Reidel, J. T., Appleby, J. F., and Schloemer, R. W. (1967). Seasonal Variation of the Probable Maximum Precipitation East of the 105th Meridian for Areas from 10 to 1000 Square Miles and Durations of 6, 12, 24, and 48 Hours. Hydrometeorological Rep. 33, U.S. Weather Bureau, Washington, D.C.

Roehl, J. W. (1962). Sediment Source Areas, Delivery Ratios and Influencing Morphological Factors. Publ. 59, IASH Comm. on Land Erosion, pp. 202–213.

Smith, R. V. and Stewart, D. A. (1977). Statistical models of river loadings of nitrogen and phosphorus in the Lough Nengh system. *Water Res.*, *11*, 631.

Snyder, F. F. (1938). Synthetic unit graphs. *Trans. Amer. Geophys. Union*, *19*, 447–454.

Soil Conservation Service (SCS) (1972). *Hydrology*. National Engineering Handbooks, Sec. 4, U.S. Department of Agriculture, Washington, D.C.

SCS (1975). Urban Hydrology for Small Watersheds. SCS TR-55, Washington, D.C.

Stalzer, R. B. and McArdle, G. W. (1983). A petroleum refinery stormwater run-off study. In *Proceedings of the 38th Industrial Waste Conference*, Purdue Univ., Lafayette, Ind., May 10–12, pp. 893–902.

Sylvester, R. O. and DeWalle, F. B. (1972). Character and Significance of Highway Runoff Waters. Res. Rept. 7.1, Wash. State Highway Dept. Res. Program, Olympia, Wash.

U.S. Army Corps of Engineers (1956). *Snow Hydrology*. North Pacific Div., Portland, Oregon.

U.S. Army Corps of Engineers (1974). Storage, Treatment, Overflow, Runoff Model "STORM". Generalized computer program users' manual, USACE Hydrologic Eng. Ctr., Davis, Calif.

Environmental Protection Agency (1974a). Feasibility of Overland-Flow Treatment of Feedlot Runoff. Office of Res. and Devel. Rept., EPA-660/2-74-062, Washington, D.C., Dec.

EPA (1974b). Urban Stormwater Management and Technology: An Assessment. EPA 670/2-74-040, Cincinnati, Ohio.

EPA (1977). Process Control Manual for Aerobic Biological Wastewater Treatment Facilities. EPA 430/9-77-006, Washington, D.C.

EPA (1980). Innovative and Alternative Technology Assessment Manual. EPA 430/9-78-009, Cincinnati, Ohio.

EPA (1984). Report to Congress: Nonpoint Source Pollution in the U.S. Office of Water Program Operations, Water Planning Div., Washington, D.C.

Viessman, W., Knapp, J. W., Lewis, G. L., and Harbaugh, T. E. (1977). *Introduction to Hydrology*. Harper & Row, New York.

Wanielista, M. P. (1978). *Stormwater Management—Quantity and Quality*. Ann Arbor Science Publishers, Ann Arbor, Mich.

Wang, L. K. and Mahoney, W. J. (1989). Treatment of storm runoff by oil-water separation, flotation, filtration and adsorption. Part A: Wastewater treatment. In *Proceedings of 44th Industrial Waste Conference*, Purdue Univ. Lafayette, Ind., May, Vol. 44, pp. 655–666.

Wang, L. K., Wang, M. H. S., and Mahoney, W. J. (1989). Treatment of storm runoff by oil-water separation, flotation, filtration and adsorption. Part B: Waste sludge management. In *Proceedings 44th Industrial Waste Conference*, Purdue Univ., Lafayette, Ind., May, Vol. 44, pp. 655–666.

Wang, L. K. and Wang, M. H. S. (1990). Decontamination of groundwater and hazardous industrial effluents by high rate air flotation processes. In *Proceedings of Great Lakes '90 Conference*, Hazardous Materials Control Res. Inst., Silver Springs, Md.

Weschmeier, W. H. and Smith, D. D. (1965). Predicting Rainfall-Erosion Losses from Cropland Costs of the Rocky Mountains. Agricultural Handbook 282, U.S. Dept. of Agriculture, Washington, D.C.

Whipple, W., Grigg, N. C., Grizzard, T., Randall, C. W., Shubinski, R. P., and Tucker, L. C. (1983). *Urbanizing Areas*. Prentice-Hall, Englewood Cliffs, N.J.

Wilber, W. G. and Hunter, J. V. (1975). Contributions of metal resulting from stormwater runoff and precipitation in Lodi, New Jersey. In *Urbanization and Water Quality Control*, American Water Resources Assoc., June, pp. 45–54.

Williams, J. R. and Berndt, H. D. (1972). Sediment yield computed with universal equation. *J. Hydraul. Div. ASCE, 98* (Hy 12), 2087.

Wu, J. S. and Ahlert, R. C. (1978). Assessment of methods for computing storm runoff loads. *Wat. Resour. Bull., 14*, 429–439.

3

Treatment of Metal Plating and Finishing Wastes

Mark Davis and Tom Sandy

CH2M HILL, Denver, Colorado

1. ORIGIN AND CHARACTERISTICS OF METAL PLATING AND FINISHING WASTES

1.1 Introduction

Metal plating and/or finishing generally encompasses one or more of the following unit operations:

Cleaning
Surface treatment
Electroplating
Electroless plating
Rinsing
Fume/exhaust scrubbing

A variety of chemical solutions and procedures are used in these operations to produce the desired metal coating and/or finish of different materials. These coatings and finishes serve the purpose of protecting, enhancing, and strengthening the material to which they are applied.

The chemical solutions or baths used in these operations have a limited life and must be replaced to maintain the desired coating or finish. These solutions, as well as the rinsewaters, are the primary sources of waste from metal plating and finishing operations.

Characteristics of each waste are dependent on the bath or solution from which it originated and the type(s) of metal being treated by the bath. Wastes typically are alkaline or acidic in nature and contain one or more toxic heavy metals.

1.2 Cleaning

Prior to finishing or plating, the metal part generally requires cleaning. Some typical cleaning operations are described below. The compositions of representative cleaning baths are listed in Table 1. Since most baths have proprietary formulations, only the gross makeup concentration of each bath is listed. Each of the baths listed will become more dilute than the makeup concentration listed over time because of drag out on the parts being cleaned.

Alkaline Cleaners

Metal surfaces are cleaned in an alkaline solution that contains surfactants and other proprietary cleaners that remove any undesired materials that might interfere with the plating or finishing operation. These cleaners typically use alkaline salt solutions at temperatures between 120 and 200°F to remove oil, grease, and organic soils from the metal surface.

Acid Cleaners

Low acid concentrations in combination with a wetting agent or detergent are also used for the removal of oil, grease, organic soils, and light metal oxides or to produce light phosphate coatings (for those acid cleaners containing phosphoric acid) prior to the finishing of the part from metal surfaces. Organic materials are removed by the etching, wetting, emulsification, or dissolution that can take place in the acid cleaning bath.

Degreasers

Metals that have had any machining or metal working performed on them usually contain larger amounts of oil and grease that exceed the capabilities of an alkaline cleaner. These materials require organic solvent cleaning in either the liquid or vapor phase to remove any oily materials that might interfere with further processing in plating or surface treatment baths. Vapor degreasing is a common method of oil and grease removal and is faster than liquid-phase solvent degreasing. Solvent degreasing is also effective in removing varnish and paint films. The solvents can be purified and recovered by a distillation

Metal Plating and Finishing Wastes

apparatus. Solvents are boiled to a vapor base and condensed and recovered in a still. The wastes or solvent sludges are left as still bottoms and manifested and discarded as hazardous wastes.

Electrocleaners

Electric current, either direct, reverse, or periodic reverse, can be used in conjunction with an alkaline cleaning bath to remove soil and smut and activate the metal surface of a part. The metal parts are either used as the cathode or anode and placed in a bath with properties similar to those of an alkaline cleaning bath. These cleaning baths can be operated at either ambient or elevated (100–150°F) temperatures.

Pickling

The immersion of metals into a strong acid solution will remove an oxide coating or scale on the metal surface. Most plating and surface treatments of metals require that the metal surface be free of oxides or rust in addition to oil. Most conventional alkaline or acid cleaners will not remove oxide coatings on metal surfaces and thereby require pickling. Depending on the strength of the acid solution, the pickling bath may also be referred to as an acid dipping or descaling bath.

1.3 Surface Treatments

Surface treatment of a metal part is performed to render the part more corrosion-resistant or make it more amenable to other coating operations by altering the outer layer. A few common surface treatment bath compositions and the waste by-products of the baths are described in Table 2.

Anodizing

Anodizing is a term referring to the electrochemical application of an oxide coating to aluminum parts. This oxide layer is applied by immersing the aluminum metal part, configured as the anode, in an electrolytic solution under an electrical current. The thickness and degree of aluminum oxidation are a function of the electrolyte composition, temperature, and the applied current. Anodizing is performed extensively in the aerospace industry for the increased corrosion and wear resistance of aluminum parts and for making the part more amenable to further surface coatings. The electrolyte generally consists of either chromic, oxalic, phosphoric, or sulfuric acids depending on the desired oxide coating.

Table 1 Typical Cleaning Bath Compositions

Cleaning Bath Type	Chemical Component(s)	Makeup concentration at 70°F (oz/gal)	Soluble metal concentrations at bath dump (mg/L)
General alkaline	Alkaline salts (NaOH, KOH), carbonates, dispersants, sequestering agents, silicates, and surfactants	8–32	<500 of treated metal
Copper or steel alkaline (Groshart, 1989)	Sodium hydroxide Sodium carbonate Trisodium phosphate Wetting agent	5–8 10–15 8–15 0.5	<500 of Cu or Fe
Acid	Emulsifying agents, organic wetting agents, and water-miscible solvents	10–15	5,000–20,000 of treated metal
	Inorganic acids (sulfuric, phosphoric) and organic acids (gluconic)	10–15% by vol.	5,000–20,000 of treated metal
Detergents	Buffering salts, dispersants, inhibitors, sequesting agents, surfactants, or wetting agents	6–12	<100 of treated metal
Degreasers (Johnson, 1989)	Methylene chloride Maximum water content	160–176 <100 mg/L	<500 of treated metal
	Perchloroethylene	192–224 <100 mg/L	<500 of treated metal
	1,1,1-Trichloroethane	160–176 <100 mg/L	<500 of treated metal
	Trichlorethylene	176–192 <100 mg/L	<500 of treated metal

Metal Plating and Finishing Wastes

Table 1 (Continued)

Cleaning Bath Type	Chemical Component(s)	Makeup concentration at 70°F (oz/gal)	Soluble metal concentrations at bath dump (mg/L)
Steel electrocleaner (Groshart, 1989)	Sodium hydroxide Trisodium phosphate Glutamic acid (sodium glutanate) Surfactant	10–20 5–10 4–8 As required to limit foaming	<1,000 of treated metal
Copper pickling (for heavy scales) (Groshart, 1989)	Nitric acid Hydrofluoric acid	25% by vol. 1% by vol.	10,000–50,000
Aluminum pickling (Groshart, 1989)	Sodium dichromate Sulfuric acid Ammonium bifluoride	7.5% 10% by vol. 1.75	30,000–40,000 of Al 10,000–20,000 of Cu
Copper pickling (for heavy scales) (Groshart, 1989)	Sulfuric acid Nitric acid	32% by vol. 12% by vol.	20,000–30,000 of Cu
Copper pickling (for light scales) (Groshart, 1989)	Sulfuric acid	10% by vol.	20,000–30,000 of Cu
Steel or copper pickling	Hydrochloric acid	55% by vol.	30,000–40,000 of Fe <4,000 of Cu

Table 2 Typical Surface Treatment Bath Compositions

Surface treatment bath type	Chemical component(s)/ operating conditions	Makeup concentration (oz/gal)
Sulfuric acid anodizing	Sulfuric acid 60–80°F	13–31
Chromic acid anodizing	Chromic acid 90–100°F	6–14
Hard anodizing	Sulfuric acid 32–38°F	21–22
Chromate conversion coatings	Chromic acid Activators (i.e., acetate, chloride, fluoride, formate, nitrate, phosphate, sulfate, or sulfamate acids)	—
Aluminum electropolishing (Mason and Tosterud)	Sulfuric acid Hydrofluoric acid 140°F	1–60% by vol. 0.2–1.5% by vol.
Nickel electropolishing	Orthophosphoric acid Sulfuric acid	15–70% by vol. 15–60% by vol.
Coloring (blueing steel) (Hall, 1989)	Sodium thiosulfate Lead acetate	8 2
Coloring (browning steel) (Hall, 1989)	Copper sulfate Mercuric chloride Ferric chloride Nitric acid Denatured alcohol	2–3 0.5–0.6 3–4 19–20 70% by vol.
Iron phosphate coating	Phosphoric acid and accelerator (i.e., chlorates, nitrates, nitrites, or peroxides) 60–160°F	0.5–2
Cadmium stripping (from steel)	Ammonium nitrate Nonelectrolytic Ambient temperature	16
Copper stripping (from nickel)	Sodium cyanide Sodium hydroxide Steel cathodes Ambient temperature	12 2

Chromate Coatings

A chromate conversion coating is a complex film of chromium compounds that is a result of the chemical or electrochemical reaction of hexavalent chrome with a metallic surface in the presence of certain anions or activators. These coatings are usually referred to as "conversion" coatings because they "convert" the metal surface of the part when metal ions at the surface of the part bond with the chromium compounds. Specific concentrations of chemicals are not listed in Table 2 since the bath formulations are application-specific and proprietary. Chromate conversion coatings are generally applied to aluminum, cadmium, copper, magnesium, silver, and zinc metal surfaces to impart corrosion resistance.

Electropolishing

A small outer layer (2.5–65 µm) of a metal part is removed via an electrochemical reaction resulting in a brightening effect of the part. The metal part is usually aluminum, carbon steel, copper, nickel, or stainless steel. Carbon steel is the least common electropolished metal because of the variations in its composition. Unlike electroplating, the parts are anodic. Chemical solutions used for electropolishing are acidic or alkaline in nature and usually have a very long life.

Metal Coloring

A variety of colors can be applied to metal via chemical or electrochemical coloring solutions. Colors such as the blueing or browning on gun barrels are a result of the immersion or application of an oxidizing chemical solution. Baths used for the coloring of metals sometimes contain a sulfide compound (i.e., sodium thiosulfate).

Phosphate Coatings

This conversion coating results in a phosphate film being integrally bonded to the metal when a metal part is immersed into a mineral acidic solution containing a soluble metal phosphate. The acid maintains the solubility of the metal phosphate solution and promotes a light pickling of the metal part surface. Nitrate anion accelerators are generally added to the solution to speed up the coating reaction rate. The primary function of phosphate coatings is to make metal surfaces more amenable to painting or plastic coating. This preconditioning is performed primarily on aluminum, cadmium, ferrous, magnesium, and zinc metal parts.

Stripping Plated Metals

Metals that have been plated and rejected due to defects such as insufficient metal thickness or roughness are usually stripped of the metal and replated. Stripping and recovery of the metal are usually practiced when the parts and/or the metal plated are valuable. Chemical immersion stripping solutions are the most common and are usually acidic in nature. Electrolytic strippers permit the metal to be stripped under anodic conditions and recovered on the cathode via applied electrical current. Baths for these electrochemical stripping may be acidic or alkaline, depending on the composition of the basis metal and the metal plate. Metals that may be stripped from aluminum, copper nickel, or steel base metals include brass, bronze, cadmium, chromate, chromium, copper, gold, lead, nickel, and tin.

Immersion Plating

The deposition from a metal salt solution, of a metal onto a surface of a base metal with a higher oxidation potential, without the use of reducing agents or electrical current, is known as immersion plating. Copper can be plated on steel or aluminum by the immersion of either metal into a heated, nonalkaline copper salt solution. There are numerous metals that can be plated by immersion plating. These metals include brass, bronze, cadmium, copper, gold, lead, nickel, palladium, platinum, rhodium, ruthenium, silver, tin, and zinc. Immersion coating may be carried out on aluminum, copper alloys, zinc, and steel metal surfaces. Table 3 lists a few of the common immersion bath's chemistry. Immersion plating is limited to thin-film metal surface coatings, since the displacement of metals onto the base metal ceases when the entire surface has been covered.

1.4 Electroplating

The electroplating of various metals such as brass, bronze, cadmium, chromium, copper, gold, indium, iron, lead, nickel, nickel-iron, palladium, platinum, rhodium, ruthenium, silver, tin-lead, tin, and zinc onto the surface of other conductive base materials is the most common plating process. Metals are reduced on the conductive part (the cathode) from an electrolyte solution by the application of an electric current. Table 4 lists some of the typical metals that are electroplated and the approximate composition of the baths. Electroplating baths can be acidic or alkaline and generally require some heating and/or agitation.

Table 3 Typical Immersion Plating Bath Composition

Surface treatment bath type	Chemical component(s)/ operating conditions	Makeup concentration (oz/gal)
Cadmium immersion plating (cadmium on aluminum) (Hirsch and Rosenstein, 1989)	Cadmium sulfate Hydrofluoric acid (70%) Ambient temperature	0.5 9.0 fl. oz/gal
Nickel immersion plating (nickel on copper) (Hirsch and Rosenstein, 1989)	Nickel sulfate Ammonium nickel sulfate Sodium thiosulfate 100–150°F	8 8 16
Tin immersion plating (tin on steel) (Hirsch and Rosenstein, 1989)	Stannous sulfate Sulfuric acid 180–212°F	0.2 0.9 fl. oz/gal

1.5 Electroless Plating

The plating or chemical reduction of soluble metal ions onto a base substrate in the presence of a catalyst, reducing agent(s), complexing agent(s), and stabilizer(s), without the use of electrical current, is referred to as electroless plating. Electroless plating requires specific aqueous metal concentrations as well as temperature and pH levels. A list of some of the typical components of electroless plating baths and/or wastes is provided in Table 5 (Henry, 1989). Electroless plating can be performed on nonmetal as well as metal substrates.

1.6 Rinsewaters

The rinsing of material before or after a plating or coating operation is necessary to remove any plating bath solution that may be left on the material. Water used in rinsing can consist of deionized, demineralized, or tap water, depending on what the finishing operation will tolerate. Rinsing operations produce the largest volumes of wastes from metal plating or finishing operations. Rinsewaters ultimately become contaminated to varying degrees with whatever bath chemistry is left on the material. Depending on the type of rinsing scheme, rinsewaters can contain varying concentrations of heavy metals. There are several types of rinsing schemes that can be used in metal plating and finishing operations; they are discussed below. Production specifications and quality control may limit the use of only one or two of the

Table 4 Typical Electroplating Bath Compositions

Surface treatment bath type	Chemical component(s)/ operating conditions	Makeup concentration (oz/gal)
Cadmium (alkaline cyanide) (Marce and Marrow, 1989)	Cadmium Cadmium oxide Sodium carbonate Sodium cyanide Sodium hydroxide 60–100°F	2.7 3.0 4.0–8.0 13.5 1.9
Cadmium (acid sulfate) (Marce and Morrow, 1989)	Cadmium oxide Sulfuric acid 60–90°F	1.0–1.5 4.5–5.0
Chromium (McCullen, 1989)	Chromic trioxide Sulfuric acid 90–150°F	26–54 0.2–0.54
Copper (sodium cyanide) (Sato and Barauska, 1989)	Copper cyanide Sodium cyanide Sodium carbonate Sodium hydroxide Free sodium cyanide 75–160°F	4–10 6.0–13.0 2–4 0.5–3.0 1.5–3.0
Copper (pyrophosphate) (Sato and Barauskas, 1989)	Copper pyrophosphate Potassium pyrophosphate Potassium nitrate Ammonium hydroxide 110–140°F	7.0–11.2 26.8–46.5 0.4–0.8 0.4–1.1% by vol.
Copper (standard acid) (Sato and Barauskas, 1989)	Copper sulfate Sulfuric acid Chloride	26–33 4–10 5–120 ppm
Gold (24K or English) (Weisberg, 1989)	Potassium gold cyanide Free potassium Dipotassium phosphate 140–160°F	0.2–0.3 1.0 2.0
Nickel (chloride) (DiBari, 1989)	Total nickel Nickel chloride Boric acid 130°F	10 40 4

Metal Plating and Finishing Wastes

Table 4 (Continued)

Surface treatment bath type	Chemical component(s)/ operating conditions	Makeup concentration (oz/gal)
Nickel (sulfate) (DiBari, 1989)	Total nickel Nickel sulfate Boric acid 130°F	9.3 44 4
Platinum (sulfate) (Morrisey, 1989)	Platinum as H_2 Pt $(NO_2)_2SO_4$ Sulfuric acid 104°F	5.0 2.0
Silver (Blair, 1989)	Potassium silver cyanide Potassium cyanide Potassium carbonate 70–85°F	0.7–6.0 1.6–16 2.0
Tin-Lead (standard 60/40 solder) (Hirsch and Rosenstein, 1989)	Stannous tin Lead Fluoroboric acid Boric acid Peptone 60–80°F	7.0–8.0 3.0–4.0 13.0–20.0 3.0–5.0 0.6–0.9
Zinc (cyanide) (Geduld, 1989)	Zinc cyanide Sodium cyanide Sodium hydroxide 70–110°F	8.1 5.6 10.0
Zinc (chloride) (Geduld, 1989)	Zinc chloride Potassium chloride Boric acid 65–115°F	9.0–12.0 25.0–32.0 3.5–5.5

schemes. Some of the rinsing operations have advantages over the others in regard to water conservation and waste minimization.

Flowing Rinses

This rinse scheme consists of a single tank with a dirty rinsewater overflow effluent point and a clean water makeup source. Water is added as required (usually continuously) to purge the rinse tank of contaminants. Sometimes the

Table 5 Typical Electroless Plating Bath Components

Electroless plating bath type	Chemical component(s)/operating conditions
Copper	Metal salt(s): copper acetate, copper carbonate, copper formate, copper nitrate, and copper sulfate
	Reducing agent(s): dimethylamine borane (DMAB), formate, formaldehyde, hydrazine, sodium hypophosphite
	Complexing agent(s): ammonium hydroxide, ethylenediamine-tetracetate (EDTA), potassium tartate, pyridium-3-sulfonic acid, and rochelle salt
	Stabilizer(s): mercaptobenzotiazone (MBT), sodium cyanide, thiodiglycolic acid, thiourea, and vanadium oxide
	pH adjuster(s): hydrochloric acid, potassium hydroxide, sodium hydroxide, and sulfuric acid
Nickel	Metal salt(s): nickel chloride and nickel sulfate
	Reducing agent(s): DMAB, hydrazine, sodium borohydride, sodium hypophosphate
	Complexing agent(s): citric acid, glycolic acid, proprionic acid, sodium citrate, sodium acetate, and succinic acid
	Stabilizer(s): fluoride compounds, heavy metal salts, MBT, oxy anions, thiourea, thioorganic compounds
	pH adjuster(s): ammonium hydroxide and sulfuric acid

Source: Henry (1989).

addition of water is controlled by a conductivity-sensing probe that activates a solenoid-actuated control valve, which introduces fresh rinsewater when the conductivity of the rinse bath becomes too high. Controlling water inflow by rinse bath conductivity generally fails due to the conductivity probe "fouling" or calibration difficulties. Flowing rinses are the largest producers of dilute wastewaters in most plating and finishing wastewaters.

Metal Plating and Finishing Wastes

Stagnant Rinses

A stagnant rinse is usually used when a high-quality rinse is not necessary. In this scheme, the rinse tank is filled with water at the beginning of each shift or day and then drained at the end of the day. Water may be added as necessary to maintain a required level and replace the volume lost by drag out or evaporation. Some stagnant rinses are replaced very infrequently. These rinses are generally used for "reworking" materials that were rejected because of poor quality. If the plating operation can tolerate this less efficient rinsing technique, stagnant rinses do produce a significantly smaller volume of wastewater at higher concentrations than flowing rinses.

On-Demand Rinses

This rinse scheme uses a manually actuated valve to initiate a rinse spray and an automatic shutoff. Figure 1 (Davis and Sandy, 1989) depicts one configuration of an on-demand rinse scheme. Automatic shutoffs generally are activated by a timer, and the manual start can be a foot pedal or hand switch. On-demand rinses are useful in manual plating lines where the production of parts is intermittent. Spray rinse chambers do not always follow a flowing rinse, but this configuration does provide better rinsing efficiency. The start and stop sequence of on-demand rinsing prevents rinsewater from being left on while

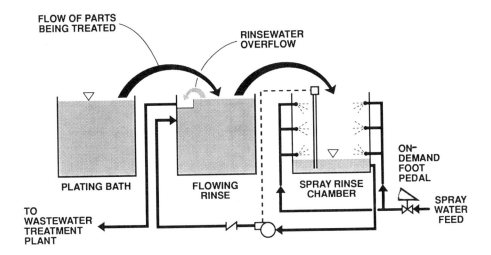

Figure 1 On-demand spray rinsing schematic. (From Davis and Sandy, 1989.)

the rinse bath is not being used. Wastewater volumes produced by this rinse scheme can be very low.

Cascade Rinsing

In this rinsing sequence, rinsewaters flow countercurrently to the movement of metal parts with water makeup to the last rinse stage and rinsewater discharge from the first rinse stage. Water makeup to the cascade rinse bath is usually continuous. Cascade rinses, at a minimum, have two stages but more commonly three. A three-stage cascade rinse is shown in Fig. 2 (Davis and Sandy, 1989). The use of this rinse scheme results in the discharged rinsewater being higher in metal concentration and lower in volume than a single-tank flowing rinse system. Cascade rinsing can reduce discharged rinsewater volume by a factor of 100 to 1000 times compared to a single-tank system. Metal concentrations in the discharged rinsewater increase by a similar factor, and metal concentrations in the final rinse are usually less than 1 mg/L.

1.7 Fume/Exhaust Scrubbing

Air pollution control equipment for plating and metal finishing exhaust systems is required by most regulatory agencies. Wet collectors are usually chosen

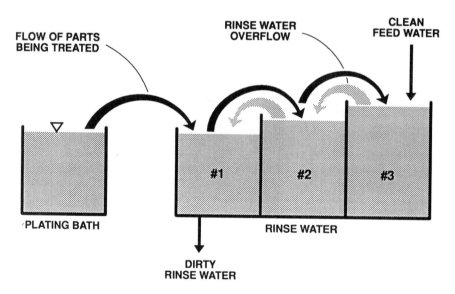

Figure 2 Countercurrent rinsing schematic. (From Davis and Sandy, 1989.)

Metal Plating and Finishing Wastes

for removing the water-soluble contaminants and gases that are released to the exhaust system during normal plating and finishing operations. A simple spray chamber or packed-bed scrubber (Figs. 3 and 4, respectively) is generally used in plating and finishing operations. The spray chamber scrubber is the most common type of collector and results in the largest consumption of water.

A lip exhaust system with activated carbon cylinders is generally used for removing solvents from vapor-degreasing operations. The activated carbon columns are usually operated in a "lead-lag" mode of operation. Solvents from the carbon can be recovered or desorbed from the carbon by steam stripping and distillation.

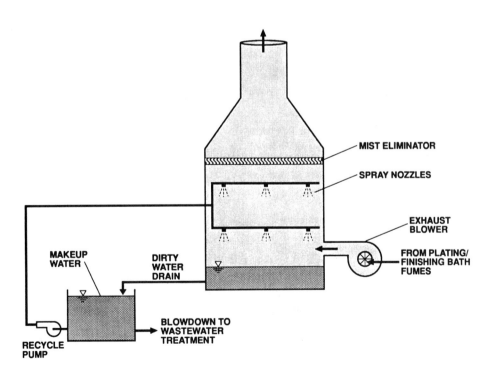

Figure 3 Spray chamber scrubber.

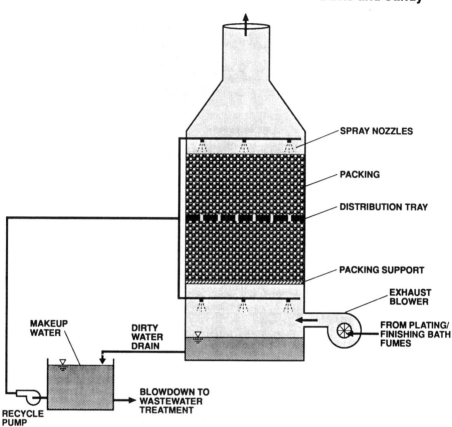

Figure 4 Packed-bed scrubber.

2. WASTE MINIMIZATION

There has been a growing trend toward the minimization of waste from manufacturing to reduce the amount of wastes being landfilled and conserve diminishing resources. Waste minimization in the metal finishing industry may involve the recovery of metal from waste streams while providing a reduction in wastewater volume and, consequently, requiring smaller treatment equipment. There are numerous methods to reduce the amount of waste discharges. Some of these methods are aimed at the plating bath while the majority are targeted for rinsewaters, since they constitute the majority of the waste volume discharged from most plating and finishing operations.

Metal Plating and Finishing Wastes

2.1 Rinsewater Reduction

On-demand, cascade, and stagnant rinsing as previously discussed are excellent methods to reduce the volume of rinsewater discharged. Since rinsing must not cause the loss of quality to the product being plated, a compromise must be met between rinse volume and rinsewater contamination limits for production standards.

The drag in D and flow Q through the rinsing tank can be calculated from the following equation (Mohler, 1989):

$$D \times Cb = Q \times Cr \tag{1}$$

where

D = drag in to the rinse
Cb = metal concentration of the plating bath
Q = flow through the rinsing tank
Cr = metal concentration in the rinsing tank

Equation (1) can be rearranged and modified (Hanson and Zabbon, 1959) to

$$Q = \frac{D}{t} \times n \times \frac{C_b}{C_r} \tag{2}$$

where

Q = flow through the rinsing tank
D = drag in to the rinse
t = time between batches as loads processed
n = number of countercurrent rinses
Cb = metal concentration of the plating tank
Cr = metal concentration in last stage of the rinsing tank

This equation accounts for bath production and the number of countercurrent rinse baths. It is useful in determining the required rinse flow rate to meet production standards and minimize rinsewater waste discharge.

The rinsing ratio R can be obtained by dividing the dissolved solids concentration in the plating bath by the dissolved solids concentration in the rinsing tank.

$$R = \frac{C_b}{C_r} \tag{3}$$

Rinsing ratios are useful in determining the dilution factors required for effective rinsing. A conductivity meter can be used to measure the dissolved solids concentrations in the plating bath and rinse tank. Drag in (or drag out) can be estimated from the rinse flow and the gain in conductivity in the rinse tank (the difference in conductivity in the rinse tank without flow to it before and after processing parts). If the flow is unknown, drag in can be calculated iteratively using Eq. (1) or the following modified formula:

$$Q = R \times D \tag{4}$$

By monitoring the conductivity of the rinsewaters and calibrating or setting the rinse flows according to the required rinse ratio, the minimization of rinse discharges can be accomplished.

Another way to reduce rinse volumes is to put maximum rate-of-flow valves or restriction orifices on the rinsewater feedline. These devices have been successful in preventing an operator from unnecessarily increasing the rinse flow. Air agitation enhances rinsing and can help to reduce the volume of discharged rinsewater.

Drag in (or drag out) can be minimized by using air knives and hanging the plating racks above the plating bath for a period of time to permit the parts to drain back into the plating tank. In the case of programmable automated hoists, the vertical lifts can be programmed to stall over the plating bath to automatically allow plated parts to drain before proceeding to the rinse cycle.

2.2 Bath Dump Reduction

In-bath refurbishing systems can be used to remove impurities that build up in metal finishing or plating baths. Bath refurbishing can be done at the tank side by electrodialytic and electrolytic processes. These processes will be discussed in detail in the following waste treatment sections.

The recovery of acid(s) lost to drag out into a stagnant or the last, most concentrated stage of a cascade rinse system can be accomplished using a proprietary ion-exchange-type resin developed and marketed by Eco-Tec, Ltd. The resin is regenerated with water, and the regenerant, composed of acid(s) lost to drag out, can be returned to the plating bath. This proprietary resin as well as other ion-exchange resins have numerous applications in concentrating the drag out lost to the rinsewater and returning it to the bath free of impurities.

Both the recovery of drag out from rinsewaters and the refurbishing of baths are effective ways to reduce the amount of bath dumps or waste discharges.

Metal Plating and Finishing Wastes

2.3 Evaporation

The evaporation of concentrated rinses and baths is a simple and effective way to reduce waste discharges. In this treatment process, water is evaporated from the solution, resulting in an increase in concentration of solute(s). The recovery of desired chemical components, such as metals, is often not possible because the rinse and bath contain contaminants.

One of the most common ways to concentrate waste baths and rinses is by an atmospheric evaporator. Figure 5 shows a schematic of a typical atmospheric evaporator used to concentrate drag out in a countercurrent rinse bath.

Atmospheric evaporators use an air blower, a recirculating pump, and a mist eliminator to concentrate rinsewater contaminants. The air exhaust is dis-

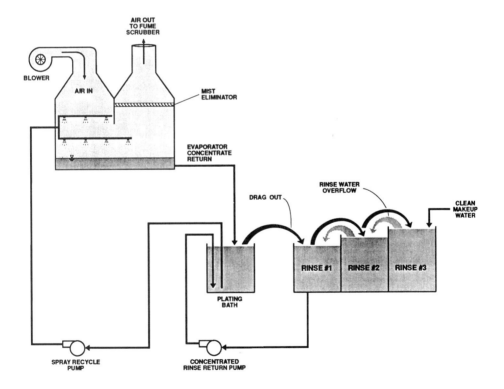

Figure 5 Atmospheric evaporator.

charged to a scrubber and the concentrate is returned to the bath or waste treatment system.

Single- and two-effect vacuum evaporators also may be used for a concentrated effluent stream and a high-purity distillate stream. The boiling point of the waste liquid is lowered by operating under a vacuum and heat is applied, causing water to evaporate. The resulting water vapor is condensed and returned to the rinse tank, and the concentrate is returned to the plating bath or waste treatment system. Figure 6 shows a two-effect evaporator. Although the capital cost is greater with this type of evaporator, it will concentrate the waste more efficiently from an energy utilization standpoint.

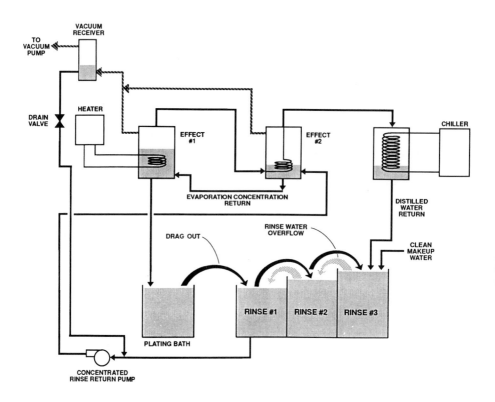

Figure 6 Two-effect evaporator.

Metal Plating and Finishing Wastes

3. CONVENTIONAL TREATMENT

Heavy metal removal from wastewater traditionally has been accomplished by precipitation of the metals, followed by collection and dewatering of the sludge. The sludge is then disposed of in a landfill. With the implementation of restrictions on the disposal of sludges containing heavy metals, however, new technology has been and continues to be developed to recover metals in reusable form, usually as a pure metal. Nevertheless, conventional, sludge-producing treatment is still a viable alternative when coupled with recent developments in sludge stabilization or other methods that allow sludges to meet landfill restrictions.

3.1 Heavy Metals

Most heavy metals can be precipitated directly by the addition of lime or caustic to produce metal hydroxide precipitants or by the addition of sulfide compounds to form metal sulfide precipitants. An exception is hexavalent chrome, which exists as chromate or dichromate anion in solution and must be reduced to trivalent chrome before hydroxide precipitation can occur.

A number of reducing agents, including sulfur dioxide, sodium bisulfite, sodium metabisulfite, and ferrous sulfate, may be used to reduce hexavalent chrome. Gaseous sulfur dioxide is one of the more commonly used reducing agents. The reduction proceeds according to the following reactions:

$$3SO_2 + 3H_2O = 3H_2SO_3 \tag{5}$$

$$3H_2SO_3 + H_2CrO_4 = Cr_2(SO_4)_3 + 5H_2O \tag{6}$$

Reactions with other sulfur-reducing agents are similar. Low pH improves the kinetics of these reactions. Normally, a pH of 2 to 3 is used for complete reduction. Above pH 4 or 5, the reaction rate is considered too slow to be practical.

Chrome reduction is usually accomplished continuously in a series of two stirred-tank reactors. Each tank has a retention time of about 45 min and is equipped with oxidation reduction potential (ORP) and pH probes. Gaseous sulfur dioxide is added to the tanks based on the ORP level, and sulfuric acid is added based on pH. The first tank in the series is usually maintained at a pH between 3 and 4, whereas the second tank is maintained at a pH of about 2. Figure 7 illustrates a typical chrome reduction process.

After reduction is complete, the wastewater, which now contains predominantly trivalent chrome, can be combined with other metal-bearing

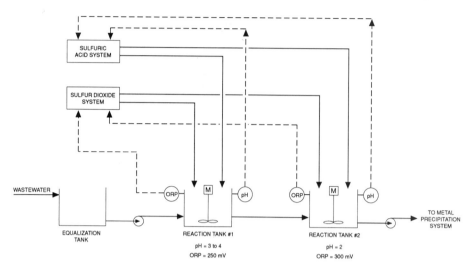

Figure 7 Sulfur-dioxide chrome reduction.

wastewaters for further treatment by precipitation and sedimentation. Any hexavalent chrome that is not reduced will not be removed by hydroxide precipitation.

For dumped baths containing high concentrations of hexavalent chrome, a batch treatment system may be used. Batch treatment systems may also have two pH reduction steps to allow the tighter control of acid addition. An alternative to batch treatment is to collect dumped baths and then bleed them into the continuous treatment system used for low-concentration wastewaters.

Hexavalent chrome may also be reduced by electrochemical reduction. In this process, current passed through iron electrodes results in the production of ferrous ions, which reduce chromate while the iron is oxidized to the ferric form. The iron electrodes are typically placed in a closed reaction vessel; at least two vessels in parallel are used to allow one vessel to be taken out of service to replace iron electrodes while the other remains in service. The voltage applied to the electrodes can be varied (as dictated by the chrome concentration) to control the rate of iron electrode dissolution.

Caustic or lime addition is required prior to the precipitation reactor to increase the pH to at least 7 or 8. After passing through the precipitation reactor, chrome and ferric hydroxide remain as insoluble solids. The precipitated solids are then settled in a clarifier. Besides chrome removal, this electrolytic processing has also been successfully used for the treatment of zinc, copper, and other heavy metals.

3.2 Chelate and Complex Breakers

Chelated and complexed metals are often used in plating operations. Because the chelating or complexing agent may prevent the metal from being precipitated, special chemical additives are sometimes required to break the chelated or complexed metal. One chemical commonly used is sodium dimethyldithiocarbamate (DTC). DTC may be restricted from discharge into some local wastewater treatment systems. Many other kinds of additives are available, but these are usually in proprietary formulations. For this reason, it is best to contact chemical suppliers for information and samples of additives for laboratory tests on the wastewater being treated.

Treatment chemicals for chelated metals tend to be expensive. Segregation of the wastewater containing chelated metals will reduce the volume of wastewater requiring costly forms of treatment, resulting in the decreased consumption of additives and lower treatment costs. Typically, a small batch system consisting of a cone-bottomed tank and mixer is used for the treatment of chelated metals. For treatment systems that handle only a few batches a day, chelate breakers may be added manually to the treatment tank. After treatment is completed, the effluent may require further treatment by conventional precipitation; usually, however, the effluent may be discharged. Sludge that accumulates in the treatment tank must be dewatered and stabilized prior to disposal.

3.3 Hydroxide Precipitation

Hydroxide precipitation is commonly used for the treatment of wastewaters containing a variety of heavy metals. With the exception of hexavalent chrome and chelated or complexed metals, virtually all heavy metals present in plating wastewater can be removed by hydroxide precipitation.

Figure 8 shows the theoretical solubilities of various metal hydroxides vs. pH. This figure illustrates the importance of maintaining proper pH during precipitation of chrome because its solubility increases above and below a pH of 7.5. Also, the removal of more than one metal may be difficult in a single stage if low effluent concentrations are required. As an example, if cadmium and chrome are both present in high concentration and the discharge requirements are less than 1 mg/L for both metals, a single-stage precipitation system

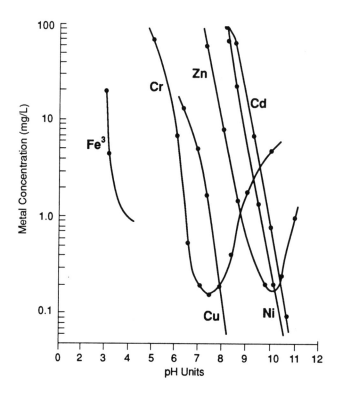

Figure 8 Metal hydroxide solubilities. (From Cherry, 1982.)

would be inadequate because the solubilities of both metals are never below 1 mg/L at the same pH.

Hydroxide precipitation processes include the following steps: pH adjustment, flocculation, and sedimentation. Settled solids are dewatered and stabilized for disposal. Either lime or sodium hydroxide is usually used to adjust wastewater pH. These chemicals may be added in a rapid-mix tank or directly to the sedimentation device. If lime is used for pH adjustment, it is normally added as a slurry.

Many different types of equipment may be used for sedimentation, including conventional clarifiers, inclined plate clarifiers, gravity settling tanks, lagoons, or filters. Coagulating agents (polymers) are usually added just ahead of the sedimentation step to improve the settling of solids.

Metal Plating and Finishing Wastes

If more than one metal (such as chrome and cadmium) is present, more than one precipitation system may be required if different pHs are needed to achieve the required degree of removal. For multiple stages, the first stage is operated at the lower pH, as shown in Fig. 9. After solids removal from the first stage, the pH is increased in the second stage. It is important to not carry over suspended solids from the first stage to the second because they may redissolve.

Besides maintaining the proper pH to achieve low solubilities, it is critical that the precipitated solids be prevented from carrying over into the effluent of the sedimentation stage. In terms of discharge limits, metals in the form of a suspended solid are treated the same as metals in a dissolved state by regulatory agencies. A filtration step after the sedimentation stage is recommended to remove suspended solids carryover. Many different types of filters may be used, including gravity, pressure, and continuous backwashing filters. Of these types, pressure filters have the greatest solids removal capacity per filter

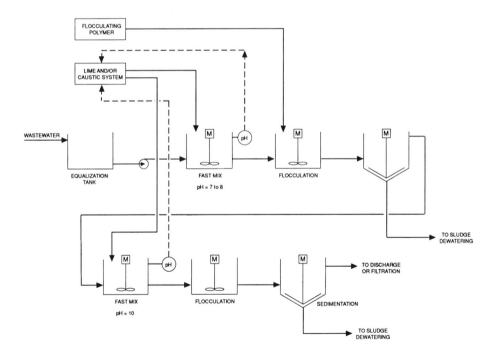

Figure 9 Metal hydroxide two-stage precipitation.

surface area, but tend to have a greater cost. Backwash from the filter may be recycled to the front of the precipitation train. A waste backwash collection tank may be necessary to prevent flow surges.

3.4 Sulfide Precipitation

Sulfide precipitation produces a sludge similar to hydroxide precipitation. Sulfide compounds are added after pH adjustment. The addition of sulfide results in the precipitation of metal sulfides that normally have much lower solubilities than metal hydroxides, as shown in Fig. 10. Because of lower solubilities, very high metal removal efficiency can be obtained.

In addition to greater removal efficiencies, sulfide precipitation has the following advantages over hydroxide precipitation:

Hexavalent chrome can be reduced and precipitated in a single step.
Most complexed metals can be precipitated without first breaking the metal complex.
Control of pH is not as critical because metal sulfides are insoluble over a wide pH range.
If more than one metal is present, only a single stage is required, in contrast to hydroxide precipitation, where a separate stage may be required for each metal.

Sulfide precipitation processes include the following steps: neutralization, sulfide addition, solids flocculation, and solids settling, flotation, or concentration in a tubular filter. Sulfide precipitates are often difficult to remove by sedimentation. The accumulated solids are dewatered and sometimes dried before disposal (Wang, 1984).

Either lime or sodium hydroxide may be used to increase wastewater pH. Lime is usually added as a slurry.

After neutralization, sulfide is added to the wastewater along with polymers or flocculating chemicals in a reactor/clarifier. Two principal methods are used to introduce sulfide: the soluble sulfide precipitation process (SSP) and the insoluble sulfide precipitation process (ICP).

In the SSP process, sulfide is added in a soluble solution such as sodium sulfide (NaS) or sodium hydrosulfide (NaHS). The amount of sulfide added is controlled by the level of sulfide in the clarifier effluent using an ORP probe or a sulfide ion-specific electrode. Overdosing can cause sulfide odors; if sulfide addition is not adequately controlled, the process tanks must be enclosed and vented to minimize odor problems. Also, it has been found that overdosing the sulfide can cause a loss of metal removal efficiency, probably from the

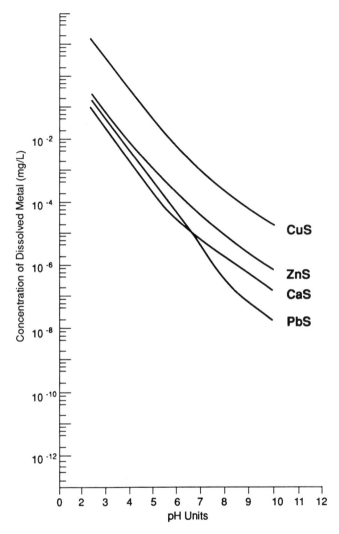

Figure 10 Metal sulfide solubilities. (From Cherry, 1982.)

production of extremely finely divided, colloidal precipitates. The reliability of the sulfide ion electrode is very important in controlling sulfide addition. The use of two or more electrodes in the reaction tank may be used to identify inaccuracies in sulfide measurement and the failure of electrodes. If an electrode fails, sulfide addition can be maintained at a low rate while maintenance is performed or a replacement installed. The routine preventative maintenance of the electrodes will help to reduce failures and the resulting operating problems.

Hexavalent chrome is reduced in the SSP process according to the following reaction:

$$2HCrO_4 + 3NaHS + H_2O = 2Cr(OH)_3 + 3S + 3NaOH \qquad (7)$$

In the ICP process, a slurry of ferrous sulfide (FeS) is used. This slurry is produced by reacting ferrous sulfate (FeSO$_4$) and sodium hydrosulfide. The FeS dissociates into ferrous and sulfide ions to the extent predicted by its solubility product. Heavy metals that have sulfide solubilities less than those of FeS will precipitate as metal sulfides, which causes additional FeS to dissociate. As shown in Table 6, most heavy metal sulfides are less soluble than FeS. FeS continues to dissociate until sulfide ions are no longer consumed by the precipitation of metal sulfides. If the reaction pH is too alkaline, the ferrous ion will precipitate as ferrous hydroxide.

Hexavalent chrome is reduced in the ISP process by the following reaction:

$$H_2CrO_4 + FeS + 2H_2O = Cr(OH)_3 + Fe(OH)_3 + S \qquad (8)$$

Table 6 Solubilities of Sulfides

Metal sulfide	K_{sp}^a at 64–77°F	Sulfide concentration (mol/L)
Manganese sulfide	1.4×10^{-15}	3.7×10^{-8}
Ferrous sulfide	3.7×10^{-19}	6.1×10^{-10}
Zinc sulfide	1.2×10^{-23}	3.5×10^{-12}
Nickel sulfide	1.4×10^{-24}	1.2×10^{-12}
Stannous sulfide	1.0×10^{-25}	3.2×10^{-13}
Cobalt sulfide	3.0×10^{-26}	1.7×10^{-13}
Lead sulfide	3.4×10^{-28}	1.8×10^{-14}
Cadmium sulfide	3.6×10^{-29}	6.0×10^{-15}
Silver sulfide	1.6×10^{-49}	3.4×10^{-17}
Bismuth sulfide	1.0×10^{-97}	4.8×10^{-20}
Copper sulfide	8.5×10^{-45}	9.2×10^{-23}
Mercuric sulfide	2.0×10^{-49}	4.5×10^{-25}

[a]K_{sp} solubility product of metal sulfide.
Source: Cherry (1982).

Metal Plating and Finishing Wastes

The primary advantage of the ISP process over the SSP process is that sulfide odors are not present because of the low solubility of FeS.

Although sulfide precipitation has a number of advantages over hydroxide precipitation, its disadvantages include the following:

1. The particle size of sulfide precipitants is much smaller than hydroxide, leading to poor settling. The proper use of flocculating polymers and coagulants is critical to producing adequate settling.

2. If excess sulfide is present in the effluent from the sulfide process, it can be released as toxic hydrogen sulfide (H_2S) gas if it mixes with acidic wastewater. For this reason, excess sulfide in the effluent may be oxidized to sulfate by air sparging in a discharge tank with about 30-min retention time.

3. If the wastewater is not neutralized properly such that sulfide reagent is added to acidic wastewater, H_2S fumes can be produced. Automatic controls that eliminate the possibility of adding sulfide to acidic wastewaters may be used to reduce the potential for H_2S generation.

4. The volume of sludge produced by sulfide precipitation is much greater than that produced by hydroxide precipitation. The removal of hydroxide precipitants after neutralization will reduce the total amount of sludge produced in the sulfide precipitation step. If the hydroxide precipitants are not removed, they will resolubize and precipitate as sulfides.

3.5 Cyanide Destruction

Alkaline Chlorination

The principle method to treat cyanide-bearing wastewaters is by alkaline chlorination. The process consists of two primary steps: conversion of cyanide to cyanate and conversion of cyanate to carbon dioxide and nitrogen.

A flow diagram of the two-step treatment process is shown in Fig. 11. In the first stage, the addition of chlorine to the reaction tank is controlled by maintaining an oxidation-reduction potential (ORP) of 350 to 400 mV. The pH of the reaction tank is controlled at 10.5 to 11 by the addition of a caustic. In the second stage, the ORP is maintained at 600 mV and the pH at 8.0. Although the second stage has a lower pH than the first, caustic addition is still required in the second stage because the reduction reaction reduces pH.

Chlorine may be added as a gas or hypochlorite. Although hypochlorite has the disadvantage of a short shelf life, higher cost, and a more complicated metering system, it has the advantage of being safer to handle and store than chlorine gas because it is an aqueous liquid.

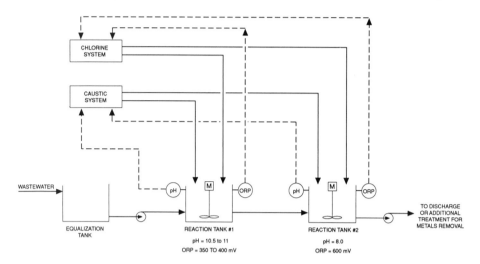

Figure 11 Cyanide destruction–alkaline chlorination.

Either batch or continuous treatment may be used. In a batch system, the pH is adjusted first before the addition of chlorine. The pH must be maintained as chlorine is added. As for continuous systems, two steps are used; ORP and pH are controlled at the same levels for the continuous process.

The addition of chlorine to wastewater that is not sufficiently alkaline can cause the release of toxic cyanogen cyanide gas or even hydrogen cyanide. Because of this potential, proper control of the system pH is critical.

Ozonation

Ozone has also been used to oxidize cyanide by the following reactions:

$$CN + O_3 = CNO + O_2 \qquad (9)$$

$$2CNO + 2O_3 = 2CO_2 + N_2 + 2O_2 \qquad (10)$$

About two pounds of ozone per pound cyanide are required for oxidation to cyanate, and about five pounds of ozone per pound cyanide are required for complete oxidation.

Ozone may be added to the reaction vessel using diffusers or a liquid eductor. Contact time is typically 5 to 10 min.

Metal Plating and Finishing Wastes

Although ozonation is an effective method for oxidizing cyanide, the capital and operating (primarily because of high energy requirements) costs are usually much greater than those for alkaline chlorination. Costs can be reduced by oxidation only to cyanate; however, although cyanate is much less toxic than cyanide, it may be considered equal to cyanide by regulating agencies. If wastewater is further dealt with by standard biological treatment, partial oxidation to cyanate may be feasible, because biological treatment will oxidize cyanate.

Ultraviolet (UV) radiation may be used in conjunction with ozonation. The primary benefit of UV radiation is that otherwise unreactive iron-complexed cyanides will also be destroyed.

4. METAL SLUDGE STABILIZATION AND SOLIDIFICATION

Heavy-metal-containing sludges produced by conventional treatment methods that are to be landfilled as the final disposal means must be treated so that the leachability of metals in the sludge are minimized to the point where the sludge passes EPA's toxicity characteristic leaching procedure (TCLP). Stabilization and solidification processes of the metal-containing sludge increase the bearing strength of the material, eliminate free liquids, and reduce the solubility, mobility, and leachability of heavy metals. The process of stabilization and solidification is relatively simple and inexpensive.

The solidification of metal sludges can be accomplished by using pozzolanic materials and/or cement. Pozzolanic materials are those materials that solidify when combined with hydrated lime and moisture. Pozzolanic materials are both naturally occurring (e.g., hydrated silicic acid) and artificial (e.g., fly ash). These materials are characterized by high amounts of silicic mineral components that can react with lime and aluminum oxide. Solidification and stabilization processes also use portland cement as a binding agent (Shively et al., 1986) or a mix of cement and pozzolanic materials. Portland cement contains a defined mixture of powdered oxides of calcium, silica, aluminum, and iron. The use of other sorbents such as bentonite will also decrease the loss of certain heavy metals. Stabilization of the soluble metal ions occurs by incorporating the metal into the solid matrix of the lime-and-pozzolan mixture or cement.

Table 7 presents estimated quantities required for solidifying and stabilizing typical plating and metal-finishing waste sludges. Treatment with commercial or waste lime alone usually will not produce a stabilized solid. Stabilization

Table 7 Estimated Solidification and Stabilization Reagent Quantities (lb/gal of waste)

Waste sludge	Commercial lime	Waste lime	Lime, fly ash, and bentonite mixture	Cement fly ash
Metal hydroxide	24	47	9	20
Copper pickle liquor	15	22	6	16
Ferric chloride pickle liquor	21	33	16	29
Sulfuric acid pickle waste (<15% H_2SO_4)	25	43	19	32

Source: Stanzyk et al. (1982).

and solidification are most effective for the treatment of cationic heavy metals. Soluble anionic species such as arsenate, chromate, selenate, and stannate are not as effectively bound within the solid matrix. The specific agents for effective solidification and stabilization should be determined by laboratory testing because waste metal types and concentrations can vary significantly with plating and metal-finishing practices. There are numerous commercial solidification and stabilization vendors in the United States that have experience stabilizing metal-containing wastes.

The encapsulation or smelting of metal-bearing waste sludges are other ways in which the metals can be stabilized before disposal. Encapsulation is more complicated and is not as universally applicable as other solidification techniques. The smelting of metal sludges is not practical for most small generators of metal sludges, but is a very effective means of stabilization.

5. NEW TECHNOLOGY

The principle alternative to sludge-producing wastewater treatment processes for heavy metal removal is a combination of ion exchange and electrowinning. In this combined process, ion exchange removes and concentrates the heavy metal, and electrowinning recovers the heavy metal as a pure, reusable metal from the ion-exchange regenerant. In addition to ion-exchange/electrowinning systems, electrodialytic equipment has been developed for the treatment of chrome, which cannot be electrowon.

Metal Plating and Finishing Wastes

5.1 Waste Segregation

Because the operating conditions (especially pH) for electrowinning heavy metals are different for each metal, it is difficult to electrowin a mixture of metals. With a mixture of heavy metals, one metal will be preferentially electrowinned while the others remain in solution. Further electrowinning of the remaining metals requires a change in the operating conditions, which complicates the operation of the electrowinning system. Also, the purity of the recovered metals is reduced, which may decrease the value of the recovered metal. For these reasons, the segregation of wastewaters according to the heavy metal present is recommended. The segregated metals can then be recovered in dedicated treatment systems.

Besides segregation by heavy metal, segregation by concentration may also be beneficial. Plating baths that contain high heavy metal concentrations may be suitable for electrowinning without a prior ion-exchange step. Ion-exchange regenerant and wasted plating baths usually may be electrowinned together.

5.2 Ion Exchange

Ion-exchange systems use resin in the form of plastic beads that contain functional groups that act as exchange sites. The exchange sites are initially located with an easily removed ion, which can be replaced by ions that will bond more strongly to the exchange sites. The loosely bound ion in cation-exchange resins (used for the removal of positive ions or cations) is usually either hydrogen or sodium. As wastewater is passed over the resin, the hydrogen or sodium in the resin in replaced by the metal ions present in the wastewater. For anion-exchange resins, the loosely bound ion is usually either sulfate or chloride.

Ion-exchange resins have a varying degree of preference for each metal ion, depending on resin type, operating conditions, and the metal being removed. This preference or selectivity is illustrated in Table 8 for IRC-718, a resin manufactured by Rohm and Haas. This resin is particularly useful for the removal of heavy metals because of the high selectivity of the resin for heavy metals over common cations, such as calcium and magnesium. This selectivity is important if wastewater is recycled after electrowinning. Otherwise, recycling would lead to a buildup of ions that would be removed by the ion-exchange resin but could not be electrowon, such as calcium. This buildup would eventually require that the treated rinse water be disposed of or treated in a separate system.

Table 8 Ion-Exchange Resin Selectivity[a]

Metal ion	Selectivity relative to calcium
Hg^{+2}	2800
Cu^{+2}	2300
Pb^{+2}	1200
Ni^{+2}	57
Zn^{+2}	17
Cd^{+2}	15
Co^{+2}	6.7
Fe^{+2}	4.0
Mn^{+2}	1.2
Ca^{+2}	1.0

[a]Rohm & Haas Amberlite IRC-718 resin. Tests were done with metals in aqueous solution at pH 4.

Selectivity is also important in determining the amount of metal that can be removed before the resin capacity is reached. If the resin had an affinity for calcium, calcium that might be present in the wastewater could be removed in competition with heavy metals; if enough calcium is present, a significant amount of the removal capacity of the resin could be consumed by removal of calcium instead of heavy metals.

Contact between the wastewater and ion-exchange resin is accomplished by passing the wastewater over beads of resin contained in one or more columns. The columns may be operated in series or in parallel. Wastewater flow can be downward or upward. For heavy metal removal, parallel operation with a downward flow is usually used. Some typical design criteria for ion-exchange columns are given in Table 9.

Usually, the only pretreatment required ahead of the ion-exchange columns is pH adjustment. A pH of about 4 to 5 is usually optimum because a lower pH will prevent the metal from loading onto the resin, and a greater pH could cause precipitation of the heavy metals. If the wastewater has an appreciable suspended solids concentration, prefiltration is recommended; otherwise, the ion-exchange column acts as a filter and must be backwashed. If organic compounds are present, an activated carbon bed ahead of the ion-exchange column may be necessary to prevent contamination and irreversible fouling of the resin.

Metal Plating and Finishing Wastes 161

Table 9 Ion-Exchange Column Design Criteria

Resin bed depth	24 in. minimum; 36–72 in. typical
Service flow rate	5 gpm/ft^2; 1–2 gpm/ft^3 resin
Number of vessels	At least one column should be on standby or in regeneration at all times. For example, if two columns are required at design wastewater flow rate, three columns should be installed.
Resin volume	As required to give a minimum of 24 hr between regenerations
Resin packing	Packed bed
Operation	Countercurrent (downflow service and upflow regeneration or upflow service and downflow regeneration)
	Concurrent (downflow or upflow service and regeneration)

The usable capacity of the resin for a particular metal will vary, depending on a number of factors. The type of resin is the most important of these factors. Resin manufacturers should be consulted for their recommendations regarding the type of resin that should be used for a specific application. Laboratory tests to determine the removal capacity of different resins are recommended if no data are available for a particular wastewater. Approximate metal removal capacities using IRC-718 are given in Table 10. The removal capacity for chromate using the anionic resin, IRA-92, is also shown. These values are based on tests conducted on wastewaters from a plating facility using a 3-in.-diameter ion-exchange column.

Usually, a single ion-exchange column will remove metal ions to below permit limits. When permit limits are particularly low, two columns in series may be needed. With this arrangement, metal ions that pass through the primary column are removed to lower concentrations in the polishing column. A three-column arrangement allows one column to be regenerated while the other two are in service. Once the primary column is exhausted, it is

Table 10 Heavy Metal Removal Rates Using Ion Exchange

Metal	Removal rate (lb/ft^3 resin)[a]
Copper	0.5–1.0
Nickel	0.8–2.0
Cadmium	0.8–2.0
Trivalent chrome	0.5–1.0
Hexavalent chrome	1.0–2.5[b]

[a]Capacities are for Rohm & Haas IRC-718 resin unless noted.
[b]Weak basic anion resin; Rohm & Haas IRA-94.

regenerated, and the polishing column becomes the primary column. The third column, which has been on standby after regeneration, is then used as the polishing column.

Ion exchange will also remove chelated metals when the ion-exchange resin has a greater affinity for the metals than the chelating agents. However, high concentrations may result in poor removal because of a slow transfer rate of metal from the chelating agent to the ion-exchange resin. Therefore, it is recommended that wastes with high concentrations of chelated metals (such as those present in dumped baths) be stored and then diluted by bleeding the chelated wastes into the waste rinsewater stream.

Once the capacity of the ion exchange is exhausted, the resin is regenerated to remove heavy metals and return the resin to its usable form. Regeneration is accomplished by passing strong acids (for cationic resins) or caustic (for anionic resins) over the resin. Sulfuric acid is usually used for regenerating cationic resins. Hydrogen ions from the acid displace the heavy metals from the resin, resulting in a regenerant waste stream with a high concentration of heavy metals. Metal concentrations from 5000 to 10,000 mg/L are typical. Other steps in the regeneration sequence include backwashing, draining, and rinse steps. A typical regeneration sequence is shown in Table 11.

Most metals present in plating wastewater can be eluted from a loaded ion-exchange bed for using sulfuric acid regeneration. For lead, however, sulfuric acid regeneration causes lead sulfate to precipitate within the ion-exchange bed. Sulfamic acid has been found to be an acceptable alternative by Dickert (1988). Hydrochloric acid may also be used for regeneration, but can

Table 11 Ion-Exchange Regeneration[a]

Step	Design critera
Backwash	3 gpm/ft^2, 10-min duration
Regenerate	4–15% acid solution (usually sulfuric); 1 gpm/ft^2 flow rate
Slow rinse	1 gpm/ft^2 flow rate, one-bed vol.
Fast rinse	5 gpm/ft^2 flow rate, 75 gal/ft^3

[a]These design criteria will vary depending on the type of resin. Consult the resin manufacturer for specific design criteria.

pose corrosion problems if electrowinning is used to recover metals from the waste regenerant.

Oxidation of the ion-exchange resin may occur if nitric acid or other strong oxidants are present in the wastewater. The potential for oxidation is greatest if a dumped bath is treated directly in the ion-exchange system without any dilution. Oxidation can result in the formation of gas, which may lead to overpressurization of the ion-exchange column. Because of this potential, ion-exchange vessels should be equipped with a pressure relief safety valve. Also, any dumped baths that contain oxidants should be stored and then bled into the waste rinsewater stream.

Conscientious operation of the ion-exchange system is critical to successful performance. It is recommended that an operating log of pertinent operating data, such as flow rates and metal concentrations, be maintained. The most important operating parameter is the metal concentration in the ion-exchange effluent. If the ion-exchange resin is regenerated before the bed is completely exhausted, the metal concentration in the regenerant will be lower, which makes metal recovery more difficult. On the other hand, if flow through the ion-exchange bed is not stopped immediately after the resin is exhausted, the effluent metal concentration may exceed permit limits. Once the breakthrough of metal begins to occur, there is a short time period (around a few hours) when the metal concentration increases up to the influent concentration. By identifying when the metal concentration just begins to increase, the resin capacity can be fully utilized without exceeding the permit limits. The operation of two columns in series avoids some of these problems.

Although on-stream analyzers are available for some metals, they tend to require a large amount of maintenance, may be unreliable, and are expensive. The monitoring of the effluent concentration with a simple Hach kit is less expensive and may be just as reliable. At least one sample per day should be analyzed by a referee method as a backup to the Hach test.

It is beneficial to maintain historical data on resin capacity that is calculated from flow rate, and influent and effluent metal concentrations. By plotting the historical trend line, the deterioration of the resin can be identified. This information can also be used to estimate the resin capacity, which can then be used along with operating data to help predict when the resin is near exhaustion. If permit limits are stringent, it may be best to regenerate the resin before the breakthrough of metal begins or operate with two columns in series. A good estimate of resin capacity is particularly important if the resin is regenerated just prior to breakthrough.

5.3 Electrowinning

Electrowinning is used to remove metal ions from wastewater by plating the metal onto the cathode of an electrochemical cell. An electrochemical cell consists of a series of anodes and cathodes immersed in the wastewater and a rectifier, which supplies power to the anodes. At the anode electrons are removed from the wastewater and travel through the external circuitry of the rectifier. These electrons combine with metal ions in solution at the cathode, causing metal to plate onto the cathode material. There are two principal types of electrowinning systems, identified by the type of cathode—parallel plate and woven mesh.

In parallel plate electrowinners, a large volume of metal may be plated on the cathodes from a metal-rich waste solution. With decreasing metal concentration, however, the efficiency (amount of metal removed per unit of power) of parallel plate electrowinners decreases to the point where it is no longer feasible to further reduce the metal concentration. An example of this decrease in efficiency is shown in Fig. 12. Because of the decrease in efficiency, additional treatment of the electrowinner effluent is usually necessary.

The approximate metal removal capacity of parallel plate electrowinners is shown in Table 12. Once this capacity is reached, the efficiency of metal removal decreases substantially, and the recovered metal plate is removed from the cathode by peeling the metal from the plate. This process can be difficult; one of the most important features that should be considered in selecting a electrowinner is the ease of metal removal. The surface condition of the plates is important. Any scratches or imperfections will be filled with the metal being electrowon, making removal more difficult.

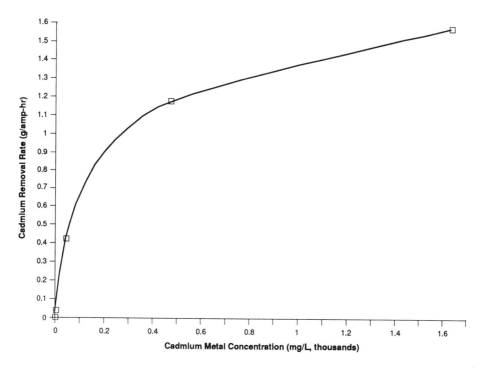

Figure 12 Parallel plate electrowinner cadmium metal removal.

The electrowinner effluent may be further treated by recycling it into the ion-exchange system. One disadvantage of recycling is that metal ions that are removed by ion exchange but not electrowinning will eventually accumulate. Depending on the ion-exchange resin, either ferrous or ferric ions could accumulate in this way. Because of the potential accumulation of ions, it is important to assess the fate of all ions present in the wastewater during laboratory or pilot plant tests. If these tests indicate that certain ions will accumulate, recycling of the electrowinner effluent may be infeasible.

An alternative to recycling to the ion-exchange system is to further treat the wastewater with a high-surface-area electrowinner. These systems use cathodes made of woven carbon or copper mesh with a much greater surface area than the plates used for parallel plate electrowinners. Because of the greater surface area, these electrowinners usually are able to remove metals to below permit limits. The amount of metal that can be removed per cathode is

Table 12 Parallel Plate Electrowinner Performance

Metal	Removal rate (lb/hr/ft^2)	Effluent metal concentration (mg/L)
Cadmium	0.04–0.06	10–100
Copper	0.04–0.20	600–1000
Nickel	0.03–0.04	1000–3000
Zinc	0.03–0.04	1000–3000

much lower, however, so that metal removal from concentrated wastewater solutions is impractical and more costly than parallel plate electrowinners. Table 13 shows the approximate removal capacity of a Model S-10 Baker Brothers HMT system.

One disadvantage of woven mesh cathodes is that the plated metal cannot be removed from the cathode material. The Baker Brothers HMT system regenerates cathodes by immersing the metal-coated cathodes in an electrorefining cell, where the metal is dissolved from the cathodes and replated on a conventional parallel plate cathode. The metal can then be stripped from the plate cathodes. Although this system reduces the cost of replacing cathodes, it does increase the capital cost of the electrowinning system.

Woven copper mesh, similar in appearance to steel wool, is frequently used for cathodes to remove copper in a high-surface-area electrowinning system. After the copper mesh has reached its capacity for copper, it can be sold as copper scrap, which helps offset the cost of the cathodes.

Electrowinners are usually operated in a batch mode. Wastewater is first collected in one of two tanks. One tank is used for collection and pH adjustment, while the other is being used for recirculation of the wastewater

Table 13 HMT Electrowinner Performance

Metal	Removal rate (lb/hr/ft^2)[a]
Cadmium	0.07
Copper	0.01–0.09
Zinc	0.02–0.03
Tin/lead	0.02–0.06

[a]Model S-10 Baker Brothers HMT.

Metal Plating and Finishing Wastes 167

through the electrowinning cell. Recirculation is necessary to supply an adequate metal concentration in the electrowinner. Without recirculation, the metal content would become depleted in the vicinity of the cathode, reducing removal efficiency.

After the wastewater has passed through the electrowinner, it is returned to the recirculation tank. Eventually, the metal concentration is reduced to the point where the electrowinner efficiency is too low for further removal to occur (parallel plate electrowinners) or the metal concentration is below permit limits (woven mesh electrowinners). At this point, the wastewater in the recirculation tank is discharged (if it is below permit limits) or recycled to another treatment system such as an ion-exchange system. After the tank is emptied, it accumulates another batch of waste and the other tank is then used as the circulation tank for the next electrowinning run.

Electrowinners are able to treat low concentrations of cyanide wastes, also, by oxidizing the cyanide to carbon dioxide and nitrogen (Water Engineering Research Laboratory, 1987). For high concentrations of cyanide, however, the conventional chlorine oxidation method of cyanide destruction is less expensive and easier to control.

A combination of electrowinning with ion exchange is common. Ion exchange serves two purposes: the concentration of metals and elimination of corrosive compounds.

The concentration of metals is particularly advantageous for waste rinsewater streams. Without concentration, only a woven mesh electrowinner would be effective, since a parallel plate electrowinner is inefficient at low concentrations (see Fig. 12). Because woven mesh systems are relatively expensive, an ion-exchange system in combination with a parallel plate electrowinner will generally be more cost-effective than a woven mesh electrowinner system alone.

Corrosive compounds, such as nitric acid and chloride compounds, can substantially increase the cost of electrowinners because cathodes and other components of the electrowinners must be made of expensive alloys, such as titanium. Ion exchange will remove metal compounds while allowing corrosive anions, such as chloride, to pass through. Regeneration with sulfuric acid produces a waste stream of metal sulfates, which are less corrosive than metal chlorides.

5.4 Electrodialytic Systems

Like electrowinners, electrodialytic systems use an electrical current between a cathode and an anode to separate metal ions from solution. In an electrodialytic system, the anode is located inside a cartridge made of a dialytic membrane. This membrane allows cations to pass through from the anode chamber to the vicinity of the cathode while repelling anions. Most heavy metals will pass through the membrane. An exception is trivalent chrome, which is oxidized to hexavalent chrome at the surface of the membrane. Since hexavalent chrome is an anion, the chrome remains in solution in the anode compartment, whereas other metals, such as copper, are removed by purging the cathode compartment. Because of this ability, one of the primary uses of electrodialytic systems is to regenerate chrome-anodizing baths. Other uses include the conversion of sodium chromate solutions to chromic acid and copper recovery from cupric chloride solutions. (The treatment of printed circuit board wastes, such as cupric chloride, will be covered in a later chapter.)

Currently, the authors are aware of only one company, Ion Sep, Inc., that markets dialytic membrane cartridge systems.

5.5 Chrome-Anodizing Bath Regeneration

Chrome-anodizing baths are used to prepare metal surfaces for plating with other metals. Over time, hexavalent chrome, which is the active ingredient in anodizing baths, is reduced to trivalent chrome. Eventually, the anodizing power of the bath becomes insufficient because the hexavalent chrome concentration is reduced below control limits. Besides the conversion of hexavalent chrome to trivalent chrome, the concentration of the metal being treated accumulates in the anodizing bath. The advent of electrodialytic systems allows these baths to be regenerated rather than discharged to a waste treatment system. Electrodialytic systems convert the trivalent chrome back to hexavalent chrome while removing other metals that have accumulated in the anodizing bath.

The regeneration of anodizing baths can be done either continuously or with batch treatment. In a continuous system, a constant flow of the anodizing bath is circulated through the electrodialytic system. The electrodialytic membrane cartridge is immersed in the recirculation tank. Metals that have accumulated in the anodizing bath pass through the membrane and are collected. Two principal methods are used to collect metals: precipitation in a catholyte solution (primarily sodium hydroxide) circulated through the cathode chamber of the cartridge and oxidation onto metal mesh collection plates.

With the latter collection method, the collection plates periodically must be cleaned off. These systems are impractical if a large amount of metal is

Metal Plating and Finishing Wastes

collected because operator time to remove metal oxides from the collectors becomes excessive.

The collection of contaminant in a catholyte solution allows a much larger amount of metal to be removed before the collected metal must be further handled. If copper is being removed, a solution with as much as 20% by weight precipitated copper can be collected. These systems include a catholyte storage tank and catholyte recirculation pumps. The storage tank sometimes will have a cone-bottom tank to aid in solids removal. After precipitated solids have built up in the catholyte tank, the solids are removed and dewatered in a filter press. The dewatered solids can then be disposed of in a landfill after stabilization or deposited directly into an electrowinner system for the recovery of pure metal. If an electrowinner is used, the solids must be dissolved in sulfuric acid before being electrowon.

The copper removal rate using a circulating catholyte solution is dependent on the wastewater copper concentration and the current applied to the electrodialytic cell. For low copper concentrations, the removal rate is between 90 and 500 mg/hr/ft^2 membrane area.

One method used for the treatment of anodizing baths is a combination waste bath treatment and waste rinsewater treatment. In this case, the bath is not regenerated and recycled; instead, it is treated along with waste rinsewater by ion exchange, followed by the electrodialytic recovery of chrome in the ion-exchange regenerant. The waste baths are first collected and bled into the waste rinsewater, then treated in cation and anion ion-exchange vessels in series. Cation ion exchange removes trivalent chrome and other metals such as copper while anion ion exchange removes hexavalent chrome. The anion ion exchanger is regenerated with sodium hydroxide, producing a solution of sodium chromate. Regeneration of the cation ion exchanger with sulfuric acid produces a solution of sulfuric acid and trivalent chrome that also contains other heavy metals that were present in the wastewater.

The trivalent chrome sulfuric acid solution is collected and batch-treated with an electrodialytic system that removes heavy metals such as copper, while converting the trivalent chrome to hexavalent chrome. After treatment, the solution is recycled to the ion-exchange system for removal of the hexavalent chrome.

The sodium chromate solution produced by regeneration of the anion ion exchanger can also be treated with an electrodialytic system to produce chromic acid. Either sodium chromate or chromic acid potentially can be sold for reuse. If sodium chromate is produced, it must be concentrated, crystallized, and then dried. Chromic acid, on the other hand, need only be concentrated by evaporation and can be shipped in tanker trucks. If chromic acid is produced, the electrodialytic system is used to remove sodium from the sodium chromate solution, producing chromic acid.

If a large volume of anodizing solution is consumed, bath regeneration will tend to be more cost-effective than chrome recovery because replacement chemical costs are substantially reduced. If anodizing bath consumption is relatively small, chrome recovery will tend to be more cost-effective (due to lower capital costs), primarily because a single system is used to treat both rinsewater and wasted baths. With bath regeneration, a treatment system is required for rinsewater and another for bath regeneration.

Electrodialytic systems are fairly easy to operate and do not require much operator attention. After filling the treatment tank with wastewater, the operator needs only to turn on the recirculation pump and the power supply to the electrodialytic cell. Periodic checks of heavy metal and trivalent chrome concentrations are required to determine if the trivalent chrome has been fully converted to chromate and heavy metals have been removed.

REFERENCES

Blair, A. (1989). Silver plating. In *Metal Finishing Guidebook and Directory*. Vol. 87, p. 279.
Cherry, K. F. (1982). *Plating Waste Treatment*. Ann Arbor Science Publishers, Ann Arbor, Mich. p. 55.
Davis, M. W. and Sandy, A. T. (1989). Zero sludge/zero hazardous waste pretreatment systems for the metal plating/finishing industry. In *Proceedings of 44th Annual Purdue Industrial Waste Conference*, West Lafayette, Ind.
DiBari, G. A. (1989). Nickel plating. In *Metal Finishing Guidebook and Directory*. Vol. 57, p. 251.
Dickert, C. (1988). Personal communication. Rohm & Haas.
Geduld, H. (1989). Zinc plating. In *Metal Finishing Guidebook and Directory*. Vol. 87, p. 309.
Groshart, E. C. (1989a). Preparation of basis metals for plating. In *Metal Finishing Guidebook and Directory*. Vol. 87, pp. 167–172.
Groshart, E. C. (1989b). Pickling and acid dipping. In *Metal Finishing Guidebook and Directory*. Vol. 87, p. 154.
Hall, N. (1989). Coloring of metals. In *Metal Finishing Guidebook and Directory*. Vol. 87, p. 436.
Hanson, N. H. and Zabban, W. (1959). *Plating, 46,* 909.
Henry, J. (1989). Electrodes (autocatalytic, chemical) plating. In *Metal Finishing Guidebook and Directory*. Vol. 87, pp. 388–392.

Metal Plating and Finishing Wastes 171

Hirsch, S. and Rosenstein, C. (1989a). Tin-lead, lead and tin plating. In *Metal Finishing Guidebook and Directory*. Vol. 87, p. 284.

Hirsch, S. and Rosenstein, C. (1989b). Immersion plating. In *Metal Finishing Guidebook and Directory*. Vol. 87, pp. 403–406.

Johnson, J. C. (1989). Vapor degreasing. In *Metal Finishing Guidebook and Directory*. Vol. 57, p. 104.

Marce, R. E. and Morrow, H. (1989). Cadmium plating. In *Metal Finishing Guidebook and Directory*. Vol. 87, p. 193.

McMullen, W. (1989). Chromium plating. In *Metal Finishing Guidebook and Directory*. Vol. 87, p. 194.

Mohler, J. B. (1989). Water rinsing. In *Metal Finishing Guidebook and Directory*. Vol. 57, pp. 136–139.

Morrisey, R. J. (1989). Platinum plating. In *Metal Finishing Guidebook and Directory*. Vol. 87, p. 272.

Sato, A. and Barauskas, R. (1989). Copper plating. In *Metal Finishing Guidebook and Directory*. Vol. 87, pp. 210–214.

Shively, W., et al. (1986). Leaching tests of heavy metals stabilized with portland cement. *J. Water Pollution Control Fed., 58*, 234–240.

Stanczyk, T. F., et al. (1982). Solidification/stabilization process appropriate to hazardous chemicals and waste spills. In *1982 Hazardous Material Spills Conference*. Government Institutes, Inc., Rockville, Md., pp. 79–84.

Water Engineering and Research Laboratory (1987). Evaluation of the HSA Reactor for Metal Recovery and Cyanide Oxidation in Metal Plating Operations. EPA Res. and Dev. Document, EPA/600/52-86/095, Cincinnati, Ohio, Jan.

Wang, L. K. (1984).

Weisberg, A. M. (1989). Gold plating. In *Metal Finishing Guidebook and Directory*. Vol. 57, p. 224.

4

Treatment of Photographic Processing Wastes

Thomas W. Bober, Thomas J. Dagon, and Harvey E. Fowler

Eastman Kodak Company, Rochester, New York

1. THE PHOTOGRAPHIC PROCESS

1.1 Exposure

A photographic material consists of a *base* made of a sheet of plastic film, glass, or paper coated with a photographic *emulsion* (consisting of a polymeric material such as gelatin containing numerous fine crystals of light-sensitive silver halide). Silver halides used in photographic emulsions include silver bromide, silver chloride, silver iodide, or mixtures of these.

Exposure of this photographic emulsion to light results in the formation of a "latent image," a four- to ten-atom speck of metallic silver on the silver halide crystal. In simplified terms, light striking molecules of silver halide (AgX) in the emulsion causes some of them to be reduced to metallic silver (Ag^o) atoms. This can be represented by the following simplified equation.

$$AgX + light \rightarrow Ag^o$$

(silver halide) (metallic silver latent image)

This latent image is invisible to the naked eye, since it consists of only a minute portion of the total amount of silver halide available in the emulsion. The number of atoms converted to silver depends on the *intensity* of the light and the *duration* (time) of exposure, as well as several other factors not discussed here.

The following discussion of development covers black-and-white images. Color development is examined in a later section. See (Duffin, 1966; Thomas, 1973; James, 1977; Mason, 1985) for a more complete description of the photographic process.

1.2 Development

In order to be useful, this nearly invisible latent image must be enhanced or "amplified" by converting other surrounding silver halide molecules to silver metal, until the metallic image becomes fully visible. This is done chemically by a process step known as *development*. The latent image is *developed* by immersing the emulsion into a solution (the *developer*) containing a mild chemical reducing agent (the *developing agent*) in water, usually at an alkaline pH. The developer may include other ingredients added to enhance the process. A chemical reaction takes place in which the developing agent furnishes an electron (e^-) to reduce the silver halide molecule (AgX) to a metallic silver atom (Ag^o), releasing halide ion (X^-) in the process. The reaction describing this step can be summarized as follows:

$$AgX + e^- \rightarrow Ag^o + X^-$$

(silver halide surrounding latent image)	(electron from developer)	(metallic silver image)	(free halide ion)

This chemical reduction proceeds as an autocatalytic chain reaction, enhanced or catalyzed by the presence of already formed metallic silver atoms. Thus, the silver halide molecules in the immediate vicinity of the latent image react more quickly than those that are farther away.

Those sites in the latent image that were originally exposed to the most intense light will contain the largest concentration of metallic silver atoms. Therefore, development will proceed faster around these areas than around other locations where weaker light had exposed fewer silver atoms. The chain reaction thus proceeds at different rates within the silver halide crystal and among crystals, producing groups of dark metallic silver "grains," with the groupings sized roughly in proportion to the amount of light that struck the area. If the emulsion were allowed to remain in the developer for a long time, eventually the reaction would proceed to completion and all the silver halide crystals would be reduced to metallic silver, leaving a fully developed black surface. Thus, *time of development* is an important factor in this step, with time (among other factors) being carefully controlled to yield a photographic image of a desired "density," "contrast," "grain," "sharpness," "resolution," and other characteristics that are discussed further in most standard photographic texts. Another very important factor is *developer temperature*. As the temperature is increased, a silver image is produced in a shorter time, but may have less sharpness and contrast than if the reaction were conducted at a lower temperature. Other factors influencing the results include pH, the measure of acidity of alkalinity of a solution (developers are usually strongly alkaline, in a pH range of 9–12); *strength* of the developer solution, with more dilute solutions producing finer developed silver particles (i.e., less grain) but at a slower rate; the *type of emulsion* used (emulsions have varying thickness and are often combined in multiple layers, with each layer containing various additives to modify the process and achieve certain desired results); *agitation* or mixing (to remove unwanted development by-products from the emulsion and allow fresh developer to access the silver halide crystals), and other factors. In the real world of the photoprocessing laboratory, *developer stability* is an important factor: Since the developing agent is a chemical reducing agent, it is easily oxidized by oxidizing agents, including air, and must be protected. Thus, certain advantages one could gain by running the process one way may be traded off for other advantages gained by operating it under slightly modified conditions.

A chemical known as a "preservative," usually sodium sulfite, is added to protect the developer from unwanted aerial oxidation; it does this by sacrificing itself and being oxidized instead, generally to sulfate. This is summarized by the reaction equation

$$SO_3^= + 1/2\, O_2 \rightarrow SO_4^=$$

(sulfite ion)   (oxygen from air)   (sulfate ion)

During development, unwanted by-products form in the developer that gradually decrease the activity of the developing agent and begin to retard development. One of these is the free halide ion (chloride, bromide, or iodide) that is released to the solution when the developer converts the silver halide to metallic silver. Another is the oxidized (used) developing agent itself: When it reduces (i.e., gives up an electron to) the silver halide molecule, it in turn becomes oxidized (i.e., has lost an electron). Thus, it can no longer enter into the development reaction and, in the case of black-and-white developer, becomes a useless, less soluble molecule that impedes the action of its unused neighboring developer molecules. It can also form an objectionable brown scum or stain on the surface of the emulsion. Fortunately, a rapid chemical reaction occurs between the oxidized developing agent and the sulfite preservative to produce a sulfonated form of the spent developing agent that is less reactive and more soluble in water. This helps prevent stain or scum formation on the emulsion by keeping the spent developer in solution and also helps reduce interference with the unused developing agent.

At some point in the development cycle, the amount of spent reaction products could build up to a point where they would begin to seriously interfere with developer activity if not controlled. This buildup of unwanted development products has historically been remedied by discarding a portion of the solution and replenishing with fresh developer.

In order to produce the highest-quality result with the desired characteristics in the final developed image, it is necessary to be able to stop the development reaction quickly. This can be done by one of several ways: quickly lowering the pH, rapidly reducing the concentration of developer, or abruptly decreasing the temperature to a low value (the latter way is usually not practical, too expensive, and therefore generally not used). pH and developer concentration can be lowered by one of two ways: with the use of an acid solution (in either an appropriately named *stop bath* or an acidic *fixer*), or by dilution with water (which is generally not as sudden or precise) using a water wash.

1.3 Stop Bath

The *stop bath* (or "stop") is generally an organic acid in water. Acetic acid is used most of the time because it is relatively inexpensive, nontoxic, commercially available, and has good buffering capability (i.e., the ability to remain at a relatively fixed acidic pH even though a certain amount of alkali may be added to it). As described above, it stops the action of the developing agent when the film is immersed in this bath. Today the stop bath is frequently combined with the next solution in the process, the fixer, to form an "acid fixer" or "acid fix" that combines the action of stopping development with that of fixing, described next. This economizes on the number of separate steps and, correspondingly, the number of processing tanks or trays needed to process the emulsion.

1.4 Fixer

After the latent silver halide image has been developed into a metallic silver image, the remainder of the emulsion still contains undeveloped silver halide. Since this is an opaque, pinkish-to-grayish material that will not pass light (and will eventually turn into metallic silver if exposed to light or heat for too long a time), it must be removed from the emulsion. The *fixer* is a solvent that will selectively dissolve silver halide molecules from the emulsion while leaving adjacent metallic silver atoms relatively untouched. It is generally composed of thiosulfate, $S_2O_3 =$ (usually the sodium, potassium, or ammonium salt, depending on certain processing considerations). This is an inexpensive, nontoxic, commercially available, quick-acting solubilizing agent (also known as a "complexing agent" or "sequestrant," since it tends to form a stable chemical complex with the metal ion it dissolves). The thiosulfate fixer dissolves the unused silver halide from the emulsion, forming a tightly bound silver thiosulfate complex, $Ag(S_2O_3)_2-3$. This is a stable, water-soluble chemical complex that has very low toxicity compared to free silver ion ($Ag+$); thus, it is relatively safe for photographic personnel to handle and does not create toxicity problems if discharged to a sewer for biological treatment.

1.5 Color vs. Black-and-White Processes

The above description adequately portrays the typical black-and-white negative process, in which the final desired image is a black metallic silver image on a transparent or white background. Color processes follow approximately the same steps with some important variations.

A black-and-white developing agent is a mild reducing agent, usually a rather simple organic molecule like hydroquinone, which has no further value to the photographic process once it has become "spent" (oxidized) in developing the silver atom. In contrast, the color-developing agent is a more complicated molecule, usually a para-phenylenediamine-based compound, that comprises one-half of a dye-forming molecule.

The emulsion of a color film or paper is also more complex than its black-and-white counterpart: in addition to the profusion of fine silver halide crystals, the gelatin matrix also contains a dispersion of tiny globules of the other half of the dye-forming molecule, known as the "coupling agent" or *coupler*. Whereas the developer half of the final dye molecule is always the same regardless of color, the coupler half of the molecule is a different type of compound, depending on the particular color (cyan, magenta, or yellow) to be formed in that emulsion layer.

When the color developing agent reduces the silver halide molecule (AgX) to metallic silver (Ag°) by supplying an electron (e⁻), it in turn becomes oxidized by losing an electron, as described above. However, instead of becoming a waste product as in the case of the black-and-white developing agent, the oxidized color-developing agent now seeks to join up ("couple") with the nearest coupler molecule, to form a dye. This dye, being a large bulky molecule and relatively insoluble in water, has no tendency to migrate from its position in the matrix but remains in place. Thus, the "image" at this point consists of tiny globules of dye sharing space with clusters of developed silver grains, with both surrounded by a dispersion of transparent, unused coupler globules and undeveloped silver halide crystals in a gelatin matrix. The reactions can be described as follows:

$$AgX + e^- \rightarrow Ag^o + X^- \quad (1)$$

(silver halide surrounding latent image) (electron from developer) (metallic silver image) (free halide ion)

$$\text{color-developing agent} - e^- \rightarrow \text{oxidized color-developing agent} \quad (2)$$

(electron lost to silver)

$$\text{oxidized color-developing agent} + \text{coupler} \rightarrow \text{colored dye image} \quad (3)$$

The emulsion now contains both developed metallic silver and undeveloped silver halide, neither of which is wanted in the final product. Fixing at this point would remove only the silver halide but not the silver metal. Thus, it is necessary to convert the silver metal back to silver halide before both can be satisfactorily dissolved from the emulsion by the same fixing step.

1.6 Bleach

The conversion of silver metal back to silver halide is accomplished by using a mild oxidizing agent known as a *bleaching agent* together with a water-soluble halide salt, such as potassium bromide, in a water solution. Together these are known as a photographic *bleach*. The bleaching agent is mild enough not to adversely affect the gelatin or dye in the emulsion, yet strong enough to take electrons from the silver metal in the presence of the halide, thus converting the silver back to silver halide. Typically, iron complexes such as iron EDTA (ethylenediamine tetraacetic acid) or ferricyanide are used since they can supply the proper bleaching activity without harming the emulsion. These are both relatively nontoxic salts that are cheap, commercially available, and safe to handle [both are used in food products: EDTA is used in bread, baked goods, and pharmaceuticals (Merck Index, 1989), whereas ferricyanide is used to prevent table salt and foods from caking and as a blue pigment in cosmetics, inks, and paints (Merck Index, 1989; American Cyanamid Co., 1953)]. The reaction between the bleach and the metallic silver halide can be described in simplified fashion as follows:

$$Fe^{+3} + Ag^o + X^- \rightarrow AgX + Fe^{+2}$$

(ferric iron salt) (metallic silver) (halide ion) (silver halide) (ferrous iron salt)

After the metallic silver is converted back to silver halide, the entire emulsion can be fixed to remove all the silver, leaving only an image of finely divided colored dye globules in a transparent matrix.

1.7 Bleach-Fixes

In some processes, particularly paper processes that are more easily bleached than film processes, the bleach and fixer can be combined into a single solution known as *bleach-fix* (also commonly known in trade jargon as a "blix" or a "bleach-fixer"). Some chemical synergy is achieved by mixing the two solutions; therefore, the concentration of each can be lowered slightly to achieve the same photographic effect. The gentle oxidizing action of the bleach component is insufficient to damage the fixer component; therefore, they are able to survive together in a single solution. This single solution saves time in processing and simplifies the processing machine. It may also save money in shipping processing chemicals and may produce environmental benefits since it may result in less chemical usage. Bleach-fixes were first introduced commercially in the late 1960s for color paper, specifically for their environmental features as well as reduced processing steps.

1.8 Washes

Before finally drying a photographic emulsion, residual chemicals from processing solutions and reaction by-products must be eliminated to avoid future interactions that would limit the life of the product. A final water wash is most often used for both color and black-and-white products. Also, at certain critical junctures in various processes, it may be desirable to introduce an intermediate wash to remove residual chemicals and/or alter the pH or chemical balance before entering the next solution. Water washes generally contain lower concentrations of the same chemicals found in the preceding tank.

In the past, all water washes were usually discarded. In recent years, because of environmental and energy concerns as well as economic considerations, many schemes for purifying and reusing wash waters have been proposed. A number of these have been successfully implemented to accomplish at least partial recycling and reuse. In some of these cases, when an additional chemical may be needed to treat the water or rejuvenate a purification bed, disposal of the treatment chemical may pose a separate problem.

1.9 Stabilizers

At the end of the process, the gelatin emulsion, having undergone a series of swelling and shrinking cycles as it passed from one processing solution to another, may have lost some of the hardness and physical strength it originally had, which could make it susceptible to scratching or damage. Also, some of the newly formed dyes in the emulsion may need to be further chemically protected against aging and light fading. Both of these tasks can be performed by treating the emulsion with a *stabilizer*, usually the last solution in a color process.

Some processes attempt to save or eliminate water, such as "washless minilab" processes designed to provide processing for a customer in 1 hr or less. These may be located in department store or storefront locations not having sewers. In these cases, the stabilizer may also serve the function of a wash, by eliminating residual chemicals in the emulsion prior to the drying step, which is necessary for image stability upon long-term keeping.

Throughout much of the history of color films and papers, the most common and effective stabilizer has been a water solution of formaldehyde, sometimes containing additional ingredients such as citric acid. However, in recent years, because of heightened medical concerns over the handling of formaldehyde

Photographic Processing Wastes

plus its annoying lachrymatory odor, attempts have been made to substitute other materials.

1.10 Solution Carryover and Replenishment

If each of the above steps could be carried out under ideal, pristine conditions, there would be few unwanted reactions or by-products and therefore waste would be minimal. Unfortunately, in actual practice this is not the case.

Oxygen from the air is the primary cause of unwanted reactions. It slowly oxidizes components such as the developing agent and fixer upon long-term standing or solution agitation, both of which tend to promote dissolving of air, and during attempts to reuse solutions or recover silver. As previously mentioned, this oxidation necessitates adding preservatives such as sulfite and other ingredients needed to counter the effects of oxidation. These preservatives are also eventually consumed by oxidation, thereby forming by-products of their own.

Solution carryover is the second major cause of chemical loss, since a solution is carried on the surface and within the saturated emulsion from one tank to the next, thereby losing the solution from the first tank and contaminating the second. To protect against the undesirable effects of contamination, each succeeding solution must be chemically bolstered to contain more of the active ingredient than might be needed strictly to react with components in the film or paper, if carryover did not occur.

Squeegees are important devices for retarding carryout on the surfaces of photographic materials, and extra washes can be important means of reducing the carrying of unwanted contaminants. However, the fact remains that some unwanted material, even if it is water, will always be trapped within an emulsion and taken into the next processing tank. In addition, small amounts of some chemicals that were originally incorporated into the film or paper, including gelatin, will leach out into the solutions during processing. As previously stated, some chemicals produce by-products by reacting with oxygen from the air. Finally, some portion of the processing chemicals will always have reacted with the emulsion, producing reaction by-products (e.g., halides) that are released in one processing tank or another. The total effect of releasing chemicals from all of these sources into the solutions during processing is known as "seasoning."

Therefore, some means of replenishing the lost components and removing the unwanted components or neutralizing their effects is necessary to operate a continuous process. This means that some waste will always be generated and needs to be treated in some fashion.

2. PHOTOPROCESSING EFFLUENT CHARACTERISTICS

2.1 Introduction

The photoprocessing industry is very diverse. It includes photofinishing laboratories, X-ray processing at medical and dental facilities and industrial sites, professional photographic operations, motion picture laboratories, processing systems for scientific uses such as astronomy and geology, aerial mapmaking and satellite photography, microfilm processors, graphic arts operations, and others.

Photographic effluents vary in composition because there are many different types of photographic processes and no two processing laboratories operate in the same manner. Processing laboratories vary greatly in size, wash water usage, daily operating time, volume of effluent, and the use of chemical recovery systems.

The actual effluent characteristics (Thomas, 1973; Versar, 1981; Petschke, 1988) for any photographic processing laboratory can best be determined by collecting a representative sample of the photoprocessing wastewater and having it analyzed by a certified analytical laboratory. However, although concentrations will vary, the effluent from most photoprocessing laboratories will be similar in chemical composition.

It is not within the scope of this section to provide the actual processing effluent characteristics of every photographic process. Tables 1 through 3 represent typical effluent concentration ranges for conventional color processes, plumbingless color processes and black-and-white processes. The chemical (and other environmental parameter) concentrations of plumbingness processes (i.e., systems that do not utilize a conventional wash cycle) are usually quite high, but the loading of these ingredients in pounds or kilograms per day will be similar to photographic processes using a conventional wash cycle. The plumbingness process is designed for use by small processing operations; consequently, the total loading from these operations will likewise be small.

2.2 Environmental Parameters

The following sections discuss parameters that might be expected to occur in typical municipal sewer codes.

Table 1 Conventional Color Photoprocessing Effluent Characteristics

Parameter	Typical concentration, mg/L
Temperature	80–110°F
pH	6.5–9.0 units
BOD_5	200–3000
COD	400–5000
TDS	300–3000
TSS	<5 to 50
Phenolic compounds	none
Flammable; explosive	none
Detergents	minimal
Oils and grease	none
NH_3-N	20–300
TKN	30–350
Thiosulfate	100–1000
Sulfate	50–250
Metals	
Silver[a]	<0–1–5
Cadmium	<0.02
Chromium	<0.05
Iron	10–100
Lead	<0.02–0.1
Mercury	<0.0002
Nickel	<0.02–0.1
Zinc	<0.02–0.75

[a] After silver recovery (see the fourth subsection in Sec. 4.3). If iron replacement techniques are used for silver recovery, higher iron levels may be present.

Temperature

The temperature of some of the most widely used photographic processes is in the 80 to 110°F (26.7–43.3°C) range. This temperature range should not present a problem to municipal sewer systems.

Table 2 Plumbingless Color Photoprocessing Effluent Characteristics

Parameter	Typical concentration, mg/L
Temperature	<95°F
pH	6.5–9 units
BOD_5	5000–14,000
COD	30,000–36,000
TDS	60,000–90,000
TSS	10–50
Phenolic compounds	none
Flammable; explosive	none
Detergents	minimal
Oils and grease	none
NH_3-N	6000–10,000
TKN	8000–13,000
Thiosulfate	20,000–25,000
Sulfate	3000–4000
Metals	
Silver[a]	<5–100
Cadmium	<0.3
Chromium	<2
Iron	1400–2000
Lead	<1
Mercury	<0001
Nickel	<2
Zinc	<2

[a] After silver recovery. See note at bottom of Table 1.

Five-Day Biochemical Oxygen Demand and Chemical Oxygen Demand

The five-day biochemical oxygen demand (BOD_5) test measures the quantity of oxygen that the effluent, chemical, or solution will consume over a five-day period through biological degradation. The BOD_5 concentration of effluent from a photographic processing laboratory will vary widely, depending on the amount of washwater used, the composition of the processing solutions, and the varying combinations of processing and nonprocessing waste. The BOD_5

Table 3 Black-and-White Photoprocessing Effluent Characteristics

Parameter	Typical concentration, mg/L
Temperature	<80–110°F
pH	6.5–9.0 units
BOD_5	300–5000
COD	2000–20,000
TDS	1500–30,000
TSS	<5–50
Phenolic compounds	<0.2
Flammable; explosive	none
Detergents	minimal
Oils and grease	none
NH_3-N	350–4300
TKN	400–4500
Thiosulfate	1000–13,000
Sulfate	100–300
Metals	
Silver[a]	<0.1–5
Cadmium	<0.02
Chromium	<0.05
Iron	<0.5
Lead	<0.5
Mercury	<0.0002
Nickel	<0.05
Zinc	<0.2

[a] After silver recovery. See note at bottom of Table 1.

of effluent from various conventional photographic processing laboratories has been found to be in the range of 300 to 3000 mg/L (see Tables 1–3).

In the chemical oxygen demand (COD) test, the chemical or sample in question is refluxed with potassium dichromate and concentrated sulfuric acid for 2 hr. The COD of effluents from various photographic processing laboratories has been found to be in the 400 to 5000 mg/L range. The COD of photographic processing wastes is generally larger than the BOD_5, but the two analyses do not completely correlate on all samples because they do not measure the same oxygen-demanding chemicals. The COD test is much faster and more reproducible than the BOD_5 test.

Total Suspended Solids

Total suspended solids (TSS) are undissolved solid material carried in effluent. Photographic processing effluent is typically very low in suspended solids (less than 50 mg/L) and therefore should not present a problem to municipal treatment plants.

pH

This measurement is an indication of how acidic or alkaline (basic) the solution is, and almost every sewer code contains restrictions on the minimum and maximum pH of mixed effluent discharges. Most are in the range of 6.0 to 10.0. The pH of individual processing solutions may range from as low as 4 to as high as 12. However, the pH of overall photographic effluent usually does not present a problem to waste treatment systems since it is generally in the range of 6.5 to 9.0.

Heavy Metals

Materials classed as heavy metals are commonly regulated by local sewer authorities. They are usually defined as those metals with a specific gravity greater than 5.0. This includes metals such as antimony, arsenic, cadmium, chromium, cobalt, copper, gold, iron, lead, manganese, mercury, molybdenum, nickel, silver, and zinc.

The concentration of some heavy metals in an effluent may be regulated because of the toxicity of these metals or their compounds. The toxicity can vary with the particular metal or compound and with the form in which the metal exists, e.g., free ion, complex, or precipitate. Some metals may be quite toxic in one form while relatively nontoxic in another form; this property for metals to exist in various forms or species is known as "speciation." In the past, nearly all regulations were based on total metal concentration in an effluent; it has only been within the past few years that scientists have begun to realize the great importance of speciation, and that the species of a metal, plus its ability or inability to convert readily between species, should be the basis for regulation.

Some photo-sensitive photographic products do contain small amounts of metals that may appear in the effluent in addition to silver. Heavy metals frequently found in photographic processing effluent and that are commonly regulated include the following.

Silver. Silver compounds are the basic light-sensitive material used in most of today's photographic films and papers. Neither elemental silver nor silver compounds are used as ingredients of packaged processing solutions.

During processing, particularly in the fixing bath, silver is removed from the film or paper and is carried out in the solution or wash overflow, usually in the form of a silver thiosulfate complex. Unlike free silver ion (Ag^+), which is toxic to microorganisms, the silver thiosulfate complex is relatively nontoxic and has no detrimental effect on the operation of a secondary waste treatment plant. Based on tests using fathead minnows, the silver thiosulfate complex has been found to be at least 17,500 times less toxic than the free silver ion. When this complex reaches a waste treatment plant, it is converted to insoluble silver sulfide (Ag_2S) through chemical or biological action and is collected as a solid sludge. Tests performed at Eastman Kodak laboratories have shown that a concentration of more than 300 mg/L of silver, present as silver sulfide in the activated sludge, does not interfere with the rate of normal biodegradation of photographic processing wastes (Dagon, 1973). This level of silver at which tests were conducted is much higher than would be expected in a typical municipal treatment plant (Cooley et al., 1988).

Even though much of the silver will be removed in the secondary treatment plant and is not harmful to its operation, recovering the silver before discharge is a sound economic practice and therefore recommended. Not only does silver recovery have environmental benefits and conserve a valuable natural resource, but selling the recovered silver becomes a source of revenue to a photographic processing laboratory. When effective silver recovery practices are used, the residual amount of silver discharged in an effluent is not an environmental concern. (See "Silver Recovery.")

Chromium. Certain bleaches used in black-and-white reversal processes, as well as process systems cleaners, contain chromium compounds. Hexavalent chromium (Cr^{+6}) is regulated by federal and state agencies as a harmful compound. Some municipal sewer codes have specific limits for Cr^{+6}, $^{+3}$ (trivalent chromium), and total chromium. Cr^{+6} can be reduced by on-site treatment in the laboratory to Cr^{+3}, which is less hazardous. (See "Chemical Recovery: Chromium.")

Iron. Iron compounds are commonly used in color photographic processing bleaches or bleach-fixes. Also, if silver recovery cartridges (steel wool metallic replacement cartridges) are used for silver recovery, the photographic effluent will contain iron. Iron is not a typical component of processing solutions for black-and-white processing.

The iron concentration in effluent is commonly regulated because it affects the appearance and taste of water and readily oxidizes to the reddish-colored ferric (Fe^{+3}) form that precipitates, causing rust stains. The iron in photographic effluent normally does not represent a toxicity problem since it is usually present only in the form of stable iron complexes.

Zinc. Zinc is present in the effluent of a few photographic color processes. It is a necessary nutrient for human and animal life. Since there is some evidence that it may be toxic to fish at relatively low levels, and because it tends to impart an astringent taste to water, its concentration is usually regulated by local sewer authorities. The concentrations of zinc in photoprocessing effluents are usually below 1 mg/L.

Cadmium. Cadmium is toxic and therefore is regulated at low levels in many sewer codes. Cadmium has been removed from most films in recent years and is present in only a few specialized films. Small amounts of cadmium may leach out during processing and be detected in processing effluents.

Other Heavy Metals

Trace amounts of other heavy metals may sometimes be detected in photographic effluents. Although very small amounts are sometimes used in photographic emulsions, the more likely source of these heavy metals may be the processing or mixing equipment, plumbing, from impurities in processing chemicals, or in the incoming water supply. The detection of these heavy metals in photographic effluents may also be due to sampling and/or analytical interferences. In any case, their concentrations will occur at very low levels.

Phenolic Compounds

Phenol is not present in photographic effluents. Although some developing agents may be included by some administrative definitions in the broad classification of "phenolic compounds," they do not behave as phenol. They do not impart the undesirable taste and smell to water that phenolics do following chlorination. Developing agents can cause interference with the phenol wet chemical test method and a false positive response; however, they should not be considered phenol.

Cyano Complexes (Hexacyanoferrates)

Bleaches containing hexacyanoferrates (very stable complexes of iron and cyanide) are used in only a few photographic processes. These bleaches contain both ferri- and ferrocyanide, but the action of hypo and other chemicals in the effluent reduces most of the ferricyanide to ferrocyanide. A "total cyanide" analysis measures both of these ions, but they should not be confused with simple (free) cyanides. Unlike free cyanide, hexacyanoferrates have a low level of toxicity and are used in many common human applications, such as cosmetics (blue eye-shadow), paints, and laundry bluing (Merck Index, 1989; American Cyanamid Co., 1953). Cyano complexes are not typically found in black-and-white processing effluents. (See "Ferrocyanide Precipitation.")

Thiocyanate

Thiocyanate is used in a few photographic processes. Thiocyanate should not be confused with cyanide. It is a different chemical substance with different properties, including much lower toxicity. Thiocyanate will biologically degrade in a secondary wastewater treatment plant.

Hydroquinone

Hydroquinone is commonly used in many black-and-white photographic developers as the reducing (developing) agent for silver. Hydroquinone, a chemical that occurs in nature, can be toxic to some organisms at relatively low concentrations. When discharged in mixed photographic processing effluents, it is present in low concentrations and readily biodegraded to innocuous products by biological treatment.

Ammonia Nitrogen

Relatively high concentrations of ammonia nitrogen may result from the use of ammonium fixers. Although some of the nitrogen-containing ions may be oxidized in a waste treatment plant, some of them may be carried through the plant and into the receiving body of water. If the waste has a pH greater than 8, some of the ammonium ion will be converted to free ammonia. Some stream standards and sewer codes do have limits on ammonia content. Ammonium ions, when in a stream for long periods of time, will oxidize to nitrates through normal biological processes.

Phosphates and Nitrates

Metaphosphates are used as sequestering agents in some processing solutions to minimize sludging due to calcium in water. A few processing solutions contain trisodium phosphate or other phosphates as buffers. Nitrates are present in only a few processing solutions and at levels that are not significant for treatment plants.

Detergents

Although detergents may be commonly used for cleaning purposes in processing laboratories, they are rarely used in processing solutions.

Color and Odor

Mixed processing wastes will generally have a very slight color and a scarcely detectable odor and, therefore, are not usually affected by effluent discharge regulations.

Flammable and Explosive Materials

Conventional photographic processing solutions or effluent are not flammable or explosive.

Volatiles

Volatile compounds in industrial effluents are of concern to regulatory agencies because sufficient concentrations of vapors in an enclosed area can present a hazard, either as irritants to the skin, eyes, lungs, and other mucous membranes or (in the case of flammables) because they can be readily ignited. With the exception of small amounts of chemicals used in some film lubricants, soundtrack applicators, or film cleaners and lacquers, most solutions and chemicals used in photographic processing laboratories are not considered volatile.

Some solutions do contain ingredients such as acetic acid, formaldehyde, or ammonia that are volatile. Formaldehyde emissions are regulated by the OSHA formaldehyde standard. At a pH above 8.0, solutions containing ammonium compounds can give off ammonia. When chemicals such as these are used, adequate ventilation is needed.

States and localities may have codes that restrict the amount of volatile organic emissions which can be discharged to the air; exhausting process fumes to the outside air may require an air pollutant source emission permit.

If large amounts of lacquers, lubricants, or cleaners are being used, the photoprocessing laboratory should check with the appropriate local or state agency to determine what limitations are imposed and what safeguards are required.

3. TREATING PHOTOPROCESSING WASTE

3.1 Introduction

Regardless of whether a waste issues from photoprocessing or any other source, all waste treatment technology can be broken down into one of three basic techniques: neutralization, oxidation/reduction, or separation.

Photographic Processing Wastes

Neutralization is the process of nullifying or dissipating the chemical effects or characteristics of a material while leaving it in place, without removing it from a chemical system, by adding one or more other ingredients. Examples would be pH adjustment with acid or alkali, solidification by mixing with a concretion additive, dilution with water, complexing of calcium or other metals by adding a sequestrant, blending of two waste streams to dilute and cancel out the effects of their individual ingredients, etc.

Oxidation/reduction is the process of chemically changing the character of a material by supplying electrons to or removing electrons from the chemical structure, breaking down chemical bonds or forming new ones to create new compounds. Examples would be biological (secondary) treatment in which the waste acts as food for microorganisms, incineration, electrolytic silver recovery, chemical oxidation with permanganate, peroxide or ozone, chemical reduction of chromium +6 to +3 with bisulfite, burning of gasoline in an internal combustion engine, photolysis, corrosion and natural decay, etc.

Separation is the process of physically separating one or more materials from a chemical system, generally without a chemical reaction but sometimes coincidental with it. Examples include filtration, activated carbon adsorption, reverse osmosis, ion exchange, settling, distillation and evaporation, freeze drying, solvent extraction, centrifuging, and numerous others.

Neutralization has limited use in waste treatment technology, except for the very critical use of pH adjustment. Oxidation/reduction and separation methods, either in individual processes or various sequential combinations, are very effective and in widespread use.

3.2 Secondary (Biological) Treatment

Biological waste treatment processes are a combination of oxidation/reduction and separation steps, assisted by appropriate pH adjustment. Microorganisms are intentionally encouraged to grow on the waste components, consuming some as food/fuel and assimilating others for cellular growth. Precipitation of materials occurs at various points of this cycle and at certain stages may be deliberately enhanced through the use of settling basins, skimmers, or filters to remove certain components as solids from the aqueous medium. Some waste components may tend to inhibit oxidation/reduction by being toxic to the microorganisms above a threshold concentration; others may form toxic or objectionable by-product gases if not limited in concentration or chemically attenuated. The appropriate management of these parameters, including the use of sewer codes to limit input concentrations, analyses to detect the presence and/or concentration of components as well as end-products, and equipment to conduct various process engineering operations as they are needed, allow the successful operation of such waste treatment processes.

Biodegradation Testing

Extensive studies were conducted in the late 1960s and early 1970s at Eastman Kodak to evaluate the biodegradability of chemicals in typical photographic processing effluents (Dagon, 1973). The results indicated that these effluents could be treated biologically and that it is safe for processing laboratories to send their waste solutions to municipal secondary waste treatment plants, after practicing silver recovery and regeneration techniques that make sense economically and photographically. Of course, this recommendation depends on the concentration of the photographic effluent, size of the waste treatment plant, and nature and quantity of other wastes being treated. These biological treatability studies were conducted using 5.5- and 55-L activated sludge units. The general results of the test were that as long as not more than 10% by volume of conventional photographic processing effluent was treated in an activated sludge unit having a retention time of between 4 and 6 hr, no adverse effect would be seen on the activated sludge unit, and the photographic effluent would be adequately treated (Dagon, 1973).

Similar tests were conducted by Hydroscience, Inc. (1974). They undertook a study to determine the environmental effects of 45 selected photoprocessing chemicals and a typical photoprocessing effluent. This testing program was sponsored by the National Association of Photographic Manufacturers (NAPM), which represents major segments of the photographic industry. The conclusions of these tests were the following:

1. Photoprocessing wastes do not present a hazard to biological systems encountered in conventional wastewater treatment schemes or to aquatic organisms indigenous to natural receiving waters. The assumption inherent in the above statement is that a photoprocessing waste is not directly discharged to a natural receiving water without adequate treatment before discharge.

2. The results indicate that photoprocessing wastes are amenable to biological treatment, with a removal efficiency approximately equivalent to that encountered with domestic wastewaters. The concentrations expected to occur in municipal or regional sewerage systems and in a receiving water upon treatment will not adversely affect natural biological systems.

3. The 45 photoprocessing chemicals investigated in this study have no significant impact on the activity or efficiency of the biomass of a conventional biological wastewater treatment system, at the concentrations realistically

attainable in municipal or regional sewerage systems. The chemical concentrations expected to occur in the discharge from a photoprocessing laboratory may usually be estimated by applying a 100-fold dilution to the working solution concentrations.

Additional studies were also conducted by the J. B. Scientific Corporation (1977) in the late 1970s. This investigation was also sponsored by the NAPM. These studies reaffirmed that the use of biological treatment to handle photographic processing effluents is a preferred technique.

Large-Scale On-Site Biotreatment

Conventional Activated Sludge. The use of biological treatment facilities operated by municipalities is a very effective way of handling photographic processing effluents. Tests by Eastman Kodak previously cited showed that it was feasible to use biological treatment to treat only photographic processing effluents (Dagon, 1973). These tests used a retention time of approximately 24 hr, indicating that extended aeration was necessary for treating exclusively photographic processing effluents without blending in domestic or other municipal waste. Further testing also showed that it was possible to pretreat photographic processing effluents using on-site treatment with biological techniques.

In addition to smaller bench-scale set-ups, an extended aeration activated sludge unit having a capacity of 76,000 L/day was operated by Kodak on photographic processing effluents. Processing effluents were collected from six different color processes. The process solutions treated included effluents from Ektaprint R and Ektaprint 3 chemicals, and from the E-4, C-22, CRI-1, and K-12 processes. Once adequate mixed liquor suspended solids (MLSS) had built up in the aeration tank, the average weekly BOD_5 reductions ranged from 78 to 91%. During this time period, the MLSS ranged from slightly over 1100 to as high as 4600 mg/L. To obtain optimum efficiency, automatically backflushed sand filters were used to remove the very finely divided suspended solids that were produced in the system.

Although the use of activated sludge systems can provide an effective method for treating photoprocessing effluents, their operation and the subsequent handling of solids produced are both time-consuming and labor-intensive tasks. These can be difficult systems to operate on-site by a photoprocessing laboratory, unless it is very large and has a well-trained, permanent maintenance staff.

Rotating Biological Contactor. An alternative treatment technique is the rotating biological contactor. Small-scale testing has indicated that it is very effective and may be more economical than an activated sludge unit. Initial tests at Eastman Kodak indicated that with an effluent having a BOD_5 of approximately 600 mg/L, retention times of between 4 and 24 hr resulted in from 83 to 94% reductions in BOD_5 (Dagon, 1978). Additional testing with rotating biological contactors was conducted by Lytle (1984). In these tests, photoprocessing effluents having BOD_5 concentrations between 440 and 1600 mg/L (with COD levels ranging from 1010–4120 mg/L) were effectively treated using retention times of 8 and 16 hr. BOD_5 reductions of 72 to 95% and COD reductions between 55 and 72% were attained. Soluble silver levels were reduced from an initial 3 to 10 mg/L level to a final level of 0.15 to 0.75 mg/L. The precipitated silver was harvested with the sludge.

Further tests were conducted on-site in a production microfilm processing plant by Petche (1989). Black-and-white microfilm developer and fixer were treated in a 3200 L/day rotating biological contactor. The concentrations in the input solutions were as follows: BOD_5 292 to 1800 mg/L, COD 1090 to 8580 mg/L, and chlorine demand 80 to 2200 mg/L. Treatment produced BOD_5 reductions averaging 85%, COD reductions averaging 61%, and chlorine demand reductions of 98%.

Impact of Biotreatment on Silver

In addition to reducing the oxygen demand of photographic processing effluent, biological treatment also affects silver speciation. Depending on the size of the photoprocessing operation and complexity (and efficiency) of the silver recovery equipment, silver concentrations of between 0.5 and 10.0 mg/L may be found in photoprocessing effluents. The silver present in these effluents is in the form of a silver thiosulfate complex, which is approximately 17,500 times less toxic than free silver ion. Studies by Dagon (1978), Cooley et al. (1988), J. B. Scientific Co. (1977), and Bard et al. (1976) showed that in secondary waste treatment plans, silver thiosulfate complex is converted by microorganisms into insoluble silver sulfide, with a small amount of metallic silver also formed. Both of these insoluble species are substantially removed from the secondary treatment plant effluent as insoluble sludge during the settling step.

In a 1974 study, Ericson and LaPerle (1974) analyzed the effluent from the 76,000 L/day activated sludge unit operated at Kodak. Total silver concentration in the effluent after biological treatment was measured at 0.95 mg/L. This effluent was filtered through a 0.45-µ filter, then through an ultrafilter with a 1000 mol. wt. cut-off. The amount of insoluble silver was found to be 0.94 mg/L, primarily in the form of silver sulfide with some metallic silver also being present. The filtrate contained less than 0.01 mg/L soluble silver.

More recently, the Environmental Protection Agency (EPA) sponsored a study of publicly owned treatment works (POTWs) (Feiler et al., 1981). In this study, hundreds of influent, effluent, and sludge samples were analyzed for various chemicals, including silver. These analyses showed that almost all of the effluent silver was incorporated into the sludge.

In 1981, Lytle (1985) carried out an analytical survey of six POTWs to evaluate speciation and fate of silver. In this study, the sources of silver were identified as municipal, photographic, and industrial. Samples of influent and effluent to and from the POTWs were analyzed for total silver and free silver ion. Regardless of the source of the silver, no significant amounts of free silver ion could be detected. Silver removal efficiencies were very high at all of the treatment plants studied. Those POTWs receiving photoprocessing effluents showed silver removal efficiencies equal to or greater than 90%. The concentration of free silver ion present in the effluents of the POTWs receiving photoprocessing waste was 4 parts per trillion. Lytle's study thus confirmed findings previously recorded by the EPA: regardless of its source, silver is rapidly and almost completely incorporated into the sludge during secondary waste treatment.

In summary, clear scientific evidence has shown that silver in photoprocessing effluent is present as the thiosulfate complex, and this complex can be satisfactorily removed during biological treatment, such as in a municipal treatment plant. A very high percentage of the complex is converted to insoluble species, primarily silver sulfide and metallic silver, most of which is removed as sludge in the settling step. Little if any soluble silver will be discharged in the liquid effluent from the treatment facility.

Impact of Biotreatment on Specific Photoprocessing Chemicals

Recommendations from the photographic industry have always been that photographic processing effluents should never be discharged directly to a receiving body of water, but rather should be either treated on-site and/or discharged to a municipal secondary waste treatment plant. If untreated, certain constituents of photographic processing effluents can indeed be toxic. Chemicals such as hydroquinone used in black-and-white developers and color-developing agents used in color developers have high levels of toxicity associated with them, if they are not adequately treated (Terhaar et al., 1972).

Work conducted by Watson et al. (1984) showed that while Kodak color-developing agent CD-3 was toxic at levels of approximately 1 mg/L to fathead minnows and *Daphnia* if untreated, significantly higher concentrations of this chemical could be safely discharged if it were first subjected to biological treatment. Levels of CD-3 as high as 30 mg/L were not toxic to aquatic species after such treatment. In addition, studies by Harbison and Belly (1982) showed that hydroquinone is readily biodegradable and would not be of concern after treatment in a secondary waste treatment system.

3.3 Other Oxidation/Reduction Treatment Technologies

Other than biological treatment, there have been few instances in which oxidation/reduction technologies have been effectively used on any large, continuous scale to treat photoprocessing effluents for direct discharge to receiving bodies of water. Some have been used as pretreatment methods prior to discharge into sewer systems. Some of the more popular past methods, as well as methods currently undergoing renewed investigation, are described below.

Ozone

Ozone is an extremely strong oxidizing gas that reacts with many organic compounds and some inorganic compounds to break them down to less environmentally objectionable materials. Experiments to evaluate the ability of ozone to decompose a number of photoprocessing chemicals were conducted in 1969 through 1973 by the Authors (Bober and Dagon, 1975).

As the result of this testing, the following conclusions were reached: (1) Ozone is particularly useful in breaking down many photographic chemicals, including developers, to more easily biodegradable materials; (2) ozone is effective in lowering the BOD_5, COD, and chlorine demand of individual and combined photographic solutions; (3) ozone has no significant effect on acetic acid, acetate ion, or glycine and appears to degrade ferricyanide ion too slowly to be of practical importance; (4) thiosulfate, sulfite, formaldehyde, benzyl alcohol, hydroquinone, and ethylene glycol can all be degraded effectively to innocuous end-products with ozone.

Four case histories that actually used ozone in production-scale photographic processing operations have been discussed by Hendrickson (1975). Although shown to be very effective in some applications, the use of ozone treatment alone to degrade photographic processing chemicals can be extremely expensive. Its use in combination with other techniques may be much more economically desirable.

Recent experiments have also shown that the combination of ozone and ultraviolet light treatment can be effective in decomposing some processing chemicals. The economics and practical operating problems with this technique have not yet been established.

Chemical Oxidation/Reduction

Other oxidation techniques, such as the use of hydrogen peroxide (Knorre et al., 1988), permanganate, persulfate, bromine, chlorine (U.S. Patents, 1971 and 1973), perchlorate, hypochlorite, and numerous other chemical oxidants, have been investigated. Many of these have been used for the regeneration of photographic bleaches, either commercially or experimentally, and in the process were discovered to degrade trace components carried into the bleach solution. Concerns in modern times with these agents, in addition to safety and cost of handling as well as general effectiveness, include the possibility of introducing yet another pollutant into the waste stream (e.g., manganese from permanganate) or forming by-products that are even more toxic, if the oxidation reaction is only partial (e.g., forming chloramines by the partial chlorination of organic amines). Currently, about the only chemical oxidants used on any significant commercial scale in photographic processing are hydrogen peroxide and "bromine sticks" (bromochlorodimethylhydantoin), the latter used in a unique application (Kreiman, 1984) to destroy residual thiosulfate while simultaneously curbing unwanted biological slime during a washwater recycling operation.

Chemical reduction has been commonly used in a number of operations in the past. As previously stated, dithionite and bisulfite have been used to reduce chromium +6 to chromium +3 in residues from systems cleaners or black-and-white reversal bleaches. Sodium borohydride has been used in certain special cases to precipitate silver as elemental silver metal from overflow fixer solutions, or other trace metals from a few processes. Sodium dithionite has been used in combination with ferrous sulfate to precipitate ferrocyanide from bleach wastes as insoluble ferrous ferrocyanide (Bober and Cooley, 1972; Kleppe and Vacco, 1979). (See "Silver: Chemical Precipitation" and "Ferrocyanide Precipitation.")

Electrooxidation/Reduction

Electrooxidation, the process of oxidizing dissolved compounds at the anode of an electrolytic cell, is a technology that many researchers have investigated for waste treatment applications, with varying results. Appropriate catalysts have been shown to enhance efficiency; however, there is difficulty in keeping the most effective of these from becoming poisoned, thus reducing their practical life. Heat build-up in the cell, cross-cell reactions whereby a component may be oxidized at one electrode and promptly reduced at the other for no net gain, gas formation, and undesired side reactions can all present practical operating problems. Currently, insufficient data exist to make any decision on the practical feasibility of such techniques for use as a waste treatment technology with photoprocessing effluents.

Electrolytic cells have been used very successfully by the photographic industry for many years to reoxidize spent ferricyanide bleach. (See "Bleach Regeneration.")

In 1975, Kodak researchers publicly disclosed a method for generating sulfide ion upon demand by the deliberate, direct electrooxidation (or alternatively, thermal treatment) of the waste silver-bearing fixer or bleach-fix (Bober and Leon, 1975). This excess sulfide ion was then used to precipitate silver from the remainder of the solution in the same apparatus. The method had the advantage of not needing to prepare and store solutions of sodium sulfide, which can produce toxic as well as very odorous hydrogen sulfide (see "Electrochemical Sulfide Precipitation").

Electrolytic reduction is most commonly used for recovering silver and other metals from solution, known as electroplating or electrowinning. This method is discussed extensively in the forthcoming section on silver. Electrochemical reduction has found no other widespread applications in the treatment of photographic wastes to date.

Incineration

Despite unfortunate and often misleading environmental publicity, and resulting opinions expressed by the public, properly conducted incineration remains one of the most viable, universally applicable, and environmentally sound waste treatment methods known to mankind.

In 1989, Eastman Kodak contracted with the Mechanical Engineering Department of the University of Wisconsin at Madison to run batch incineration tests on waste solutions from the Kodak Flexicolor C-41 process, both combined and individual processing solutions (Ragland et al., 1990; Holm, 1990). The results represent only initial information from only a few solutions among a vast array of possible tests. However, the data were very encouraging: No significant residual organics were detected in the emissions when the solutions were incinerated with excess air at or above 800°C (1472°F), which was the lowest temperature tested. This is a reasonable operating condition for commercial incinerators. A second phase of these tests has begun.

Wet-Air Oxidation

Wet-air oxidation is the "pressure-cooking" of waste with air at high pressure and moderately high temperatures. Experiments were funded by Eastman Kodak in 1989 for the Zimpro Passavant Corp. to conduct wet-air oxidation studies on Kodak Flexicolor C-41 process wastes at temperatures up to 280°C and pressures of 1500 to 2500 psi (Zimpro Passavant, 1989). As in the case of incineration tests above, the information obtained to date is only preliminary. However, as of this writing, the data indicate that the treatment can achieve a high degree of destruction of certain developer components and can considerably increase the biodegradability of the developer.

3.4 Separation Methods

Evaporation

Evaporation experiments at atmospheric pressure to eliminate waste from waste photoprocessing solutions were first performed at Eastman Kodak in 1970 and have continued since, both at that company and others. Water was successfully removed, and sludges or slurries containing nearly all the residual chemicals occupied only 10 to 15% of the original volume. It was established that at higher temperatures near the boiling point of water, certain ingredients such as thiosulfate would decompose, liberating noxious gases such as sulfur dioxide or hydrogen sulfide, especially under acidic conditions. However, it was further discovered that an upper temperature limit of 160°F along with a pH adjusted to near neutral would minimize such gas formation, yet still allow reasonably rapid water removal. Suggestions since that time in response to inquiries from prospective manufacturers or users of evaporation equipment have been to limit the upper temperature of 140°F, thereby including a safety factor to cover cases of extreme solution aging or unforeseen equipment malfunctions.

Some manufacturers have overcome the temperature problem while maintaining a maximum evaporation rate by conducting the evaporation under vacuum, which lowers the boiling point of the solution to below 100°F. This, along with an appropriate pH adjustment, appears to virtually eliminate the formation of sulfurous gases. A few other manufacturers have proposed conducting atmospheric evaporation at higher temperatures, near the boiling point of water, and trapping the evolved gases in various absorptive or scrubbing media, but in most cases this appears to be a cumbersome and impractical answer since the media would then need to be treated or discarded.

In recent years, ammonium salts have become popular ingredients in fixers and bleaches because of their rapid penetration into gelatin emulsions, allowing shorter process times and less overall chemical consumption for significant environmental benefits. However, if the pH is too high, the evaporation of wastes containing ammonium salts can liberate ammonia gas. The best compromise, between the liberation of ammonia at alkaline pH's and sulfurous gas formation at low acidic pH's, appears to be to limit the pH for most combined photoprocessing waste solutions to a range between 6.0 and 6.8, aiming for about 6.2. This successfully ties up most of the ammonium salts in the residual sludge. In a few specific instances, such as high-ammonium X-ray fixers that had not aged significantly, pH's as low as 5.5 have been successfully used without significant sulfurous gas formation.

Freezing and Freeze Drying

Freeze and freeze drying techniques have been investigated to purify and separate components of photoprocessing waste, but no methods are currently in commercial use.

Ion Exchange

Ion exchange is a process in which "ions held by electrostatic forces to charged functional groups on the surface of a solid are exchanged for ions of a similar charge in a solution in which the solid is immersed" (Weber, 1972). Ion exchange is most commonly used to soften water by utilizing a cation-exchange resin to remove calcium and magnesium ions. In photoprocessing applications, anion-exchange resins are widely used for removing halide ions for developer regeneration, recovering silver thiosulfate complexes from dilute solutions, and recovering hexacyanoferrates from dilute solutions. Cation-exchange resins have been used to remove color-developing agents from stop baths.

Precipitation

Precipitation, as described in this chapter, is the formation of insoluble solids through the reaction of two or more water-soluble species. Although most photographic chemicals are deliberately chosen to be soluble in water so they do not precipitate spontaneously during processing, in several instances a reagent can be found to solidify certain components for recovery. (See "Silver Recovery; Chemical Recovery Methods.")

Photographic Processing Wastes

Settling and Decanting

After precipitation, methods must be found to separate the precipitate from the solution. Generally in photoprocessing applications, the solids are often either very finely divided or they form very colloidal, gelatinous suspensions. In either case, the materials are difficult to filter and often "blind" the fine-pore filters that would be required for adequate separation. The alternative is often to add a flocculating agent, allow the material to settle over a period of time, and then carefully decant the supernatant liquid to a point just over the sediment. By removing most of the water this way and allowing the mass to settle for a time, which sometimes promotes the growth of larger crystals or particles (known as "ripening"), direct separation by decantation may be possible or often the resulting mixture becomes easier to filter.

Filtration

Several different types of filters are used in photoprocessing applications. Woven, pleated paper or particulate cartridge filters are often used on the processing machine itself to remove carried-in dirt or in regeneration operations. Sediments and precipitates that are formed either in waste mixtures or intentionally during recovery operations may need more elaborate filters. These should be oriented toward higher solids loading in production quantities and preferably be reusable. Filter types used in the past for the treatment of photoprocessing chemicals have included filter presses, vacuum belt filters, pressure leaf filters, sand, carbon or other particulate filters that can be backwashed, and others. To assist filtration, flocculants and filter aids have been used when necessary (Bober and Cooley, 1972). (See "Silver Recovery, Chemical Recovery.")

Centrifuging

Centrifuging is the process of separating two or more materials of dissimilar specific gravity by a rotary spinning (centrifugal) action. Centrifuges are more expensive capital investments than filters, although they generally produce drier cakes of solids and may require less operating labor. In photographic processing, centrifuges have found limited use in Kodachrome process operations for recovering color coupler. (See "Coupler Recovery.")

Reverse Osmosis

Reverse osmosis is the process of applying pressure to a salt solution on one side of a membrane, to force water through while retaining the salts on the original side (in effect "squeezing" fresh water out of salty water). The original stream is known as the "feed," the fresh water product as the "permeate," and the rejected concentrated solution as the "brine" or "concentrate." The process works best on dilute solutions such as wash water, where more than 90% of the original feed can be reclaimed before the salt buildup in the brine begins to hamper recovery. The advantages of the process are the very low energy needed to reclaim water as compared to distillation and the lack of need for water treatment chemicals. The disadvantages include high capital costs, the size of equipment needed to obtain sufficient flow to keep up with a process, and frequent maintenance of the membrane system and pumps. Reverse osmosis has been tested for photoprocessing applications since the mid-1960s but to date has not found widespread application. In the future, increased equipment reliability and better membranes may widen its use.

Ultrafiltration

Ultrafiltration appears similar to reverse osmosis superficially, but rather than relying on osmotic principles for removing dissolved materials, it is a true filtration process for removing finely divided particles using a membrane filter. Ultrafiltration also has not found widespread use in photoprocessing to date, namely because (1) simple wound cartridge filters are cheaper and more convenient to use for ordinary process dirt, and (2) situations where solids formation is deliberately induced for waste treatment reasons (see "Chemical Recovery Methods") need filters that do not blind easily and can handle large quantities of gelatinous solids. Possible applications continue to be explored.

3.5 Silver Recovery

Introduction

As previously discussed, silver from photoprocessing operations is much less toxic than free silver ion. Silver is generally removed from photographic products during processing in the form of silver thiosulfate complex, $Ag(S_2O_3)_2^{-3}$. This complex has a dissociation constant of 5×10^{-14}; thus, it is virtually impossible for free silver ion (Ag^+) to be present at any significant concentration levels in photoprocessing effluents (Cooley et al., 1988).

However, there are good economic and regulatory reasons for processing laboratories to recover silver. In black-and-white products, since the final image is metallic silver, the amount of silver removed during processing will depend on the amount of exposed image area. In color products, processing removes virtually all the silver from the emulsion. Although primarily found in fixers and bleach-fixes, small quantities of silver will also have dissolved in the developer and bleach, and some will be carried over into washwaters following the fixer or bleach-fix.

All except the very smallest processing operations have adequate economic and regulatory justification to recover silver from their exhausted fixers and bleach-fixes. Whether the same justification applies to washwater depends on the size of the operation, treatment alternatives, and applicable regulations. The amount of silver found in washwaters usually represents about 2 to 5% of the potentially recoverable silver in the process.

The two most commonly used methods of silver recovery are electrolytic recovery and metallic replacement using steel wool cartridges (Eastman Kodak, 1979). Electrolytic silver recovery can be successfully applied to concentrated fixer and bleach-fix solutions, in either an in-line or a terminal application, but not generally to washes.

Metallic replacement is normally used with fixers and bleach-fixes in a terminal application, known as "tailing," to recover silver from solutions destined for the drain. However, it has also been used for in-line desilvering of bleach-fixes, in which the addition of dissolved iron to the solution is a benefit and has been successfully used to treat washes or washes that have been combined with fixer and bleach-fix overflows.

Some large photoprocessing installations use ion-exchange methods to recover silver from washwater or mixtures of washwater, fixers, and bleach-fixes. Certain precipitation methods can also be applied to fixers and bleach-fixes; however, these are infrequently used at present. The destructive oxidation of bleach-fixes and fixers with chemicals such as hydrogen peroxide also causes silver to precipitate as silver sulfide, which can be removed by settling or filtration.

If electrolytic silver recovery and/or metallic replacement are effectively used, a photoprocessing laboratory should be able to recover on-site between 90 and 99% of the potentially recoverable silver. The effluent silver concentration from such a laboratory would be in the 1 to 5 mg/L range, a concentration that will have no adverse impact on a secondary waste treatment plant or a receiving body of water after the effluent is biologically treated. If ion exchange is used to treat the overall discharge, the treated effluent will contain between 0.5 and 2 mg/L silver. If the fixer and bleach-fix overflows are pretreated by electrolysis followed by ion-exchange treatment of the overall effluent, the effluent silver concentration can be reduced to the 0.1 to 0.5 mg/L range. The overall silver recovery efficiency can be 98 to 99+% of the potentially recoverable silver (Quinones, 1985).

The EPA contracted with Versar, Inc. to provide a guidance document describing the control of water pollution in the photographic processing industry. This document, published in 1981, described the results of sampling effluents from 48 photoprocessing laboratories. The maximum silver concentration in the effluents averaged over a 30-day period was 1.1 mg/L for laboratories using conventional silver recovery methods. Laboratories using conventional silver recovery plus the ion-exchange treatment of washwater averaged 0.4 mg/L over the period. Maximum single-day concentrations were 3.7 and 1.3 mg/L, respectively. The Versar report stated that more than 99% of the photoprocessing facilities it surveyed (over 1100 plants) discharged wastewater to POTWs (Versar, 1981).

In a 1980 study, total silver concentrations in effluents from two POTWs were reported as 0.2 and 0.004 mg/L, respectively, while both effluents contained free ionic silver concentrations of only 4×10^{-6} mg/L (Lytle, 1985).

Metallic Replacement

Metallic replacement has been an important means of recovering silver from fixers, bleach-fixes, and washwaters for many years. It can be used as a primary method of recovery or a secondary method following the electrolytic treatment of fixers or bleach-fixes.

Metallic replacement involves passing a silver-bearing solution through a vessel containing a more active metal in elemental form, usually iron. The reaction of dissolved silver with metallic iron is

$$Fe^{o} + 2\,Ag(S_2O_3)_2^{-3} \rightarrow Fe^{+2} + 2\,Ag^{o} + 4\,S_2O_3 =$$

If conditions were ideal and no other reactions were involved, 1 g of steel wool would recover 3.86 g of silver. In actual practice, usually less than 1 g of silver is recovered per gram of steel wool. Most fixers, such as those used in X-ray and graphic arts processing, are acidic; therefore, a competing reaction that consumes much of the iron is the acidic dissolution of the steel wool. Spontaneous oxidation of iron with air (i.e., rusting) also occurs, particularly upon long-term standing under moist conditions. Insoluble iron hydroxide compounds are also formed when the solutions are at higher pH values, above 6 to 7.

The most common size of steel wool cartridge (also known as a "chemical recovery cartridge" or "CRC"), made by Kodak, contains about 8 lb of steel wool. However, many other manufacturers in the United States and overseas produce such devices in various shapes, configurations, and sizes, with various types of steel wire filling.

Although a number of active metals (e.g., aluminum, manganese, zinc) can replace silver, steel wool has been shown to be the best choice from several technical, economic, safety, and environmental points of view. Certain metals higher than iron in the electromotive series, particularly aluminum, can react so violently with an acidic fixer that copious amounts of sulfurous gases, which are odorous and noxious, would be given off. Iron reacts more slowly and thus does not form such gases.

There are a number of advantages associated with metallic replacement, including low initial cost, simple nonelectrical installation, small size and low weight, little maintenance, and high efficiency of silver recovery if properly monitored. One disadvantage is the high shipping and refining costs after exhaustion, compared with those for silver flake; this may offset the lower initial cost of CRCs. Another disadvantage is the high iron concentration in the cartridge effluent, which has been measured as high as 3000 mg/L. This precludes the reuse of fixer after silver recovery with CRCs, produces a colored effluent, and could cause problems in meeting local sewer codes.

High silver-recovery efficiency can best be ensured if two CRCs are used in series. When the first cartridge shows exhaustion, it is removed. The cartridge in the second position is moved to the first position and a fresh cartridge placed in the second position.

Studies have been conducted by Cooley (1988) to determine how to optimize the use of CRCs. These showed that the optimum pH in actual laboratory practice is between 4.5 and 5.5. At lower pH values, the acidity will consume too much of the steel wool, whereas at higher values the reaction becomes slower since considerable amounts of iron hydroxide are formed that may obscure the surface of the steel wool. Silver deposited on the steel wool may also restrict the mass transfer of silver to the iron surface at higher pH's, while below pH 5 the iron may be continuously acid-etched to provide a fresh reaction surface.

Electrolytic Silver Recovery

Electrolysis is the most widely used and universally applicable method for silver recovery in the photoprocessing industry. An electrolytic silver-recovery cell consists of a cathode and an anode. Oxidation occurs at the anode (positive electrode) and reduction at the cathode (negative electrode). Silver deposits on the cathode during electrolysis when a direct current is passed through the silver-bearing photoprocessing solution. After sufficient silver has been plated, the cathode is removed from the system and the silver stripped off (Cooley and Dagon, 1976; Hickman et al., 1933).

The primary reaction occurring at the cathode is

$$Ag(S_2O_3)_2^{-3} + e^- \rightarrow Ag^\circ + 2\,S_2O_3^=$$

If the cathode voltage is allowed to become too high, thiosulfate could be reduced at the cathode as shown in the following equation:

$$S_2O_3^= + 8e^- + 8\,H^+ \rightarrow 2\,HS^- + 3\,H_2O$$

The production of sulfide is undesirable, since it will react with the silver complex to produce insoluble silver sulfide (known as "sulfiding"). Although from a recovery standpoint, a small amount of silver sulfide can be tolerated, too high a level will result in a poor plate (Cooley, 1986). Additionally, if in-line fixer desilvering were being done, silver sulfide formation would contaminate the fixer and could damage the photographic product. Therefore, it is necessary to compromise on the voltage applied to obtain optimum current efficiency while minimizing sulfide production.

Electrolytic silver recovery requires a larger capital expenditure than the use of CRCs and also necessitates an electrical connection. However, it has the advantage of yielding nearly pure silver, resulting in lower refining and shipping costs. A primary advantage from an environmental viewpoint is that it allows fixer reuse for many processes since it does not contaminate the fixer when properly controlled.

There are essentially two ways in which electrolytic silver recovery can be applied (Cooley and Dagon, 1976). One involves its use in a terminal manner, and one concerns its application for the in-line desilvering of fixer. When used in a terminal manner, the silver-bearing solution is passed through the electrolytic cell to recover the silver and the desilvered solution is slowly discharged to the drain, perhaps through a secondary metallic replacement cartridge for additional low-level silver recovery. An alternate terminal approach is to mix the electrolytically desilvered solution with silver-containing washwaters and pass the mixture through an ion-exchange system for further silver recovery.

Finally, part of the desilvered fixer or bleach-fix, if not mixed with other solutions or otherwise contaminated or altered during silver recovery, may be reused in making fresh replenisher, thus minimizing the environmental impact.

It is also possible to operate electrolytic equipment for the in-line desilvering of the fixer solution. The equipment is set to function so that the silver in the fixer tank is constantly maintained in the 0.5 to 1 g/L range. (This compares with typical silver concentrations in fixer tanks of 5 to 10 g/L when well seasoned.) Although careful control is essential to preclude the formation of silver sulfide, this method offers several environmental benefits. Depending on the process, the fixer replenishment rate can be reduced from 50 to 70% compared to the standard rate. Additionally, the lower silver level in the tank means that significantly less silver (only about 5–10% as much) will carry over to the washwaters, thus assuring that, overall, more silver is recovered and less is lost.

Several factors are involved in choosing and operating an electrolytic silver recovery unit (Cooley, 1982 and 1984a). The amount of current that a device delivers is important: Low-current density units can be used for desilvering fixers, but high-current density is needed for bleach-fixes. Some method of agitation is required to keep the fresh silver-containing fixer in contact with the cathode, but too much turbulence that produces a vortex will whip air into the solution, consuming sulfite preservative and promoting sulfiding. A rotating cathode unit provides its own agitation, whereas a pump or impeller may be needed for a stationary cathode.

Some method for controlling the current is also important. Several methods are available, including timers, selective ion electrodes for on-line monitoring, and constant voltage operation including the more complicated use of potentiostatic control with IR compensation (Branch, 1988; Cooley, 1984b). The current density relative to the solution silver concentration should be high enough to desilver the solution in a reasonable time, yet low enough to prevent sulfiding. Well-designed controls step the current down as silver is depleted from the solution.

In addition to the time, voltage, and current, pH is an important factor affecting electrolytic silver recovery. Tests have shown that the optimum pH for desilvering fixers is approximately 6.2, whereas the optimum for desilvering bleach-fixes is approximately 8 to 8.5. How high a pH can be used for desilvering bleach-fixes is limited by the evolution of ammonia. As the pH of the bleach-fix increases, a side reaction involving the reduction of iron is inhibited and the electrolytic silver-recovery efficiency increased (Krauss, 1987).

Ion Exchange

If silver must be recovered from dilute solutions, ion exchange is the method of choice. This method can be applied to washwaters, mixtures of desilvered fixers and/or bleach-fixes with washwaters, and mixtures of silver-bearing fixers and/or bleach-fixes with washwaters. The lowest effluent silver concentrations can be attained when washwaters alone are treated, since input thiosulfate and silver levels are lowest. Treating a solution of desilvered fixer and/or bleach-fix mixed with washwater will produce the next lowest effluent silver levels.

The basic principle involves exchanging ions from solution with similarly charged ions bound to the resin, described as

$$R^+(S_2O_3)^{-2} + Ag(S_2O_3)_2^{-3} \rightarrow 2\ S_2O_3^= + R^+AgS_2O_3^-$$

where R^+ represents the ion-exchange resin.

A weak-base anionic resin is generally used for silver recovery. Rohm & Haas Amberlite IRA-68, a gel-type acrylic resin, has been shown to be the most effective choice in various silver-recovery operations. Two approaches to recovering silver from dilute photoprocessing solutions are conventional ion exchange and in-situ precipitation.

With conventional ion exchange, the silver-bearing solution is pumped through a vessel containing the resin (Mina, 1980). Normally, two columns are used in series to minimize the loss of silver when the first column reaches exhaustion. Exchange occurs between the silver thiosulfate anion in the solution and the anion on the exchange site of the resin, which is usually thiosulfate. Silver will continue to be removed by the resin until the silver thiosulfate concentrations on the resin and in the solution reach equilibrium. The duty cycle is stopped when the effluent silver level reaches an undesirably high value, as the resin bed nears exhaustion. A concentrated thiosulfate solution is then pumped through the column as a resin regenerant. The silver thiosulfate on the exchange sites of the resin is replaced by the fresh, more concentrated thiosulfate; simultaneously, nearly all the silver previously on the resin has now been transferred to the thiosulfate regenerant, from which it can be removed electrolytically since it is now more concentrated. After a water rinse (which itself can be added to the washwater overflow stream for treatment, after being filtered), the system is then ready to recover additional silver. The bed can be reused for over 150 cycles, although eventually the resin deteriorates and must be replaced with fresh material. The electrolytically desilvered regenerant can also be reused for the next regeneration step after make-up chemicals have been added, to conserve chemical usage.

This technique has a very high degree of efficiency, about 98 to 99+%. Depending on the solutions treated, effluent silver levels in the 0.1 to 0.5 mg/L range can be attained, although levels in the 1 to 2 mg/L range are more common. The silver is recovered as high purity flake. Because thiosulfate is the only anion used, there is no contamination from other anions which would otherwise need to be rinsed out and discarded. The regenerant solution and resin can be used for multiple cycles; therefore, operating costs are moderate.

Photographic Processing Wastes

However, the initial capital cost for the ion-exchange and electrolytic silver-recovery equipment is high. Additionally, this technique as described above has been used successfully only by a few high-production laboratories with good maintenance capabilities. The most common operational problem is biological decomposition of the silver thiosulfate complex inside the column, which causes silver sulfide to form within the resin beads. Silver sulfide cannot be dissolved off the resin by the thiosulfate regenerant solution.

The in-situ precipitation technique takes advantage of this natural tendency of the thiosulfate complex to decompose. It involves deliberately precipitating silver as silver sulfide inside the resin, with no attempt to regenerate (Quinones, 1984; Lorenzo, 1988). Rohm & Haas Amberlite IRA-68 resin is also used with this method. After the silver thiosulfate complex has adsorbed on the resin, dilute sulfuric acid is then pumped through the column, intentionally converting the silver thiosulfate to silver sulfide. This acid is then neutralized and the column rinsed before the next cycle is run. The resin swells in size considerably to accommodate large amounts of silver sulfide, but it can be reused until its ultimate adsorption capacity is reached. The spent resin is finally removed from the column and incinerated to recover the silver. Effluent silver concentrations as low as 0.1 mg/L can be achieved using this method; however, levels of 1 mg/L are more common. The recovery efficiency attained is usually in the 98 to 99.5% range. This technique is generally recognized as the most efficient way of recovering silver from dilute photoprocessing solutions, since fewer steps, less labor, and simpler equipment are involved, and less silver is lost.

Sulfide Precipitation

Precipitation methods have also been shown to be very effective for silver recovery (Eastman Kodak, 1979). In particular, precipitation with sodium sulfide has been used for desilvering both fixers and washwaters since silver sulfide is one of the most insoluble species known, having a solubility product in water of 10^{-51}. This method involves first adjusting the pH of the normally acidic fixer with sodium hydroxide to an alkaline pH, to prevent the liberation of toxic hydrogen sulfide gas when the reactants are combined. Sodium sulfide is then added and the precipitated silver sulfide allowed to settle, after which it can be removed by decanting and/or filtration. This method has been used for many decades; however, although highly effective for removing silver, it is not very popular because of the "rotten egg" smell of sulfide and the potential hazard of forming hydrogen sulfide gas if personnel are not adequately trained. Silver levels as low as 0.01 mg/L have been attained in the laboratory.

In actual practice, the efficiency of this method is highly dependent on the filtration step.

This sulfide precipitation technique was automated in the early 1970s by Laperle (1976). The process uses automatic pH and specific ion electrodes to control the pH and add specific reagents as needed. An enclosed reaction tank and filtration system with automatic pumping cycles, as well as an emergency override actuated by a hydrogen sulfide gas detector, virtually eliminate the safety hazards and much of the bad odor usually associated with handling sodium sulfide.

Electrochemical Sulfide Precipitation

This technique, for generating sulfide ion directly from waste thiosulfate by electrolytic oxidation and using it to precipitate silver, was disclosed to the public by Kodak researchers in 1975, in lieu of seeking a patent (Bober and Leon, 1985). In this method, an electrolytic cell is deliberately operated in such a manner as to produce excess sulfide ion directly from a small side stream of the silver-bearing fixer or bleach-fix solution containing thiosulfate. The presence of silver or iron helps somewhat to catalyze this reaction. This sulfide-laden side stream is then recombined with the parent solution and reacted with the remaining silver (silver thiosulfate complex) to precipitate silver sulfide, in the same manner as if the sulfide had been added as a separate solution. The silver sulfide is then precipitated and collected by settling and/or filtration. This method has the advantage of not having to mix or store odorous sulfide solution, but instead generates it only on demand from a portion of the waste-processing solution with the flip of an electric switch.

This electrochemical sulfiding reaction takes advantage of the well-known electrochemical phenomenon, sulfiding, which has long been an operational problem for persons attempting to recover silver by electrolytic plating. However, very careful, continuous control is necessary that may involve the use of expensive instrumentation. The pH must be kept neutral, since the same precautions concerning liberation of toxic hydrogen sulfide gas under acid conditions apply (see "Sulfide Precipitation" above), and ammonia can be generated at alkaline pH's if ammonium ion is present. In addition, the silver sulfide formed is a very fine precipitate that requires good filtration; the blinding of normal filters was routinely encountered during laboratory experiments. As this publication goes to press, there is currently no known commercial device using this technology specifically for silver recovery.

Photographic Processing Wastes

Other Chemical Precipitation of Silver

Sodium borohydride can also be used to precipitate silver according to the reaction

$$BH_4^- + 2\,H_2O + 8\,Ag^+ \rightarrow 8\,Ag^o + 8\,H^+ + BO_2^-$$

This method can be used for bleach-fixes, fixers, and washwaters (Eastman Kodak, 1979). Significantly more than the stoichiometric amount of borohydride is required to complete the reaction, and it must be performed under alkaline conditions. The silver recovered has a purity in the 90 to 95% range. Borohydride can be quite dangerous to handle, since it can explosively liberate hydrogen gas under acid conditions; therefore, this technique should only be used by trained personnel.

Other chemical reducing agents have been tested experimentally for the precipitation of silver but are not in commercial use, and therefore, they will not be discussed here.

3.6 Solution Regeneration Techniques

Introduction

The collection of processing tank overflows and subsequent regeneration of solutions for reuse can reduce the quantity of chemicals discharged by between 40 to 90%, depending on the solution. In addition to process and product modifications that have resulted in the use of fewer processing solutions, lower concentrations of chemicals in the solutions, and reduced replenishment rates (Cribbs and Dagon, 1987), regeneration and reuse have significantly decreased the quantity of chemicals discarded from photographic processing operations. The following section discusses the considerations that impact regeneration and reuse techniques for color developer solutions, desilvered fixers and bleach-fixes, and bleaches.

Whether or not a specific processing solution can be regenerated and/or reused depends on a number of factors. First, reuse requires that the reclaimed solution is or can be made photographically acceptable. That is, sensitometric measurements, generated from photographic test materials processed with the reclaimed solution, must meet specific quality standards.

Second, the practicality of regeneration techniques will depend on the size of the photographic processing operation and the consumption rate of the solutions. If sophisticated equipment and techniques are required, a small photo-processing operation will most likely not have the technical expertise or be able to afford the capital expenditure needed to regenerate certain solutions.

In most photographic processes, fresh replenisher is added to each solution tank at a predetermined, fixed rate while the exposed product moves through the process, to maintain a certain minimal concentration of each required chemical in the processing solution. However, individual chemical constituents may be depleted at varying rates due to differences in exposure, size, or photographic characteristics of the photographic films or papers, differences in the rate they are fed through the process, effects of oxidation or carry-in of other solutions, etc. While the replenisher formula attempts to correct for this as well as possible to maintain a chemical balance, the need to discard unwanted by-products means that overflows from processing tanks will still contain large quantities of good chemicals that can potentially be reused. The exact techniques chosen for regeneration depend on the balance of unwanted vs. wanted components and their chemical nature.

Color Developer Regeneration

Color developer solutions become exhausted through the oxidation of developing agent and the increased concentration of reaction products, which may significantly reduce the activity of the solution. In most cases, the limiting factor for reuse is the increased halide concentration. Therefore, to reuse color developers, the halide concentrations must be controlled. Two approaches are employed to control the halide level.

First, the undesirable reaction products such as halides can be removed. A strong-base anion-exchange resin such as Rohm & Haas IRA-400 is usually chosen to remove bromide and chloride ions from the solution. If proper techniques are used, this treatment may also remove other constituents such as color-developing agents and oxidized color-developing agents, but only to a very small extent. This regeneration method has been applied to color developers from the color paper, reversal color paper, conventional color film, Kodak Ektachrome reversal film (in limited applications), Kodak Kodachrome reversal film, and Kodak Eastman Color motion picture processes (Dagnault, 1977; Allen, 1979; Kleppe, 1979a and 1979b; Meckl, 1979; Bard, 1980; Burger and Mina, 1983).

After the halides have been removed, the purified solution is then collected and analyzed, and needed make-up chemicals and water are added to bring it to replenisher strength. Previously, these regeneration techniques required considerable analytical capabilities. This limited the use of developer regeneration to only large laboratories having such facilities. However, during the past several years, developer regeneration kits have become available that minimize analytical demands, usually requiring only pH

Photographic Processing Wastes

measurement and therefore making the technique available to smaller laboratories.

Second, some color developers can be reclaimed using the reconstitution technique. The solutions are collected and then diluted until an appropriate halide level is reached. Chemicals are then added to bring the solution to replenishment strength. This method has been used in some color processes including Eastman Color motion picture films and more recently for color paper processing. In the later case, a slightly elevated temperature is used during development to offset the reduced activity due to higher halide content. The benefit of this approach over ion exchange is that minimal equipment is needed, desirable chemicals such as a color-developing agent are not adsorbed by the ion-exchange resin, and there is no need to discard regenerant solutions containing high salt concentrations.

Besides significantly reducing the concentration of color-developing agents discharged to sewer systems, the regeneration of color developer solutions can produce substantial COD reductions. Use of developer regeneration for color paper presents minimal risks since the paper can always be reprinted if not satisfactory. However, application to color negative or, especially, color reversal films represent a much greater risk since a customer's film may be ruined if the reused developer is out of specification.

Bleach Regeneration

As previously described, bleaches contain oxidizing agents that oxidize metallic silver to silver ion. Today most photographic processing bleaches use the selective oxidizing ability of ferric iron in a chelated form, often as an iron EDTA complex (Dagon, 1976). Bleaches used in color negative film and some color paper processes are usually regenerated. In most instances, the bleach overflow is collected and the ferrous ion oxidized to ferric by simple aeration, then makeup chemicals are added to bring the solution back to replenisher strength. Bleach regeneration will significantly reduce the amount of iron, chelating agent, and COD discharged to the environment.

Although most modern processes use alternative bleaches, several processes such as the Kodachrome and aerial Ektachrome film processes still use ferricyanide as their bleaching agent of choice. A ferricyanide bleach option is also available for processing Eastman Color motion picture film.

In ferricyanide bleaches, ferricyanide ion is the oxidizing agent, which becomes reduced to ferrocyanide upon use. Together, these ions and related forms of the iron-cyanide complex are known as hexacyanoferrates. The concentrations of hexacyanoferrates in an effluent can be minimized by a reliable regeneration method for the beach overflow. This requires collecting the overflow and treating it with a strong oxidizing agent. Options include persulfate, peroxide, bromine, ozone, and electrolysis (Cooley, 1976). When persulfate is used (Hutchins and West, 1957), the specific gravity of the solution builds up due to the formation of sulfate by-product. Eventually, after several regeneration cycles, the sulfate concentration can grow high enough to reduce the bleaching activity. This is usually remedied by discarding between 5 and 10% of the overflow. However, a precipitation technique can be used to prevent the wasted material from entering the sewer (see "Chemical Recovery: Ferrocyanide Precipitation").

An alternative bleach oxidant to persulfate is ozone (Bober and Dagon, 1972a and 1972b). The use of ozone requires a fairly significant capital investment in an ozone gas generator and contact system, and certain safeguards are needed to minimize risk to personnel since ozone is a toxic and unstable gas. However, the specific gravity build-up problem attributable to persulfate is eliminated (see "Ozone").

Another technique having many of the advantages of ozone without the risk of a toxic gas is electrolytic bleach regeneration (Kleppe and Nash, 1978). Ferrocyanide is oxidized to ferricyanide at the anode of an electrolytic cell. Because of the reduction reaction that occurs simultaneously at the cathode, the cell must be divided by some type of semipermeable membrane. Commercial units are available (see "Electrooxidation"). This is the most widely used method today.

In some cases, it may also be necessary to remove hexacyanoferrates from washwaters following a bleach or fixer. Since the complexes will be very diluted, it is not feasible to use precipitation techniques. Several alternatives are available to concentrate these salts and allow them to be treated by precipitation.

Reverse osmosis can be used to concentrate the bleach components in washwater (Cooley, 1976). By using high pressure (300–600 psi), it is possible to produce a permeate stream containing 90% of the volume but only a small quantity of hexacyanoferrate. The smaller brine stream, although only about 10% of the flow, will contain almost all the hexacyanoferrate complex. Although this technique has been demonstrated repeatedly on a laboratory scale, operational problems have limited its use in commercial practice. (See "Reverse Osmosis.")

Another technique for removing hexocyanoferrate from washwaters is by ion exchange (Brugger, 1979). Rohm & Haas Amberlite IRA-68 resin has been used successfully. Experiments have shown that 50 to 60 g of hexacyanoferrate can be collected on 1 L of resin before the effluent exceeds 1 mg/L hexacyanoferrate. Following ion-exchange treatment, the resin can be regenerated with a dilute sodium hydroxide solution, producing a solution containing as much as 25 g/L hexacyanoferrate. (See "Ion Exchange.")

It is also possible to treat ferricyanide solution by breaking down the hexacyanoferrate to innocuous products, by severe oxidation methods. Hendrickson and Daignault (1973) have discussed the destruction of ferricyanide of hexacyanoferrate solution by chlorination and ozonation. Although the chemical destruction of hexacyanoferrate solution by oxidation is possible, it is generally not economical.

Fixer and Bleach-Fix Reuse

As previously stated, electrolytic recovery and metallic replacement are used to desilver concentrated fixer and bleach-fix (see "Silver Recovery"). Steel wool cartridges have also been used to treat washes alone or washes combined with previously desilvered fixer or bleach-fix. Some photoprocessing installations use ion exchange to recover silver from washwater. Additionally, a few processing laboratories use chemical precipitation methods. Of these options, only electrolysis is used to any significant degree to recover silver from fixers and bleach-fixes when the solution is to be subsequently reused.

The primary factors limiting the direct recycling of fixers and bleach-fixes are the build-up of silver, halide ions, and, in some cases, oxidized developer products that can stain the product. However, with appropriate chemical treatment virtually every fixer or bleach-fix can be reused, although the degree of reuse possible varies from solution to solution.

When silver is recovered from the fixer or bleach-fix by plating at the cathode, the sodium sulfite preservative is consumed in the anode reaction, causing a pH decrease. In addition, the fixer or bleach-fix will have been diluted by water or carried-in products, usually including developer oxidation products, from the preceding tanks. To counter these effects, the unwanted components must be at least partially removed and makeup chemicals added to rejuvenate the desilvered overflow to replenisher strength.

The first step in regenerating either a fixer or bleach-fix involves collecting the overflow and desilvering the solution by electrolysis to a silver level of between 0.5 and 1.0 g/L. Lower levels are usually not attempted because of the possibility of sulfiding, which would irreversibly contaminate the solution.

A certain percentage of the desilvered overflow is saved for reuse. This can vary from a low of approximately 50% for C-41 (due to iodide build-up in this solution) to as high as 75% with Kodachrome fixer. Makeup chemicals are added, and the fixer or bleach-fix can then be reused. The portion that cannot be reused should be further desilvered to a low silver level prior to disposal.

A more common technique used by many processing laboratories is in-line desilvering, previously discussed. With this method, the electrolytic desilvering cell is plumbed directly to the fixer tank. The tank solution is continuously desilvered to silver levels in the range of 0.5 to 1.0 g/L. This permits lowering of the replenisher rate and effectively reduces the amount of fixer used per unit of film or paper processed. Again, reductions in fixer use of 50 to 75% can be obtained, depending on the process (Cooley and Dagon, 1976).

In some black-and-white processes, a considerable amount of developer is carried into the fixer by the film. During electrolytic desilvering the developer is oxidized at the anode and, if allowed to build up, could form objectionable stain on the product. In at least one X-ray film process, a technique has been devised to pass the fixer through an ion-exchange cartridge after desilvering. The ion-exchange resin effectively removes the staining material so that at least 50% of the fixer can be reused. A similar method has been used to remove excess bromide or iodide.

The regeneration method for desilvering and reusing bleach-fix varies slightly from the method for reusing fixer. The bleach-fix is collected and the pH adjusted to over 7 to facilitate electrolytic silver recovery. After desilvering to between 0.5 and 1.0 g/L silver, a certain portion is saved for reuse. Aeration may be performed to reoxidize the ferrous salt back to ferric (although in many cases, the solution spontaneously reaerates itself simply upon standing or when being pumped to and mixed in the processing tank). Make-up chemicals are added together with water to bring the solution back to replenisher levels. As much as 45% of the bleach-fix can be reused. Reuse of fixer and bleach-fix effectively reduces the ammonium ion, thiosulfate, BOD_5, and COD concentrations in the effluent (Krauss, 1987).

3.7 Chemical Recovery Methods

Introduction

Certain chemicals can be recovered individually from waste solutions even when the overall solution can no longer be salvaged. Often these are primary ingredients of a spent processing bath; in other situations, they may be foreign ingredients carried in from a preceding tank. Occasionally, it may be technically possible and economical to purify them in the laboratory and reuse them directly in a process. Other times the recovery may be done simply to extract a

Photographic Processing Wastes

waste material that cannot be discharged to a sewer and isolate it for separate disposal.

Recovery of Color-Developing Agents from Stop Baths

If color-developing agents are carried over into bleaches or bleach-fixes, they are irreversibly oxidized and can no longer be reused. However, when a stop bath follows the color developer solution, the color-developing agents remain essentially in their original form and can be recovered. A frequently used recovery method is to collect the stop bath and pass it through a column containing an ion-exchange resin such as Rohm & Haas XAD (Burger et al., 1985). The color-developing agents will be adsorbed on the resin. A method developed by Linkopia in Sweden then uses alkaline color developer solution to strip the color-developing agents off the resin for reuse. This technique is useful when a processing laboratory must meet very stringent discharge limitations for color-developing agents.

Coupler Recovery

In the Kodachrome film process, cyan, magenta, and yellow couplers are present in the three separate color developer solutions. (In all other processes, they are incorporated in individual layers within the color film or paper.) The Kodachrome couplers are all soluble in an alkaline solution but will precipitate at neutral or acid pH. Therefore, recovery becomes simple: The developer overflows are individually collected, pH is lowered (usually with dilute sulfuric or acetic acid or carbon dioxide gas), and the precipitated coupler is then removed by centrifuging or filtration and dried for storage. When needed, the dried coupler can be resolubilized in alkali (usually preceded by a grinding or homogenizing step to make mixing easier) and reused. Analytical facilities are required for this technique (Eastman Kodak, 1979).

Ferrocyanide Precipitation and Recovery

Although ferricyanide and ferrocyanide (hexacyanoferrates) have relatively low toxicities, to the degree that they are even used as ingredients in foods, they can be of concern because they have the capability of being degraded slowly through photolysis, especially by strong sunlight, to form free cyanide (Terhaar et al., 1972). Therefore, the discharge of hexacyanoferrate from a photoprocessing operation must be controlled. Any overflow concentrates that cannot be regenerated and reused, and sometimes washes or fixers containing hexacyanoferrate, should be treated. A very effective technique has been developed for removing hexacyanoferrate complexes from an effluent by precipitation (Bober and Cooley, 1972). The ferricyanide in solution is reduced to ferrocyanide by the addition of dithionite or a similar reducing agent. The ferrocyanide is then precipitated by adding ferrous sulfate, to form insoluble ferrous ferrocyanide. This technique is generally used to precipitate ferrocyanide from either fixers or washes following bleaches. The precipitate is removed by filtration or centrifuging. After washing off by-product salts, the precipitate can sometimes be reused in new bleach solutions by redissolving in alkali and reoxidizing it, although this latter method is usually economically feasible only for a very large production operation having good analytical capabilities (Kleppe and Vacco, 1979).

Chromium Precipitation

As previously stated, chromium is a relative rare metal in today's photographic processing laboratories, although it is used in a few black-and-white reversal bleaches as well as some process system cleaners. When a dichromate bleach is mixed with other processing solutions that are alkaline, and with solutions containing reducing agents such as thiosulfate and sulfite, the Cr^{+6} is reduced to Cr^{+3} and precipitated as chromium hydroxide or other insoluble particulate matter. A chromium +3 precipitate is then removed during primary or secondary clarification at the wastewater treatment plant as a component of the sludge.

If necessary, because of sewer code discharge limits, chromium can be recovered from solution in the processing laboratory by precipitation. This is done by collecting the chromium-bearing overflow in tanks and reducing the hexavalent chromium to the trivalent form by adding bisulfite or dithionite. The chromium is then precipitated as a chromium hydroxide solid by adjusting the solution to pH 8 with an alkaline material such as dilute sodium hydroxide. This sludge can then be removed by settling and decantation, filtration, or similar techniques. The chromium hydroxide sludge will have to be managed as a hazardous waste under the stipulations of the Resource Conservation and Recovery Act.

Phosphate Recovery

During the 1970s there was great public pressure to remove phosphates from many discharge sources, primarily household detergents but including industrial wastes, because of the general absence of adequate municipal treatment facilities and the resulting eutrophication of streams. At that time, great effort was put into reformulating photographic solutions to exclude phosphates. Only small quantities remain today, generally as sequestrants to prevent calcium in hard water from sludging or crystallizing in process solutions and damaging

Photographic Processing Wastes

emulsions, or for buffering high-pH developers. These levels are generally so low that no pretreatment would be required by the average laboratory that discharges to a POTW.

However, should phosphate removal be required because of stringent discharge limitations, it can readily be precipitated by various agents, particularly lime (calcium hydroxide). Unfortunately, the resulting precipitate is a sticky mass resembling toothpaste or milk of magnesia, from which it is difficult to extract water. Successful ways of handling and dewatering this sludge include evaporation, filtration on a vacuum belt filter or in a filter press, or drying of the solution (Bober and Cooley, 1972).

3.8 Conservation Methods

A number of techniques can be applied in a photoprocessing laboratory to help reduce chemical discharges to the sewer. These methods include the use of squeegees, careful maintenance of replenishment rates, use of holding tanks and floating lids, and good housekeeping techniques.

Squeegees

Squeegees are used to reduce the carry-over of processing solutions and washes. Both environmental and economic benefits result from their use. When properly installed and maintained, they can reduce carry-over by 75% or more. This reduction can mean, in turn, that replenishment rates will be lowered significantly and regeneration efficiencies optimized. Squeegees are usually installed after all washes except the final wash in a processing machine, and before and after all solutions being regenerated. Specifications for recommended placement of squeegees should be checked for each particular process to prevent emulsion damage.

Squeegees exist as many different types, including wiper blades, vacuum squeegees, wringer-slinger, and air squeegees, etc. (Ott and Dunn, 1968; Edgcomb and Zankowski, 1970; Perkins, 1970; Boutet, 1972). The type to be used will depend on the photographic product being processed, the particular solution tank (e.g., some of the most flexible synthetic rubber squeegees used on downstream tanks could not survive the relatively harsh alkaline environment of the developer tank), the kind of processing machine, and the level of technical expertise available in the laboratory for installation and maintenance.

Replenishment Rates

The replenishment rate required for adequate processing of a photographic product will depend on many factors including the following: the nature of the photographic product, speed at which it is being transported through the machine, temperature of the solution (especially for developers), concentration of chemicals in the replenisher solution, rate of build-up of seasoning products in the solution, design of the machine (certain configurations introduce more air into the solution during operation than others), and the overall utilization (i.e., running vs. standby time) of the processing machine. Photographic product manufacturers will provide specifications for replenishment rates required for a particular product under varying processing conditions. An adequate safety factor is built into these recommendations. It is important that a processing laboratory use the recommended rates. Although too low a rate could cause adverse photographic results, too high a rate will result in a waste of chemicals and money. Therefore, all processing laboratories should periodically check and adjust processing solution replenishment rates.

Good Housekeeping

Careful operating procedures in a photoprocessing laboratory can have a significant impact on the laboratory's typical effluent characteristics as measured over a period of time. Careful mixing of processing solutions will minimize dumping of concentrated processing solutions because of an error. Frequent maintenance of squeegees, tanks, and rollers and readjustment of pH, temperatures, and replenishment rates will also aid in decreasing the waste of processing solutions and chemicals in daily operations.

Floating Lids

The use of floating lids on solution storage tanks will help to reduce the aerial oxidation of easily oxidized chemicals, as well as evaporation. As the solution level decreases, the lid stays in contact with the solution surface to eliminate any air space. It also safeguards against airborne dust and dirt and accidental contamination by other laboratory chemicals and materials. It helps protect the quality of the processed materials, helps extend solution life, and thus lessens the dumping of stored solutions gone "bad." All these factors play important roles in the environmental as well as economic health of a laboratory.

3.9 Holding Tanks

One way to lessen the potential impact of having to dump contaminated solution tanks to the sewer in batches is to use holding tanks. A holding tank should have enough capacity to contain the largest sudden processing solution dump that a laboratory might anticipate. The solution batch is pumped to the holding tank and then slowly released over an extended time interval through a

Photographic Processing Wastes

valve to the drain, where it is diluted by another normally generated processing effluent. This approach will ensure that a sudden dump will not adversely affect the microorganisms in a secondary waste treatment plant.

3.10 Alternative Off-Site Disposal and Waste Concentration Options

Some situations occur in which it is not possible to discharge processing solution overflows to a sewer. This may be due to stringent sewer codes that cannot be met or because no sewer system is available. Under these circumstances, it may be necessary to have the solutions hauled away by an off-site waste disposal company. These services will then treat the solutions at a centralized waste treatment facility. It is important that the company chosen be reputable and comply with all pertinent regulations; if not, the photoprocessing laboratory could be held liable for illegal disposal of the waste. The use of off-site disposal may require the processing laboratory to comply with its state's waste management requirements, usually under a small quantity generator program.

Before signing a contract with such a waste disposal company, the laboratory manager and his or her chief technical assistant should personally visit the disposal facility. Facility personnel should be asked to describe or demonstrate their methods for handling, storing, and disposing of waste solutions. They should be able to produce permits and licenses to operate, as well as inspection records. Copies of such documents should be furnished to the prospective customer laboratory for its permanent files. The ultimate disposal of wastes from the treatment facility should be clearly stated, and the service company should be willing to furnish a signed statement to the laboratory, after each chemical pickup, that the laboratory no longer has title to the waste. The officers and principal stockholders of the company should be known, and their names checked with the state regulatory agency to be certain that they have no past record of waste-handling violations. A Dun and Bradstreet report or the equivalent should be requested on the company to verify its financial stability, to ensure that it has sufficient funds to properly treat all wastes that may be in current storage at its treatment sites. The above requirements are subject to regulatory change; an attorney should be consulted before proceeding.

A laboratory that needs to use an off-site waste disposal company will often find it very advantageous to minimize waste volume. On-site evaporation of water from the waste is the most widely used technique. Properly designed evaporators are simple to operate, either manually or automatically, and can generally remove from 85 to 95% of the water from a photoprocessing waste, depending on the original composition. Up to 99% may be removed if only washwaters are evaporated, but this may too costly an approach for these very dilute solutions because of the energy consumed; a better approach might be to use a water purification method such as ion exchange to clean up the washwater, then simply concentrate the ion-exchange regeneration chemicals in the evaporator. Preadjusting the pH of the solution before evaporation may be necessary to prevent the formation of unwanted gases at high or very low pH's. Before purchasing an evaporator, local regulatory codes should be checked to determine whether operating permits will be required. It is also important to make sure that the waste disposal firm will agree to handle the more concentrated waste. In addition, an economic check should be made to ensure that the smaller volume of more concentrated effluent will not cost as much to dispose of as the higher volume of more dilute, unevaporated effluent, when both waste-hauling and disposal charges and the purchase and operating costs of the evaporator are included. (See "Evaporation.")

In some branches of the photographic industry (particularly medical X-ray), small solution-service companies operate to pick up silver-bearing waste from a photoprocessing customer, including overflow fixer, silver flake from electrolytic units, exhausted metallic replacement cartridges, and even scrap film and paper. The materials are processed and a credit for silver, minus the processing and refining charges, is returned to the laboratory. Sometimes, these companies will perform maintenance work on the processing machines and recovery devices and may act as dealers to supply new chemicals and film or paper to the customer during their visit. With the advent of new waste regulations, some of these companies are making additional investments in their facilities and becoming properly licensed to also haul away other wastes from the laboratory on routine pickup cycles. This frees the laboratory from having to worry about waste discharge and allows it to concentrate on processing photographic film and paper. The silver credit can help defray a significant part of the cost for such a service.

3.11 Washwater Conservation

Washwater conservation can be important for many reasons, including water shortages in some locations, energy savings associated with reduced water consumption, and hydraulic limitations and/or discharge fees imposed on effluent discharged to municipal sewer systems (Fields, 1976). A laboratory

Photographic Processing Wastes

can conserve water in several ways. These include simple steps such as checking water supply equipment for leaks, using recommended wash rates, and running washwater only during actual processing.

Several methods can also be applied to reuse washwaters. Reverse osmosis has been chosen in a few instances, but its use is not widespread because of cost, maintenance, and operational problems that to date remain largely unresolved.

Perhaps the most commonly practiced methods of reducing water use involve ion exchange. A process such as the PACEX water recycling system incorporates ion-exchange technology together with filtration and bacterial control to remove contaminants (Kreiman, 1984). An anion-exchange resin removes silver thiosulfate complex from solution. An oxidizing agent subsequently oxidizes residual thiosulfate to sulfate and a biocide minimizes biological growth. A filter removes unwanted sediment. Depending on the process, reduction in washwater usage from 50 to 80% may be possible. This method not only conserves water and recovers silver but also provides energy savings by minimizing the quantity of fresh water that needs to be heated. Other manufacturers are beginning to announce similar systems. In each case, the photoprocessing laboratory has a responsibility to ensure that any recycling of water is properly done and constantly monitored for chemical build-up, such that the quality and storage life of its customers' processed films and papers do not suffer.

Washwater reuse raises several technical concerns. Proper operation of the system with strict quality control is mandatory, or the image quality of the photographic produce can be damaged. Also, washwater reuse will cause the effluents discharged from the photoprocessing laboratory to be more concentrated. Although the total chemical loading will not be higher, the higher concentrations could result in exceeding the sewer code.

In 1981, the EPA published a "Guidance Document for the Control of Water Pollution in the Photographic Processing Industry." In its conclusions, the report states, "the agency does not recommend the use of concentration-based limitations for controlling pollutants at facilities in the industry" (Versar, 1981). The report goes on to state that limits based on concentration tend to discourage water use reductions, and that the agency encourages the reduction of wastewater quantities by various water-saving controls. Unfortunately, this recommendation has been largely ignored, and since discharge monitoring is more difficult by other methods, most municipalities continue to use concentration-based limitations. Therefore, before installing washwater reuse equipment, it is important to check its potential effect on the laboratory's continued ability to use the municipal sewer system.

As a final comment, the above discussion is relevant to the photographic industry as of mid-summer 1990. The management of any chemical waste discharges, including photographic processing wastes, is a constantly shifting target, both because of rapidly changing regulations as well as somewhat less rapid advances in waste treatment technology. Consultation with reliable engineering and legal sources should be sought by any laboratory considering the appropriate, up-to-date management of its particular waste stream.

REFERENCES

Allen, L. E. (1979). Ion-exchange recovery techniques for the reuse of color developers. *J. SMPTE, 88*, 165–167.

American Cyanamid Co. (1953). *The Chemistry of the Ferrocyanides*, Vol. VII. New York.

Bard, C. C. (1980). Recovery and reuse of color developing agents. *J. SMPTE, 89*, 225–228.

Bard, C. C., Murphy, J. J., Stone, D. J., and Terhaar, C. J. (1976). Silver in photoprocessing effluents. *J. Water Pollut. Cont. Fed. 8*, 389–394.

Bober, T. W. and Cooley, A. C. (1972). The filter press for the filtration of insoluble photographic processing wastes. *J. Photo. Sci. and Eng., 16*(2), 131–135.

Bober, T. W. and Dagon, T. J. (1972a). The regeneration of ferricyanide bleach using ozone—Part 1. *Image Tech.*, 13–16, 24, 25.

Bober, T. W. and Dagon, T. J. (1972b). The regeneration of ferricyanide bleach using ozone—Part 2. *Image Tech.*, 19–24.

Bober, T. W. and Dagon, T. J. (1975). Treating photographic processing solutions and chemicals with ozone. *J. Water Poll. Cont. Fed., 47*, 2114–2129.

Bober, T. W. and Leon, R. B. (1975). Recovering metals from waste photographic processing solutions. *Res. Disclosure, 37*, 5, Item 13702.

Boutet, J. C. (1972). Spring-loaded wiper-blade squeegees. *J. SMPTE, 1*, 792–796.

Branch, D. A. (1988). Silver recovery methods for photoprocessing solutions. *J. Imaging Technol., 14*, 160–166.

Brugger, D. A. (1979). Removal of hexacyanoferrate from selected photographic process effluents by ion exchange. *J. SMPTE, 88*, 237–243.

Burger, J. L. and Mina, R. (1983). An alternative ion-exchange regeneration system for recovery of Kodak Ektaprint 2 developer. *J. Appl. Photo. Eng., 9*, 71–75.

Burger, J. L., Fowler, H. E., McPhee, B. A., and Yager, J. E. (1985). Recovery of Kodak color developing agent CD-2 from process ECP-2A color developer and stop bath. *J. SMPTE, 94*, 648–653.

Cooley, A. C. (1976). Regeneration and disposal of photographic processing solution containing hexacyanoferrate. *J. Appl. Photo. Eng., 2*.

Cooley, A. C. (1982). An engineering approach to electrolytic silver recovery systems design. *J. Appl. Photo. Eng., 8*, 171–180.

Cooley, A. C. (1984a). A study of the major parameters for designing rotating cathode electrolytic silver recovery cells. *J. Imaging Technol., 10*, 226–232.
Cooley, A. C. (1984b). Three-electrode control procedure for electrolytic silver recovery. *J. Imaging Technol., 10*, 233–238.
Cooley, A. C. (1986). The effect of the chemical components of fixer on electrolytic silver recovery. *J. Imaging Technol., 12*, 316–322.
Cooley, A. C. (1988). Silver recovery using steel wool metallic replacement cartridges. *J. Imaging Technol., 14*, 167–173.
Cooley, A. C. and Dagon, T. J. (1976). Current silver recovery practices in the photographic processing industry. *J. Appl. Photo. Eng., 2*, 36–41.
Cooley, A. C., Dagon, T. J., Jenkins, P. W., and Robillard, K. A. (1988). Silver and the environment. *J. Imaging Technol., 14*(6), 183–189.
Cribbs, T. P. and Dagon, T. J. (1987). A Review of Waste Reduction Programs in the Photoprocessing Industry. CMA Waste Minimization Workshop Notebooks, Vol. I, C-5, Nov. 12.
Dagon, T. J. (1973). Biological treatment of photoprocessing effluents. *J. Water Poll. Cont. Fed., 45*, 2123–2135.
Dagon, T. J. (1976). Processing chemistry of bleaches and secondary processing solution and applicable regeneration techniques. *J. Appl. Photo. Eng., 2*, 42–45.
Dagon, T. J. (1978). Photographic processing effluent control. *J. Appl. Photo. Eng., 2*, 62–71.
Daignault, L. G. (1977). Pollution control in the photoprocessing industry through regeneration and reuse. *J. Appl. Photo. Eng., 3*, 93–96.
Duffin, G. F. (1966). *Photographic Emulsion Chemistry*. Focal Press, Ltd., London.
Eastman Kodak (1979). Recovering Silver from Photographic Materials. Publ. 5-10, Rochester, N.Y.
Edgcomb, L. I. and Zankowski, J. S. (1970). Molded squeegee blades for photographic processing. *J. SMPTE, 79*, 123–126.
Ericson, F. A. and LaPerle, R. L. (1974). The State of Silver Photographic Effluents After Secondary Treatment; the Effectiveness of Extended-Aeration Treatment in Removing Silver from Photographic Processing Effluents. Eastman Kodak, Rochester, N.Y. Photographic Technol., Tech. Memo. TMB 74-36.
Feiler, H. D., Storch, D. J., and Shattuck, A. (1981). Treatment and Removal of Priority Industrial Pollutants at Publicly Owned Treatment Works. Burns and Roe Industrial Services Corp., EPA Rep. 400/1-79-300, April, Washington, D.C., PB83-142414.
Fields, A. E. (1976). Reducing wash water consumption in photographic processing. *J. Appl. Photo. Eng., 2*, 128–133.
Harbison, K. G. and Belly, R. T. (1982). The biodegradation of hydroquinone. *Environ. Toxicol. Chem., 1*, 9–15.
Hendrickson, T. N. (1975). *International Symposium on Ozone for Waste and Wastewater Treatment*, Industrial Ozone Inst., pp. 578–586.
Hendrickson, T. N. and Daignault, L. G. (1973). Treatment of photographic ferrocyanide-type bleach solution for reuse and disposal. *J. SMPTE, 82*, 727–732.

Hickman, K. C. D., et al. (1933). Electrolysis of silver-bearing thiosulfate solutions. *Ind. Eng. and Chem., 25*, 202–212.

Holm, C. E. (1990). Laboratory investigation of incineration of spent photofinishing liquids. Master's thesis, Jan. Univ. of Wis. Madison.

Hutchins, B. A. and West, L. A. (1957). The preparation or regeneration of a silver bleach solution by oxidizing ferrocyanide with persulfate. *J. SMPTE, 66.*

Hydroscience, Inc. (1974). *Environmental Effect of Photoprocessing Chemicals*, Vol. 1. Nat. Assoc. Photographic Manufacturers, Harrison, New York.

James, T. H. (ed.) (1977). *The Theory of the Photographic Process*, 4th ed. Macmillan, New York.

J. B. Scientific Corp. (1977). *Pathways of Photoprocessing Chemicals in Publicly Owned Treatment Works*. Nat. Assoc. Photographic Manufacturers, Harrison, New York.

Kleppe, J. W. (1979a). The application of an ion exchange method for color developer reuse. *J. Appl. Photo. Eng., 5*, 132–135.

Kleppe, J. W. (1979). Practical application of an ion-exchange method for color developer reuse. *J. SMPTE, 88*, 168–171.

Kleppe, J. W. and Nash, C. R. (1978). A simplified electrolytic method for ferricyanide bleach regeneration. *J. SMPTE, 878*, 4.

Kleppe, J. W. and Vacco, D. (1979). Settle and decant process for ferrocyanide removal from fixer. Eastman Kodak.

Knorre, H., Maennig, D., and Stidetzel, K. (1988). Chemical treatment of effluent from photofinishing plants. *J. Imaging Technol., 4*, 154–156.

Krauss, S. J. (1987). Factors affecting the desilvering and reuse of bleach-fix. Presented at SPSE Symp. Environmental Issues in Photofinishing, Sept. 15–17, Los Angeles, Calif.

Kreiman, R. T. (1984). Photo wash water recycling system utilizes ion exchange technology. *J. Imaging Technol., 10*, 244–246.

LaPerle, R. L. (1976). The removal of metals from photographic effluent by sodium sulfide precipitation. *J. SMPTE, 85*, 206–216.

Lorenzo, G. A. (1988). In-situ ion exchange silver recovery for pollution control. *J. Imaging Technol., 14*, 174–177.

Lytle, P. E. (1984). Treatment of photofinishing effluents using rotating biological contractors (RBC's). *J. Imaging Technol., 10*, 221–226.

Lytle, P. E. (1985). Fate and speciation of silver in publicly owned treatment works. *Environ. Toxicol. Chem., 3*, 21–30.

Mason, L. F. A. (1985). *Photographic Processing Chemistry*, Focal Press, Ltd., London.

Meckl, H. (1979). Recycling of color paper developer. *J. Appl. Photo. Eng., 5*, 216–219.

The Merck Index, 11th ed. (1989). Rahway, N.J., Item 3484, p. 550, Items 8562 and 8563, p. 1361; Item 3694, p. 631; Items 8562 and 8563, p. 1361.

Mina, R. (1980). Silver recovery from photographic effluents by ion-exchange methods. *J. Appl. Photo. Eng., 6*, 120–125.

Ott, H. F. and Dunn, J. E. (1968). The rotary-buffer squeegee and its use in a motion-picture film lubricator. *J. SMPTE, 77*, 121–124.

Perkins, P. E. (1970). A review of the effects of squeegees in continuous-processing machines. *J. SMPTE, 79*, 121–123.

Petche, K. S. (1989). Meeting municipal sanitary sewer discharge permits limits for production operations at University Microfilms, International. Presented at SPSE Ann. Conf., May 19, Boston, Mass.

Petschke, D. (ed.) (1988). Disposal and Treatment of Photographic Effluent, in Support of Clean Water. Kodak Publ. J-55, Rochester, N.Y.

Quinones, P. R. (1984). In-situ precipitation as the regeneration step in ion exchange for silver recovery. *J. SMPTE, 93*, 800–807.

Quinones, P. R. (1985). Optimizing silver recovery in photofinishing operations. *J. Imaging Technol., 11*, 43–50.

Ragland, K. W., Holm, C. E., and Andren, A. W. (1990). Laboratory investigation of incineration of spent photofinishing liquids. Presented at SPSE 6th Internat. Symp. Photofinishing Technol., Feb. 20, Las Vegas, Nev.

Terhaar, C. J., Ewell, W. S., Dziuba, S. P., and Fassett, D. W. (1972). Toxicity of photographic processing chemicals to fish. *Photo. Sci. and Eng., 16*, 370–377.

Thomas, W., Jr. (ed.) (1973). *SPSE Handbook of Photographic Science and Engineering*. J. Wiley, New York, Sec. 9, 10.

U.S. Patent 3,594,157 (1971). Alkaline Chlorination of Waste Photographic Processing Solutions Containing Silver. July 20, Washington, D.C.

U.S. Patent 3,767,572 (1973). Destruction of EDTA by Alkaline Chlorination. Oct. 23, Washington, D.C.

Versar, Inc. (1981). Guidance Document for the Control of Water Pollution in the Photographic Processing Industry. EPA Rep. 440/1-81/082-9, April, Washington, D.C., PB82-177643, pp. 111-1 thru 111-62; I-3, I-5, II-1, V-1 thru V-25.

Watson, H. M., Boatman, R., and Ewell, W. S. (1984). Simulated Secondary Waste Treatment of Color Developer CD-3. Tech. Rep. ETS-TR-84-30, Health and Environmental Labs., Eastman Kodak, Rochester, N.Y.

Weber, W. J., Jr. (1972). *Physicochemical Processes for Water Quality Control*. Wiley-Interscience, New York.

Zimpro Passavant (1989). Wet-air oxidation of three photographic developing solutions from Eastman Kodak. Unpublished rept., Nov. 9, Rothschild, Wis.

5

Treatment of Soap and Detergent Industry Wastes

Constantine Yapijakis

The Cooper Union, New York, New York

1. INTRODUCTION

Natural soap was one of the earliest chemicals produced by man. Historically, its first use as a cleaning compound dates back to Ancient Egypt (Callely et al., 1976). In modern times, the soap and detergent industry, although a major one, produces relatively small volumes of liquid wastes directly. However, it causes great public concern when its products are discharged after use in homes, service establishments, and factories.

A number of soap substitutes had been developed for the first time during World War I, but the large-scale production of synthetic surface-active agents (surfactants, syndets) became commercially feasible only after World War II. Since the early 1950s, syndets have replaced soap in cleaning and laundry formulations in virtually all countries with an industrialized society. In the past 40 years, the total world production of synthetic detergents increased about 50-fold, but this expansion in use has not been paralleled by a significant increase in the detectable amounts of surfactants in soils or natural water bodies to which waste surfactants were discharged (Callely et al., 1976). This is due to the fact that primarily the biological degradation of these compounds has been taking place in the environment or in treatment plants.

The water pollution resulting from the production or use of detergents represents a typical case of the problems that followed the very rapid evolution of the industrialization that contributed to the improvement of the quality of life after World War II. Prior to that time, this problem did not exist. The continuing increase in consumption of detergents (in particular, their domestic use) and the tremendous increase in production of syndets are the origin of a type of pollution whose most significant impact is the formation of toxic or nuisance foams in rivers, lakes, and treatment plants.

1.1 Classification of Surfactants

Soaps and detergents are formulated products designed to meet various cost and performance standards. The formulated products contain many components, such as surfactants to tie up unwanted materials (commercial detergents usually contain only 10–30% surfactants), builders or polyphosphate salts to improve surfactant processes and remove calcium and magnesium ions, and bleaches to increase reflectance of visible light. They also contain various additives designed to remove stains (enzymes), prevent soil redeposition, regulate foam, reduce washing machine corrosion, brighten colors, give an agreeable odor, prevent caking, and help processing of the formulated detergent (Greek, 1990).

The classification of surfactants in common usage depends on their electrolytic dissociation, which allows the determination of the nature of the hydrophilic polar group, e.g., anionic, cationic, nonionic, and amphoteric. As reported by Greek (1990), the total 1988 U.S. production of surfactants consisted of 62% anionic, 10% cationic, 27% nonionic, and 1% amphoteric.

Anionic Surfactants

These produce a negatively charged surfactant ion in aqueous solution, usually derived from a sulphate, carboxylate, or sulphonate grouping. The usual types of these compounds are carboxylic acids and derivatives (largely based on natural oils), sulfonic acid derivatives (alkylbenzene sulfonates LAS or ABS and other sulfonates), and sulfuric acid esters and salts (largely sulfated alcohols and ethers). Alkyl sulfates are readily biodegradable, often disappearing within 24 hr in river water or sewage plants (Payne, 1963). Because of their instability in acidic conditions, they were to a considerable extent replaced by ABS and LAS, which have been the most widely used of the syndets because of their excellent cleaning properties, chemical stability, and low cost. Their biodegradation has been the subject of numerous investigations (Swisher, 1970).

Treatment of Soap and Detergent Industry Wastes

Cationic Surfactants

These produce a positively charged surfactant ion in solution and are mainly quaternary nitrogen compounds such as amines and derivatives and quaternary ammonium salts. Due to their poor cleaning properties, they are little used as detergents; rather their use is due to their bacteriocidal qualities. Relatively little is known about the mechanisms of biodegradation of these compounds.

Nonionic Surfactants

These are mainly carboxylic acid amides and esters and their derivatives, and ethers (alkoxylated alcohols), and they have been gradually replacing ABS in detergent formulations (especially as an increasingly popular active ingredient of automatic washing machine formulations) since the 1960s. Therefore, their removal in wastewater treatment is of great significance, but although it is known that they readily biodegrade, many facts about their metabolism are unclear (Osburn and Benedict, 1966). In nonionic surfactants, both the hydrophilic and hydrophobic groups are organic, so the cumulative effect of the multiple weak organic hydrophils is the cause of their surface-active qualities. These products are effective in hard water and are very low foamers.

Amphoteric Surfactants

As previously mentioned, these presently represent a minor fraction of the total surfactants production with only specialty uses. They are compounds with both anionic and cationic properties in aqueous solutions, depending on the pH of the system in which they work. The main types of these compounds are essentially analogues of linear alkane sulfonates, which provide numerous points for the initiation of biodegradation, and pyridinium compounds that also have a positively charged N-atom (but in the ring) and they are very resistant to biodegradation (Wright and Cain, 1972).

1.2 Sources of Detergents in Waters and Wastewaters

The detergent concentrations that actually find their way into wastewaters and surface water bodies have quite diverse origins. (1) Soaps and detergents, as well as their component compounds, are introduced into wastewaters and water bodies at the point of their manufacture, at storage facilities and distribution warehouses, and at points of accidental spills on their routes of transportation. (The origin of pollution is dealt within this chapter.) (2) The additional industrial origin of detergent pollution notably results from the use of surfactants in various industries, e.g., textiles, cosmetics, leather tanning and products, paper, metals, dyes and paints, production of domestic soaps and detergents, and from the use of detergents in commercial/industrial laundries and dry cleaners. (3) The contribution from agricultural activities is due to the surface runoff transporting of surfactants that are included in the formulation of insecticides and fungicides (Prat and Giraud, 1964). (4) The origin with the most rapid growth since the 1950s comprises the wastewaters from urban areas and it is due to the increased domestic usage of detergents and, equally important, their use in cleaning public spaces, sidewalks, and street surfaces.

1.3 Problem and Biodegradation

Notable improvements in washing and cleaning resulted from the introduction and increasing use of synthetic detergents. However, this also caused difficulties in sewage treatment and led to a new form of pollution, the main visible effect of which was the formation of objectionable quantities of foam on rivers. Although biodegradation of syndets in soils and natural waters was inferred by the observation that they did not accumulate in the environment, there was widespread concern that their much higher concentrations in the effluents from large industrial areas would have significant local impacts. In agreement with public authorities, the manufacturers fairly quickly introduced products of a different type.

The surface-active agents in these new products are biodegradable (called "soft" in contrast to the former "hard" ones). They are to a great extent eliminated by normal sewage treatment, and the self-purification occurring in water courses also has some beneficial effects (OECD, 1971). However, the introduction of biodegradable products has not solved all the problems connected to syndets (i.e., sludge digestion, toxicity, and interference with oxygen transfer), but it has made a significant improvement. Studies of surfactant biodegradation have shown that the molecular architecture of the surfactant largely determines its biological lability (Callely et al., 1976). Nevertheless, one of the later most pressing environmental problems was not the effects of the surfactants themselves, but the eutrophication of natural water bodies by the polyphosphate builders that go into detergent formulations. This led many local authorities to enact restrictions in or even prohibition of the use of phosphate detergents.

2. IMPACTS OF DETERGENT PRODUCTION AND USE

Surfactants retain their foaming properties in natural waters in concentrations as low as 1 mg/L, and although such concentrations are nontoxic to

Treatment of Soap and Detergent Industry Wastes

humans (Swisher, 1968), the presence of surfactants in drinking water is esthetically undesirable. More important, however, is the generation of large volumes of foam in activated sludge plants and below weirs and dams on rivers.

2.1 Impacts in Rivers

The principal factors that influence the formation and stability of foams in rivers (Prat and Giraud, 1964) are the presence of ABS-type detergents, the concentration of more or less degraded proteins and colloidal particles, the presence and concentration of mineral salts, the temperature and pH of the water. Additional very important factors are the BOD of the water that under given conditions represents the quantity of biodegradable material; the time of travel and the conditions influencing the reactions of the compounds presumed responsible for foaming, between the point of discharge and the location of foam appearance; and last but not least, the concentration of calcium ion that is the main constituent of hardness in most natural waters and merits particular attention with regard to foam development.

The minimum concentrations of ABS or other detergents above which foam formation occurs vary considerably, depending on the water medium, i.e., river or sewage, and its level of pollution (mineral or organic). Therefore, it is not merely the concentration of detergents that controls foam formation, but rather their combined action with other substances present in the waters. Various studies have shown (Prat and Giraud, 1964) that the concentration of detergents measured in the foams is quite significantly higher, up to three orders of magnitude, than that measured at the same time in solution in the river waters.

The formation of foam constitutes trouble and worries for river navigation as well. For instance, in the areas of dams and river locks, the turbulence caused by the intensive traffic of barges and by the incessant opening and closing of the lock gates results in foam formation that may cover entire boats and leave a sticky deposit on the decks of barges and piers. This renders them extremely slippery and may be the cause of injuries and drownings. Also, when the winds are strong, masses of foam are detached and transported to great distances in the neighboring areas, causing problems in automobile traffic by deposition on car windshields and by rendering the road surfaces slippery. Finally, masses of foam floating on river waters represent an esthetically objectionable nuisance and a problem for the tourism industry.

2.2 Impacts on Public Health

For a long time, detergents were utilized in laboratories for the isolation, through concentration in the foam, of mycobacteria such as the bacillus of Koch (tuberculosis), as reported in the annals of the Pasteur Institute (Prat and Giraud, 1964). This phenomenon of extraction by foam points to the danger existing in river waters where numerous such microorganisms may be present due to sewage pollution. The foam transported by wind could possibly serve as the source of a disease epidemic. In fact, this problem limits itself to the mycobacteria and viruses (such as those of hepatitis and polio), which are the only microorganisms able to resist the disinfecting power of detergents. Therefore, waterborne epidemics could also be spread through airborne detergent foams.

2.3 Impacts on Biodegradation of Organics

Surfactant concentrations in polluted natural water bodies interfere with the self-purification process in several ways. First, certain detergents such as ABS are refractory or difficult to biodegrade and even toxic or inhibitory to microorganisms, and influence the BOD exhibited by organic pollution in surface waters. On the other hand, readily biodegradable detergents could impose an extreme short-term burden on the self-purification capacity of a water course, possibly introducing anaerobic conditions.

Surfactant concentrations also exert a negative influence on the bio-oxidation of certain substances, as evidenced in studies with even readily biodegradable substances (Chambon and Giraud, 1960). It should be noted that this protection of substances from bio-oxidation is only temporary and it slowly reduces until its virtual disappearance in about a week for most substances. This phenomenon serves to retard the self-purification process in organically polluted rivers, even in the presence of high concentrations of dissolved oxygen.

An additional way in which detergent concentrations interfere with the self-purification process in polluted rivers consists of their negative action on the oxygen rate of transfer and dissolution into waters. According to Gameson et al. (1955), the presence of surfactants in a water course could reduce its reaeration capacity by as much as 40%, depending on other parameters such as turbulence. In relatively calm waters such as estuaries, under certain conditions, the reduction of reaeration could be as much as 70%. It is the anionic surfactants, especially the ABS, that have the overall greatest negative impact on the natural self-purification mechanisms of rivers.

Treatment of Soap and Detergent Industry Wastes 235

2.4 Impacts on Wastewater Treatment Processes

Despite the initial apprehension on the possible extent of impacts of syndets on the physico-chemical or biological treatment processes of municipal and industrial wastewaters, it soon became evident that no major interference occurred. As mentioned previously, the greatest problem proved to be the layers of foam that not only hindered normal sewage-plant operation, but when wind-blown into urban areas, also aided the probable transmission of fecal pathogens present in the sewage.

The first unit process in a sewage treatment plant is primary sedimentation, which depends on simple settling of solids partially assisted by flocculation of the finer particles. The stability, nonflocculating property, of a fine particle dispersion could be influenced by the surface tension of the liquid or by the solid/liquid interface tension—hence, by the presence of surfactants. Depending on the conditions, primarily the size of the particles in suspension, a given concentration of detergents could either decrease (finer particles) or increase (larger particles) the rate of sedimentation (Prat and Giraud, 1964). The synergistic or antagonistic action of certain inorganic salts, which are included in the formulation of commercial detergent products, is also influential.

The effect of surfactants on wastewater oils and greases depends on the nature of the latter, as well as on the structure of the lipophilic group of the detergent that assists solubilization. As is the case, emulsification could be more or less complete. This results in a more or less significant impact on the efficiency of physical treatment designed for their removal. On the other hand, the emulsifying surfactants play a role in protecting the oil and grease molecules from attacking bacteria in a biological unit process.

In water treatment plants, the coagulation/flocculation process was found early to be affected by the presence of surfactants in the raw water supply. In general, the anionic detergents stabilize colloidal particle suspensions or turbidity solids that, most times, are negatively charged. Langelier et al. (1952) reported problems with water clarification due to syndets, although according to Nichols and Koepp (1961) and Todd (1954) concentrations of surfactants on the order of 4 to 5 ppm interfered with flocculation. The floc, instead of settling to the bottom, floats to the surface of sedimentation tanks. Other studies, such as those conducted by Smith et al. (1956) and Cohen et al. (1959), indicated that this interference could be not so much due to the surfactants themselves, but to the additives included in their formulation, i.e., phosphate complexes. Such interference was observed both for alum and ferric sulfate coagulant, but the use of certain organic polymer flocculants was shown to overcome this problem.

Concentrations of detergents, such as those generally found in municipal wastewaters, have been shown to insignificantly impact on the treatment efficiency of biological sewage treatment plants (McGauhey and Klein, 1959). Studies indicated that significant impacts on efficiency can be observed only for considerable concentrations of detergents, such as those that could be possibly found in undiluted industrial wastewaters, on the order of 30 ppm and above. As previously mentioned, it is through their influence of water aeration that the surfactants impact the organics' biodegradation process. As little as 0.1 mg/L of surfactant reduces to nearly half the oxygen absorption rate in a river, but in sewage aeration units the system could be easily designed to compensate. This is achieved through the use of the alpha and beta factors in the design equation of an aeration system.

Surfactants are only partially biodegraded in a sewage treatment plant, so that a considerable proportion may be discharged into surface water bodies with the final effluent. The shorter the overall detention time of the treatment plant, the higher the surfactant concentration in the discharged effluent. By the early 1960s, the concentration of syndets in the final effluents from sewage treatment plants was in the 5 to 10 ppm range, and while dilution occurs at the site of discharge, the resulting values of concentration were well above the threshold for foaming. In more recent times, with the advent of more readily biodegradable surfactants, foaming within treatment plants and in natural water bodies is a much more rare and limited phenomenon.

Finally, according to Prat and Giraud (1964), the process of anaerobic sludge digestion, commonly used to further stabilize biological sludge prior to disposal and to produce methane gas, is not affected by concentrations of syndets in the treated sludge up to 500 ppm or when it does not contain too high an amount of phosphates. These levels of concentration are not found in municipal or industrial effluents, but within the biological treatment processes a large part of the detergents is passed to the sludge solids. By this, it could presumably build up to concentrations (especially of ABS surfactants) that may affect somewhat the sludge digestion process, i.e., methane gas production. Also, it seems that anaerobic digestion (Lawson, 1959) does not decompose syndets and, therefore, their accumulation could pose problems with the use of the final sludge product as a fertilizer.

Treatment of Soap and Detergent Industry Wastes

2.5 Impacts on Soil and Groundwater

The phenomena related to surface tension in groundwater interfere with the mechanisms of water flow in the soil. The presence of detergents in wastewaters discharged on soil for groundwater recharge or filtered through sand beds would cause an increase in headloss and leave a deposit of surfactant film on the filter media, thereby affecting permeability. Surfactants, especially those resistant to biodegradation, constitute a pollutant that tends to accumulate in groundwater and has been found to remain in the soil for a few years without appreciable decomposition. Since surfactants modify the permeability of soil, their presence could possibly facilitate the penetration of other pollutants, i.e., chemicals or microorganisms, to depths where they would not have reached due to the filtering action of the soil, thereby increasing groundwater pollution (Robeck et al., 1962).

2.6 Impacts on Drinking Water

From all the aforementioned, it is obvious that detergents find their way into drinking water supplies in various ways. As far as imparting odor to drinking water, only heavy doses of anionic surfactants yield an unpleasant odor (Renn and Barada, 1961), and someone has to have a very sensitive nose to smell detergent doses of 50 mg/L or less. On the other hand, it seems that the impact of detergent doses on the sense of taste of various individuals varies considerably. As reported by Cohen (1962), the U.S. Public Health Service conducted a series of taste tests which showed that although 50% of the people in the test group detected a concentration of 60 mg/L of ABS in drinking water, only 5% of them detected a concentration of 16 mg/L. Since tests like this have been conducted using commercial detergent formulations, most probably the observed taste is not due to the surfactants but rather to the additives or perfumes added to the products. However, the actual limit for detergents in drinking water in the United States is a concentration of only 0.5 mg/L; less than even the most sensitive palates can discern it.

2.7 Toxicity of Detergents

There is an upper limit of surfactant concentration in natural waters above which the existence of aquatic life, particularly higher animal life, is endangered. Trout are particularly sensitive to concentrations as low as 1 ppm and show symptoms similar to asphyxia (Callely et al., 1976). On the other hand, numerous studies, which extended over a period of months and required test animals to drink significantly high doses of surfactants, showed absolutely no apparent ill-effects due to digested detergents. And there are no instances in which the trace amounts of detergents present in drinking water were directly connected to adverse effects on human health.

River pollution from anionic surfactants, the primarily toxic ones, is of two types essentially: (1) acute toxic pollution due to, i.e., an accidental spill from a container of full-strength surfactant products, and (2) chronic pollution due to the daily discharges of municipal and industrial wastewaters. The international literature contains the result of numerous studies that have established dosages for both types of pollutional toxicity due to detergents, for most types of aquatic life such as species of fish.

3. CURRENT PERSPECTIVE AND FUTURE OUTLOOK

This section summarizes the main points of a recent product report (Greek, 1990), which presented the new products of the detergent industry and its proposed direction in the foreseeable future.

If recent product innovations sell successfully in test markets in the United States and other countries, rapid growth could begin again for the entire soap and detergent industry and especially for individual sectors of that industry. Among these new products are formulations that combine bleaching materials and other components, and detergents and fabric softeners sold in concentrated forms. These concentrated materials, so well accepted in Japan, are now becoming commercially significant in Western Europe. Their more widespread use will allow the industry to store and transport significantly smaller volumes of detergents, with the consequent reduction of environmental risks from housecleaning and spills. Some components of detergents, i.e., enzymes, will very likely grow in use, although the use of phosphates employed as builders will continue to drop for environmental reasons. Consumers shift to liquid formulations in areas where phosphate materials are banned from detergents, because they perceive that the liquid detergents perform better than powdered ones without phosphates.

In fuel markets, detergent formulations such as gasoline additives that limit the buildup of deposits in car engines and fuel injectors will very likely grow fast from a small base, with the likelihood of an increase in spills and discharges from this industrial source. Soap, on the other hand, now has become a small part (17%) of the total output of surfactants, whereas the anionic forms (which include soaps) accounted for 62% of total U.S. production in 1988. Liquid detergents (many of the LAS type), which are

Treatment of Soap and Detergent Industry Wastes 239

generally higher in surfactant concentrations than powdered ones, will continue to increase in production volume, therefore creating greater surfactant pollution problems due to housecleaning and spills. (Plus, a powdered detergent spill creates less of a problem, since it is easier to just be scooped up or vacuumed.)

Changes in the use of builders resulting from environmental concerns have been pushing surfactant production demand. Outright legal bans or consumer pressures on the use of inorganic phosphates and other materials as builders generally have led formulators to raise the contents of surfactants in detergents. Builders provide several functions, most important of which are to aid the detergency action and to tie up and remove calcium and magnesium from the washwater, dirt, and the fabric or other material being cleaned. Besides sodium and potassium phosphates, other builders that may be used in various detergent formulations are citric acid and derivatives, zeolites, and other alkalis. Citric acid causes caking and is not used in powdered detergents, but it finds considerable use in liquid ones. In some detergent formulations, larger and larger amounts of soda ash (sodium carbonate) are replacing inert ingredients due to its functionality as a builder, an agglomerating aid, a carrier for surfactants, and a source of alkalinity.

Incorporating bleaching agents into detergent formulations for home laundry has accelerated, because its performance allows users to curtail the need to store as well as add (as a second step) bleaching material. Because U.S. home laundry requires shorter wash times and lower temperatures than European home laundry, chlorine bleaches (mainly sodium hypochlorite) have long dominated the U.S. market. Institutional and industrial laundry bleaching, when done, also has favored chlorine bleaches (often chlorinated isocyanurates) because of their rapid action. Other kinds of bleaching agents used in the detergent markets are largely sodium perborates and percarbonates other than hydrogen peroxide itself.

The peroxygen bleaches are forecast to grow rapidly, for both environmental and technical reasons, as regulatory pressures drive the institutional and industrial market away from chlorine bleaches and toward the peroxygen ones. The Clean Water Act amendments are requiring lower levels of trihalomethanes (products of reaction of organics and chlorine) in wastewaters. Expensive systems may be needed to clean up effluents, or the industrial users of chlorine bleaches will have to pay higher and higher surcharges to municipalities for handling chlorine-containing wastewaters that are put into sewers. Current and expected changes in bleaching materials for various segments of the detergent industry are but part of sweeping changes to come due to environmental concerns and responses to efforts to improve the world environment.

Both detergent manufacturers and their suppliers will make greater efforts to develop more "environmentally friendly" products. BASF, e.g., has developed a new biodegradable stabilizer for perborate bleach, which is now being evaluated for use in detergents. The existing detergent material, e.g., LAS and its precursor linear alkylbenzene, known to be nontoxic and environmentally safe as well as effective, will continue to be widely used. It will be difficult, however, to gain approval for new materials to be used in detergent formulations until their environmental performance has been shown to meet existing guidelines. Some countries, e.g., tend to favor a formal regulation or law (i.e., the EEC Countries) prohibiting the manufacture, importation, or use of detergents not satisfactorily biodegradable (OECD, 1971).

4. INDUSTRIAL OPERATION AND WASTEWATER

The soap and detergent industry is a basic chemical manufacturing industry in which essentially both the mixing and chemical reactions of raw materials are involved in production. Also, short- and long-term chemicals storage and warehousing, as well as loading/unloading and transportation of chemicals, are involved in the operation.

4.1 Manufacture and Formulation

This industry produces liquid and solid cleaning agents for domestic and industrial use, including laundry, dishwashing, bar soaps, specialty cleaners, and industrial cleaning products. It can be broadly divided into two categories: (1) soap manufacture that is based on the processing of natural fat, and (2) detergent manufacture that is based on the processing of petrochemicals. The information presented here includes establishments primarily involved in the production of soap, synthetic organic detergents, inorganic alkaline detergents, or any combinations of these, and plants producing crude and refined glycerine from vegetable and animal fats and oils. Types of facilities not discussed here include plants primarily involved in the production of shampoo or shaving creams/soaps whether from soap or syndets, and of synthetic glycerine as well as specialty cleaners, polishing and sanitation preparations.

Numerous processing steps exist between basic raw materials for surfactants and other components that are used to improve performance and desirability, and the finished marketable products of the soap and detergent industry.

Treatment of Soap and Detergent Industry Wastes

Inorganic and organic compounds such as ethylene, propylene, benzene, natural fatty oils, ammonia, phosphate rock, trona, chlorine, peroxides, and silicates are among the various basic raw materials being used by the industry. The final formulation of the industry's numerous marketable products involves both simple mixing of and chemical reactions among compounds such as the above.

The following categorization system (Table 1) of the various main production streams and their descriptions is taken from federal guidelines (EPA, 1977) pertaining to state and local industrial pretreatment programs. It will be used in the discussion that ensues to identify process flows and to characterize the resulting raw waste. Figure 1 shows a flow diagram for the production streams of the entire industry.

Table 1 Soap and Detergent Categorization

Category	Subcategory	Code
Soap manufacture	batch kettle and continuous	A
	fatty acid manufacture by fat splitting	B
	soap from fatty acid neutralization	C
	glycerine recovery	
	glycerine concentration	D
	glycerine distillation	E
	soap flakes and powders	F
	bar soaps	G
	liquid soap	H
Detergent manufacture	oleum sulfonation and sulfation (batch and continuous)	I
	air-SO_3 sulfation and sulfonation (batch and continuous)	J
	SO_3 solvent and vacuum sulfonation	K
	sulfamic acid sulfation	L
	chlorosulfonic acid sulfation	M
	neutralization of sulfuric acid esters and sulfonic acids	N
	spray-dried detergents	O
	liquid detergent manufacture	P
	detergent manufacture by dry blending	Q
	drum-dried detergents	R
	detergent bars and cakes	S

Source: EPA (1977).

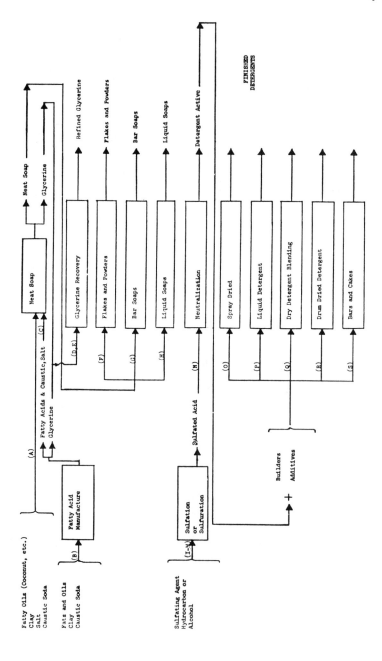

Figure 1 Flow diagram of soap and detergent manufacture. (After EPA, 1977.)

4.2 Soap Manufacture and Processing

Manufacturing of soap consists of two major operations: the production of neat soap (65–70% hot soap solution) and the preparation and packaging of finished products into flakes and powders (F), bar soaps (G), and liquid soaps (H). Many neat soap manufacturers also recover glycerine as a by-product for subsequent concentration (D) and distillation (E). Neat soap is generally produced in either of two processes: the batch kettle process (A) or the fatty acid neutralization process, which is preceded by the fat splitting process (B, C).

Neat Soap Manufacture and Waste Streams

Batch Kettle Process (A). It consists of the following operations: (1) receiving and storage of raw materials, (2) fat refining and bleaching, and (3) soap boiling. The major wastewater sources, as shown on the process flow diagram (Fig. 2), are the washouts of both the storage and refining tanks, as well as from leaks and spills of fats and oils around these tanks. These streams are usually skimmed for fat recovery prior to discharge to the sewer.

The fat refining and bleaching operation is carried out to remove impurities that would cause color and odor in the finished soap. The wastewater from this source has a high soap concentration, treatment chemicals, fatty impurities, emulsified fats, and sulfuric acid solutions of fatty acids. Where steam is used for heating, the condensate may contain low-molecular-weight fatty acids, which are highly odorous, partially soluble materials.

The soap boiling process produces two concentrated waste streams: sewer lyes that result from the reclaiming of scrap soap and the brine from Nigre processing. Both of these wastes are low-volume, high pH with BOD values up to 45,000 mg/L.

Fatty Acid Neutralization. Soap manufacture by the neutralization process is a two-step process:

fat + water → fatty acid + glycerine *(fat splitting) (B)*

fatty acid + caustic → soap *(fatty acid neutralization) (C)*

Fat Splitting (B). The manufacture of fatty acid from fat is called fat splitting (B), and the process flow diagram is shown in Fig. 3. Washouts from the storage, transfer, and pretreatment stages are the same as those for process (A). Process condensate and barometric condensate from fat splitting will be contaminated with fatty acids and glycerine streams, which are settled and skimmed to recover the insoluble fatty acids that are processed for sale. The water will typically circulate through a cooling tower and be reused.

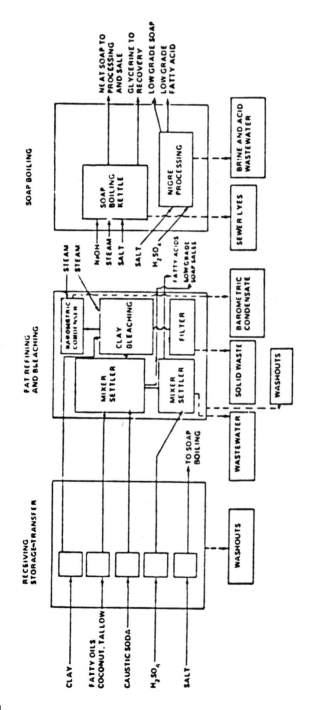

Figure 2 Soap manufacture by batch kettle (A). (After EPA, 1977.)

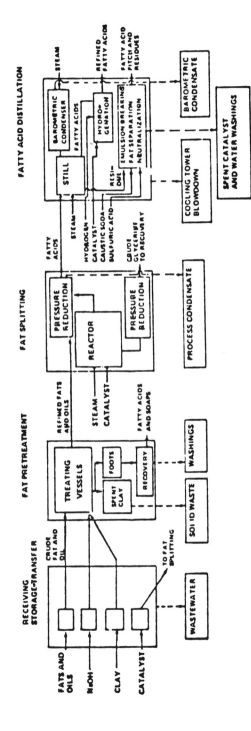

Figure 3 Fatty acid manufacture by fat splitting (B). (After EPA, 1977.)

Occasional purges of part of this stream to the sewer release high concentrations of BOD and some grease and oil.

In the fatty acid distillation process, wastewater is generated as a result of an acidification process, which breaks the emulsion. This wastewater is neutralized and sent to the sewer. It will contain salt from the neutralization, zinc and alkaline earth metal salts from the fat splitting catalyst, and emulsified fatty acids and fatty acid polymers.

Fatty Acid Neutralization (C). Soapmaking by this method is a faster process than the kettle boil process and generates less wastewater effluent (see Fig. 4). Because it is faster, simpler, and cleaner than the kettle boil process, it is the preferred process among larger as well as small manufacturers.

Often, sodium carbonate is used in place of caustic. When liquid soaps (at room temperature) are desired, the more soluble potassium soaps are made by substituting potassium hydroxide for the sodium hydroxide (lye). This process is relatively simple and high-purity raw materials are converted to soap with essentially no by-products. Leaks, spills, storm runoff, and washouts are absent. There is only one wastewater of consequence: the sewer lyes from reclaiming of scrap. The sewer lyes contain the excess caustic soda and salt added to grain out the soap. Also, they contain some dirt and paper not removed in the strainer.

Glycerine Recovery Process (D, E). A process flow diagram for the glycerine recovery process uses the glycerine by-products from kettle boiling (A) and fat splitting (B). The process consists of three steps (Fig. 5): (1) pretreatment to remove impurities, (2) concentration of glycerine by evaporation, and (3) distillation to a finished product of 98% purity.

There are three wastewaters of consequence from this process: Two barometric condensates, one from evaporation and one from distillation, plus the glycerine foots or still bottoms. Contaminants from the condensates are essentially glycerine with a little entrained salt. In the distillation process, the glycerine foots or still bottoms leave a glassy dark brown amorphous solid rich in salt that is disposed of in the wastewater stream. It contains glycerine, glycerine polymers, and salt. The organics will contribute to BOD, COD, and dissolved solids. The sodium chloride will also contribute to dissolved solids. Little or no suspended solids, oil, and grease or pH effect should be seen.

Glycerine can also be purified by the use of ion-exchange resins to remove sodium chloride salt, followed by evaporation of the water. This process puts additional salts into the wastewater but results in less organic contamination.

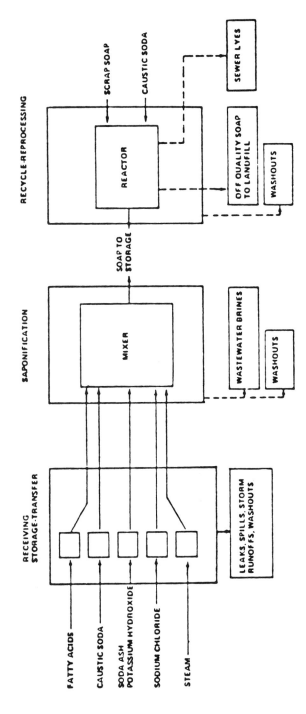

Figure 4 Soap from fatty acid neutralization (C). (After EPA, 1977.)

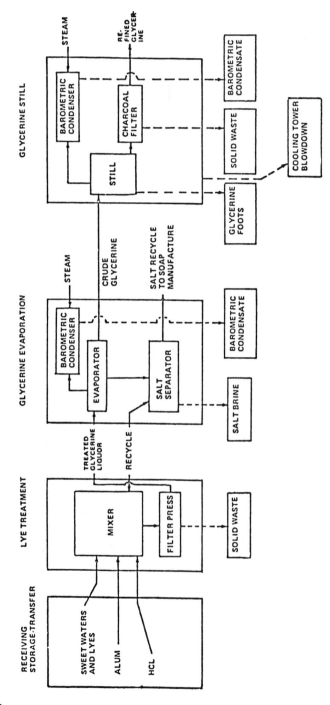

Figure 5 Glycerine recovery process flow diagram (D, E). (After EPA, 1977.)

Treatment of Soap and Detergent Industry Wastes 249

Production of Finished Soaps and Process Wastes

The production of finished soaps utilizes the neat soap produced in processes A and C to prepare and package finished soap. These finished products are soap flakes and powders (F), bar soaps (G), and liquid soap (H). See Figs. 6, 7, and 8 for their respective flow diagrams.

Flakes and Powders (F). Neat soap may or may not be blended with other products before flaking or powdering. Neat soap is sometimes filtered to remove gel particles and run into a crutcher for mixing with builders. After thorough mixing, the finished formulation is run through various mechanical operations to produce flakes and powders. Since all of the evaporated moisture goes to the atmosphere, there is no wastewater effluent.

Some operations will include a scrap soap reboil to recover reclaimed soap. The soap reboil is salted out for soap recovery and the salt water is recycled. After frequent recycling, the salt water becomes so contaminated that it must be discharged to the sewer. Occasional washdown of the crutcher may be needed. The tower is usually cleaned down dry. There is also some gland water that flows over the pump shaft, picking up any minor leaks. This will contribute a very small, but finite, effluent loading.

There are a number of possible effluents shown on the flow diagram for process F (Fig. 6). However, a survey of the industry showed that most operating plants either recycled any wastewater to extinction or used dry clean-up processes. Occasionally, water will be used for clean-up.

Bar Soaps (G). The procedure for bar soap manufacture (G) will vary significantly from plant to plant, depending on the particular clientele served. A typical flow diagram for process G is shown in Fig. 7. The amount of water used in bar soap manufacture varies greatly. In many cases, the entire bar soap processing operation is done without generating a single wastewater stream. The equipment is all cleaned dry, without any washups. In other cases, due to housekeeping requirements associated with the particular bar soap processes, there are one or more wastewater streams from air scrubbers.

The major waste streams in bar soap manufacture are the filter backwash, scrubber waters, or condensate from a vacuum drier and water from equipment washdown. The main contaminant of all these streams is soap that will contribute primarily BOD and COD to the wastewater.

Liquid Soap (H). In the making of liquid soap, neat soap (often the potassium soap of fatty acids) is blended in a mixing tank with other ingredients such as alcohols or glycols to produce a finished product, or the pine oil and kerosene for a product with greater solvency and versatility (see Fig. 8). The

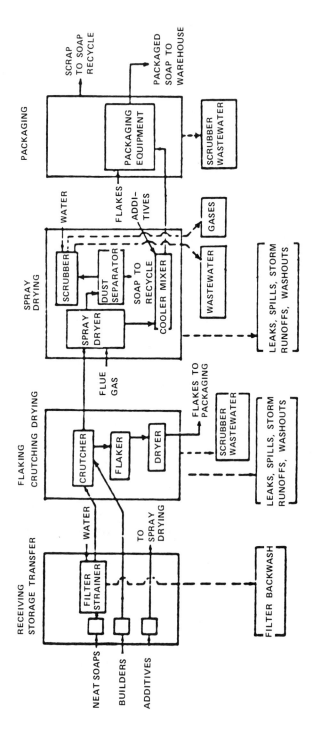

Figure 6 Soap flake and powder manufacture (F). (After EPA, 1977.)

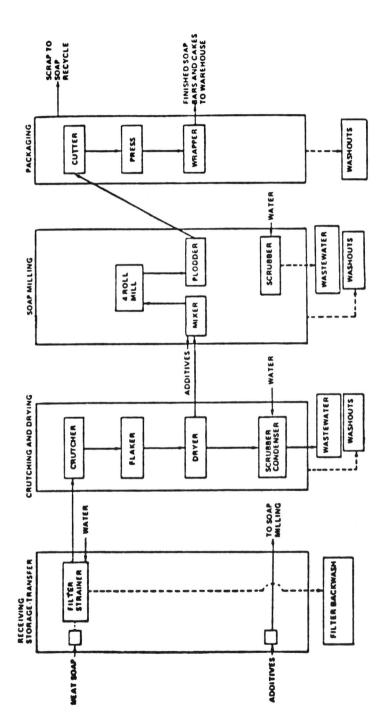

Figure 7 Bar soap manufacture (G). (After EPA, 1977.)

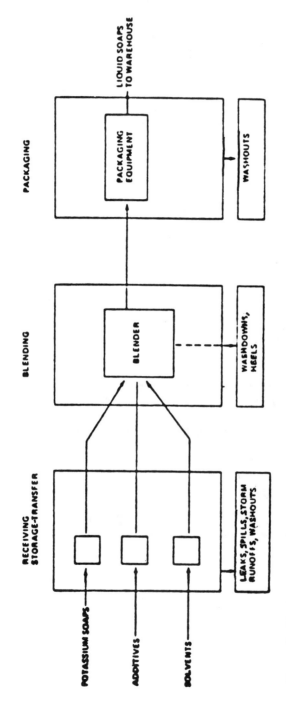

Figure 8 Liquid soap processing (H). (After EPA, 1977.)

final blended product may be, and often is, filtered to achieve a sparkling clarity before being drummed. In making liquid soap, water is used to wash out the filter press and other equipment. According to manufacturers, there are very little effluent leaks. Spills can be recycled or handled dry. Washout between batches is usually unnecessary or can be recycled to extinction.

4.3 Detergent Manufacture and Waste Streams

Detergents, as mentioned previously, can be formulated with a variety of organic and inorganic chemicals, depending on the cleaning characteristics desired. A finished, packaged detergent customarily consists of two main components: the active ingredient or surfactant, and the builder. The processes discussed in the following will include the manufacture and processing of the surfactant as well as the preparation of the finished, marketable detergent. The production of the surfactant (Fig. 1) is generally a two-step process: (1) sulfation or sulfonation and (2) neutralization.

Surfactant Manufacture and Waste Streams

Oleum Sulfonation/Sulfation (I). One of the most important active ingredients of detergents is the sulfate or sulfonate compounds made via the oleum route. A process flow diagram is shown in Fig. 9. In most cases, the sulfonation/sulfation is carried out continuously in a reactor where the oleum (a solution of sulfur trioxide in sulfuric acid) is brought into contact with the hydrocarbon or alcohol and a rapid reaction ensues. The stream is then mixed with water, where the surfactant separates and is then sent to a settler. The spent acid is drawn off and usually forwarded for reprocessing, and the sulfonated/sulfated materials are sent to be neutralized.

This process is normally operated continuously and performs indefinitely without need of periodic cleanout. A stream of water is generally played over pump shafts to pick up leaks as well as to cool the pumps. Wastewater flow from this source is quite modest but continual.

Air—SO_3 Sulfation/Sulfonation (J). This process for surfactant manufacture has many advantages and is used extensively. With SO_3 sulfation, no water is generated in the reaction. A process flow diagram is shown in Fig. 10. SO_3 can be generated at the plant by burning sulfur or sulfur dioxide with air instead of obtaining it as a liquid. Because of this reaction's particular tendency to char the product, the reactor system must be cleaned thoroughly on a regular basis. In addition, there are usually several airborne sulfonic acid streams that must be scrubbed, with the wastewater going to the sewer during sulfation.

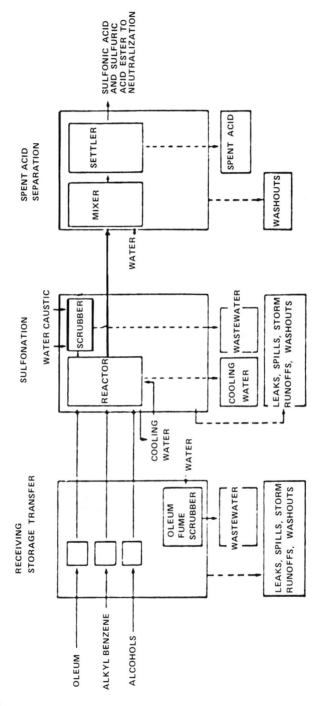

Figure 9 Oleum sulfation and sulfonation (batch and continuous) (1). (After EPA, 1977.)

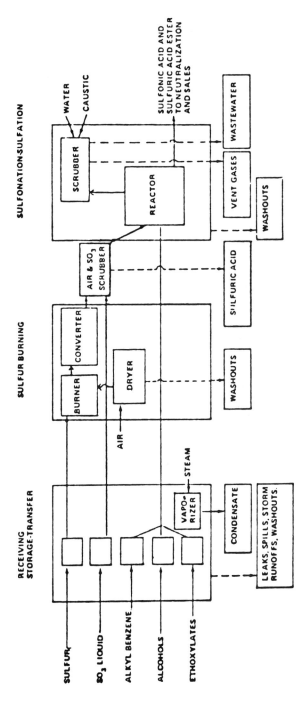

Figure 10 Air-SO₃ sulfation and sulfonation (batch and continuous) (J). (After EPA, 1977.)

***SO₃* Solvent and Vacuum Sulfonation (K).** Undiluted SO_3 and organic reactant are fed into the vacuum reactor through a mixing nozzle. A process flow diagram is shown in Fig. 11. This system produces a high-quality product, but offsetting this is the high operating cost of maintaining the vacuum. Other than occasional washout, the process is essentially free of wastewater generation.

Sulfamic Acid Sulfation (L). Sulfamic acid is a mild sulfating agent and is used only in very specialized quality areas because of the high reagent price. A process flow diagram is shown in Fig. 12. Washouts are the only wastewater effluents from this process as well.

Chlorosulfonic Acid Sulfation (M). For products requiring high-quality sulfates, chlorosulfonic acid is an excellent corrosive agent that generates hydrochloric acid as a by-product. A process flow diagram is shown in Fig. 13. The effluent washouts are minimal.

Neutralization of Sulfuric Acid Esters and Sulfonic Acids (N). This step is essential in the manufacture of detergent active ingredients as it converts the sulfonic acids or sulfuric acid esters (products produced by processes I–M) into neutral surfactants. It is a potential source of some oil and grease, but occasional leaks and spills around the pump and valves are the only expected source of wastewater contamination. A process flow diagram is shown in Fig. 14.

Detergent Formulation and Process Wastes

Spray-Dried Detergents (O). In this segment of the processing, the neutralized sulfonates and/or sulfates are first blended with builders and additives in the crutcher. The slurry is then pumped to the top of a spray tower of about 4.5 to 6.1 m (15–20 ft) in diameter by 45 to 61 m (150–200 ft) high, where nozzles spray out detergent slurry. A large volume of hot air enters the bottom of the tower and rises to meet the falling detergent. The design preparation of this step will determine the detergent particle's shape, size, and density, which in turn determine its solubility rate in the washing process.

The air coming from the tower will be carrying dust particles that must be scrubbed, thus generating a wastewater stream. The spray towers are periodically shut down and cleaned. The tower walls are scraped and thoroughly washed down. The final step is mandatory since the manufacturers must be careful to avoid contamination to the subsequent formulation.

Wastewater streams are rather numerous, as seen in the flow diagram of Fig. 15. They include many washouts of equipment from the crutchers to the spray tower itself. One wastewater flow that has high loadings is that of the air scrubber which cleans and cools the hot gases exiting from this tower. All the plants recycle some of the wastewater generated, while some of the plants

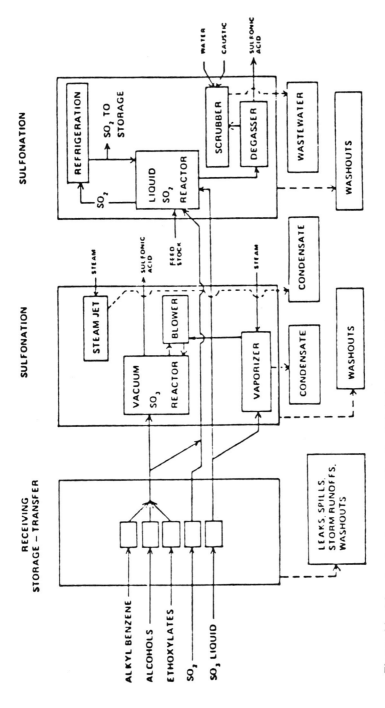

Figure 11 SO₃ solvent and vacuum sulfonation (K). (After EPA, 1977.)

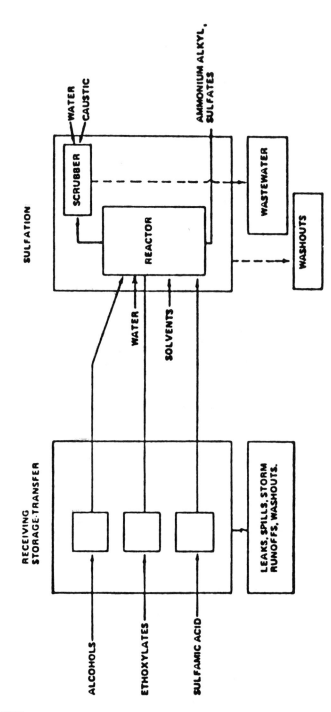

Figure 12 Sulfamic acid sulfation (L). (After EPA, 1977.)

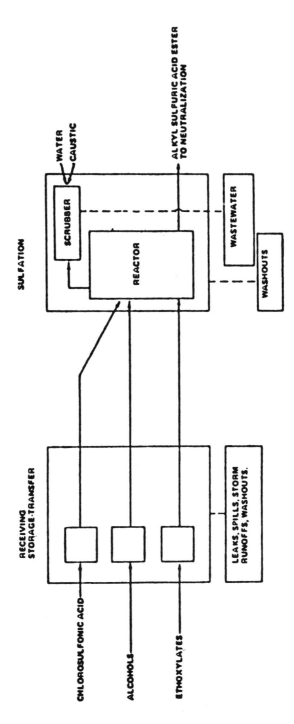

Figure 13 Chlorosulfonic acid sulfation (M). (After EPA, 1977.)

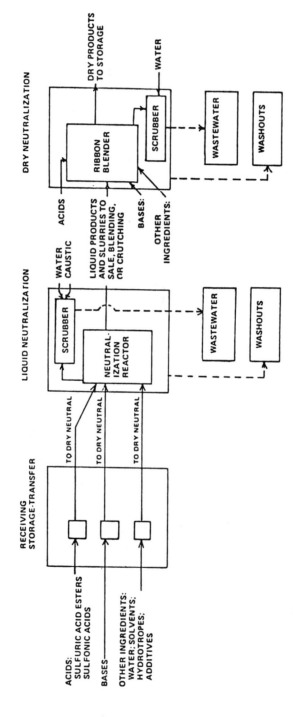

Figure 14 Neutralization of sulfuric acid esters and sulfonic acids (N). (After EPA, 1977.)

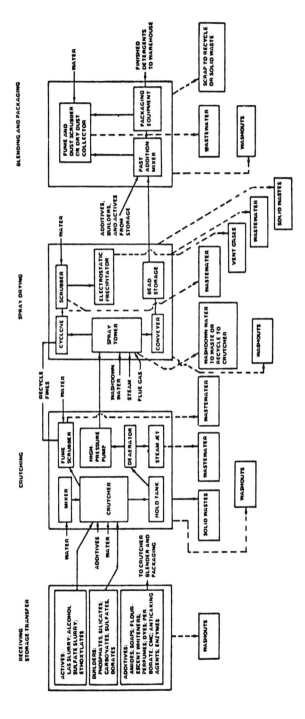

Figure 15 Spray-dried detergent production (O). (After EPA, 1977.)

recycle all the flow generated. Due to increasingly stringent air quality requirements, it can be expected that fewer plants will be able to maintain a complete recycle system of all water flows in the spray tower area.

After the powder comes from the spray tower, it is further blended and then packaged.

Liquid Detergents (P). Detergent actives are pumped into mixing tanks where they are blended with numerous ingredients, ranging from perfumes to dyes. A process flow diagram is shown in Fig. 16. From here, the fully formulated liquid detergent is run down to the filling line for filling, capping, labeling, etc. Whenever the filling line is to change to a different product, the filling system must be thoroughly cleaned out to avoid cross contamination.

Dry Detergent Blending (Q). Fully dried surfactant materials are blended with additives in dry mixers. Normal operation will see many succeeding batches of detergent mixed in the same equipment without anything but dry cleaning. However, when a change in formulation occurs, the equipment must be completely washed down and a modest amount of wastewater is generated. A process flow diagram is shown in Fig. 17.

Drum-Dried Detergent (R). This process is one method of converting liquid slurry to a powder and should be essentially free of the generation of wastewater discharge other than occasional washdown. A process flow diagram is shown in Fig. 18.

Detergent Bars and Cakes (S). Detergent bars are either 100% synthetic detergent or a blend of detergent and soap. They are blended in essentially the same manner as conventional soap. Fairly frequent cleanups generate a wastewater stream. A process flow diagram is shown in Fig. 19.

4.4 Wastewater Characteristics

Wastewaters from the manufacturing, processing, and formulation of organic chemicals such as soaps and detergents cannot be exactly characterized. The wastewater streams are usually expected to contain trace or larger concentrations of all raw materials used in the plant, all intermediate compounds produced during manufacture, all final products, co-products, and by-products, and the auxiliary or processing chemicals employed. It is desirable, from the viewpoint of economics, that these substances not be lost, but some losses and spills appear unavoidable and some intentional dumping does take place during housecleaning and vessel emptying and preparation operations.

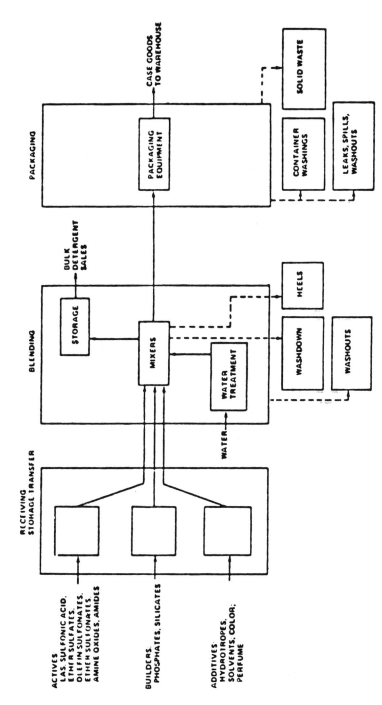

Figure 16 Liquid detergent manufacture (P). (After EPA, 1977.)

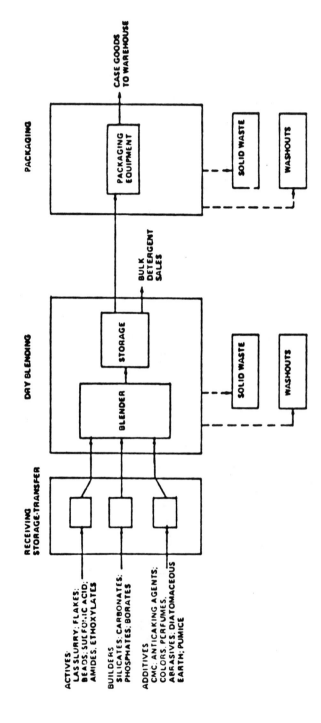

Figure 17 Detergent manufacture by dry blending (Q). (After EPA, 1977.)

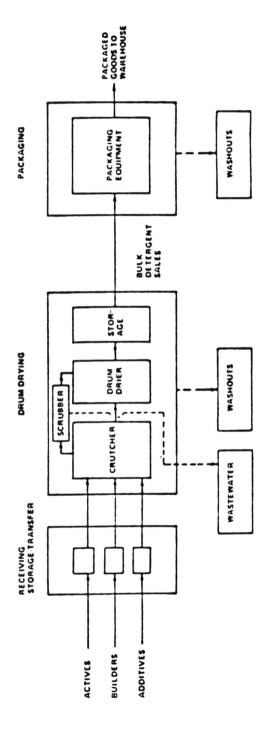

Figure 18 Drum-dried detergent manufacture (R). (After EPA, 1977.)

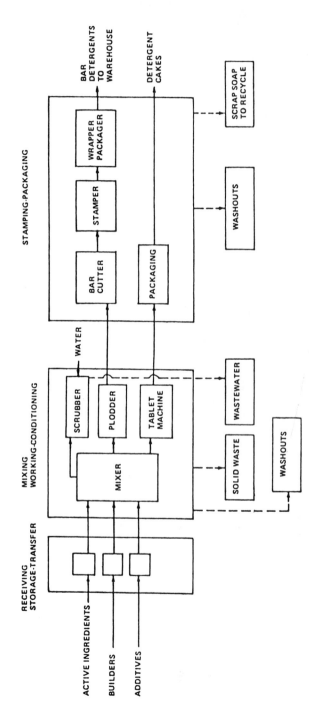

Figure 19 Detergent bar and cake manufacture (S). (After EPA, 1977.)

Treatment of Soap and Detergent Industry Wastes 267

According to a study by Zimmerman (EPA, 1974) presenting estimates of industrial wastewater generation as well as related pollution parameter concentrations, the wastewater volume discharged from soap and detergent manufacturing facilities per unit of production ranges from 0.3 to 2.8 gal/lb (2.5–23.4 L/kg) of product. The reported ranges of concentration (mg/L) for BOD, suspended solids, COD, and grease were 500 to 1200, 400 to 2100, 400 to 1800, and about 300, respectively. These data were based on a study of the literature and the field experience of governmental and private organizations. The values represent plant operating experience for several plants consisting of 24-hr composite samples taken at frequent intervals. The ranges for flow and other parameters generally represent variations in the level of plant technology or variations in flow and quality parameters from different subprocesses. In particular, the more advanced and modern the level of production technology, the smaller the volume of wastewater discharged per unit of product. The large variability (up to one order of magnitude) in the ranges is generally due to the heterogeneity of products and processes in the soap and detergent industry.

The federal guidelines (EPA, 1977) for state and local pretreatment programs reported the raw wastewater characteristics (Table 2) in mg/L concentration and the flows and quality parameters (Table 3) based on the production or 1 ton of product manufactured for the subcategories of the industry. Most soap and detergent manufacturing plants contain two or more of the subcategories shown in Table 1, and their wastewaters are a composite of these individual unit processes.

5. U.S. CODE OF FEDERAL REGULATIONS

The information presented in this section has been taken from the U.S. Code of Federal Regulations (40 CFR), containing documents related to the protection of the environment (Federal Register, 1987). In particular, the regulations contained in Part 417, Soap and Detergent Manufacturing Point Source Category, pertaining to effluent limitations guidelines and pretreatment or performance standards for each of the 19 subcategories shown in Table 1.

The effluent guideline regulations and standards of 40 CFR, Part 417 were promulgated on February 11, 1975. According to the most recent notice in the Federal Register (Federal Register, 1990) regarding industrial categories and regulations, no review is under way or planned and no revision is proposed for the soap and detergent industry. The effluent guidelines and standards applicable to this industrial category include (1) the best practicable control technology currently available (BPT); (2) the best available technology

Table 2 Soap and Detergent Industry Raw Wastewater Characteristics

Parameter (mg/L)	Batch kettle A	Fat splitting B	Fatty acid neutralization C	Glycerine concentration D	Glycerine distillation E	Flakes and powders F	Bar soap G	Liquid soap H
BOD	3600[a]	60-3600[a]	400				1600-3000[a]	
COD	4267[a]	115-6000[a]	1000					
TSS	1600-6420	115-6000	775					
Oil and grease	250[a]	13-760[a]	200[a]					
pH	5-13.5	High	High	Neutral	Neutral	Neutral	Neutral	Neutral
Chlorides	20-47m[a]							
Zinc		Present						
Nickel		Present						

Parameter (mg/L)	Oleum sulfation and sulfonation I	Air sulfation and sulfonation J	SO₃ solvent and vacuum sulfation K	Sulfamic acid sulfation L	Chlorosulfonic M	Neutral sulfuric N	Spray-dried O	Liquid detergent P	Dry blend Q	Drum-dried R	Bars and cakes S
BOD	75-2000[a]	380-520				8.5-6m[a]	48-19m[a]	65-3400[a]	Neg.		
COD	220-6000[a]	920-1589[a]				245-21m[a]	150-60m[a]	640-11m[a]			
TSS	100-3000										
Oil and grease	100-3000[a]										
pH	1-2[a]	2[a]-7	Low	Low	Low	Low					
Surfactant	250-7000							60-2m			
Boron	Present	Present	Present	Present	Present	Present	Present	Present	Present	Present	Present

[a]In high levels these parameters may be inhibitory to biological systems; m = thousands.
Source: EPA (1977).

Table 3 Raw Wastewater Characteristics Based on Production

Parameter	Batch kettle A	Fat splitting B	Fatty acid neutralization C	Glycerine concentration D	Glycerine distillation E	Flakes and powders F	Bar soap G	Liquid soap H
Flow range (L/kkg)[a]	623/2500	3.3M/1924M	258			Neg.		Neg.
Flow type	B	B	B	B	B	B	B	B
BOD (kg/kkg)[b]	6	12	0.1	15	5	0.1	3.4	0.1
COD (kg/kkg)	10	22	0.25	30	10	0.3	5.7	0.3
TSS (kg/kkg)	4	22	0.2	2	2	0.1	5.8	0.1
Oil and grease (kg/kkg)	0.9	2.5	0.05	1	1	0.1	0.4	0.1

Parameter	Oleum sulfation and sulfonation I	SO$_3$ sulfation and sulfonation J	SO$_3$ solvent and vacuum sulfonation K	Sulfamic acid sulfation L	Chlorosulfonic M	Neutral sulfuric acid esters N	Spray-dried O	Liquid detergent P	Dry blend Q	Drum-dried R	Bars and cakes S
Flow range (L/kkg)[a]	100/2740	249				10/4170	41/2084	625/6250			
Flow type	C	C	B	B	B	B&C	B	B	B	B	B
BOD (kg/kkg)[b]	0.2	3	3	3	3	0.10	0.1–0.8	2–5	0.1	0.1	7
COD (kg/kkg)	0.6	9	9	9	9	0.3	0.3–25	4–7	0.5	0.3	22
TSS (kg/kkg)	0.3	0.3	0.3	0.3	0.3	0.3	0.1–1.0		0.1	0.1	2
Oil and grease (kg/kkg)	0.3	0.5	0.5	0.5	0.5	0.1	Nil–0.3			0.1	0.2
Chloride (kg.kkg)					5						
Surfactant (kg/kkg)	0.7	3	3	3	3	0.2	0.2–1.5	1.3–3.3		0.1	5

[a] L/kkg-L/1000 kg product produced (lower limit/upper limit).
[b] kg/kkg-kg/1000 kg product produced.
B = Batch, C = Continuous, Neg. = Neglible, M = Thousand.
Source: EPA (1977).

economically achievable (BAT); (3) pretreatment standards for existing sources (PSES); (4) standards of performance for new sources (NSPS); and (5) pretreatment standards for new sources (PSNS).

For all 19 subcategories of the soap and detergent manufacture industry, there are no pretreatment standards establishing the quantity and quality of pollutants or pollutant properties that may be discharged to a publicly owned treatment works (POTW) by an existing or new point source. If the major contributing industry is an existing point source discharging pollutants to navigable waters, it will be subject to Section 301 of the Federal Water Pollution Control Act and to the provisions of 40 CFR, Part 128. However, practically all the soap and detergent manufacturing plants in the United States discharge their wastewaters into municipal sewer systems. The effluent limitations guidelines for certain subcategories regarding BPT, BAT, and NSPS are presented in Tables 4 through 10.

6. WASTEWATER CONTROL AND TREATMENT

The sources and characteristics of wastewater streams from the various subcategories in soap and detergent manufacturing, as well as some of the possibilities for recycling and treatment, were discussed Sec. 4. The pollution control and treatment methods and unit processes used are discussed in more detail in the following. The details of the process design criteria for these unit treatment processes can be found in any design handbook.

6.1 In-Plant Control and Recycle

Significant in-plant control of both waste quantity and quality is possible particularly in the soap manufacturing subcategories where maximum flows may be 100 times the minimum. Considerably less in-plant water conservation and recycle are possible in the detergent industry, where flows per unit of product are smaller.

The largest in-plant modification that can be made is the changing or replacement of the barometric condensers (subcategories A, B, D, and E). The wastewater quantity discharged from these processes can be significantly reduced by recycling the barometric cooling water through fat skimmers, from which valuable fats and oils can be recovered and then through the cooling towers. The only waste with this type of cooling would be the continuous small blowdown from the skimmer. Replacement with surface condensers has been

Treatment of Soap and Detergent Industry Wastes

used in several plants to reduce both the waste flow and quantity of organics wasted.

Significant reduction of water usage is possible in the manufacture of liquid detergents (P) by the installation of water recycle piping and tankage and by the use of air rather than water to blowdown filling lines. In the production of bar soaps (G), the volume of discharge and the level of contamination can be reduced materially by installation of an atmospheric flash evaporator ahead of the vacuum drier. Finally, pollutant carry-over from distillation columns such as those used in glycerine concentration (D) or fatty acid separation (B) can be reduced by the use of two additional special trays.

In a recent document (Overcash, 1986) presenting techniques adopted by the French for pollution prevention, a new process of detergent manufacturing effluent recycle is described. As shown in Fig. 20, the washout effluents from reaction and/or mixing vessels and washwater leaks from the paste preparation and pulverization pump operations are collected and recycled for use in the paste preparation process. The claim was that pollution generation at such a plant is significantly reduced and, although the savings on water and raw materials are small, the capital and operating costs are less than those for building a wastewater treatment facility.

Besselievre (1969) reported in a review of water reuse and recycling by the industry that soap and detergent manufacturing facilities showed an average ratio of reused and recycled water to total wastewater effluent of about 2:1. That is, over two-thirds of the generated wastewater stream in an average plant was being reused and recycled. Of this volume, about 66% was used as cooling water and the remaining 34% for the process or other purposes.

6.2 Wastewater Treatment Methods

The soap and detergent manufacturing industry makes routine use of various physico-chemical and biological pretreatment methods to control the quality of its discharges. A survey of these treatment processes is presented in Table 11 (EPA, 1977), which also shows the usual removal efficiencies of each unit process on the various pollutants of concern. According to Nemerow (1978), the origin of major wastes is in washing and purifying soaps and detergents and the resulting major pollutants are high BOD and saponified soaps (oily and greasy, alkali, and high-temperature wastes), which are removed primarily through air-flotation and skimming, and precipitation with the use of $CaCl_2$ as a coagulant.

Figure 21 presents a composite flow diagram describing a complete treatment train of the unit processes that may be used in a large soap and detergent manufacturing plant to treat its wastes. As a minimum requirement, flow equalization to smooth out peak discharges should be utilized even at a production facility that has a small-volume batch operation. Larger plants with integrated product lines may require additional treatment of their wastewaters for both suspended solids and organic materials' reduction. Coagulation and sedimentation are used by the industry for removing the greater portion of the large solid particles in its waste. On the other hand, sand or mixed-bed filters used after biological treatment can be utilized to eliminate fine particles. One of the biological treatment processes or, alternatively, granular or powdered activated carbon is the usual method employed for the removal of particulate or soluble organics from the waste streams. Finally, as a tertiary step for removing particular ionized pollutants or total dissolved solids (TDS), a

Table 4 Effluent Limitations for Subpart A, Batch Kettle

	Effluent limitations	
Effluent characteric	Maximum for any 1 day	Average of daily values for 30 consecutive days shall not exceed
(a) BCT	Metric units (kg/1000 kg of anhydrous product)	
BOD_5	1.80	0.60
COD	4.50	1.50
TSS	1.20	0.40
Oil and grease	0.30	0.10
pH	a	a
(b) BAT and NSPS		
BOD_5	0.80	0.40
COD	2.10	1.05
TSS	0.80	0.40
Oil and grease	0.10	0.05
pH	a	a

[a]Within the range 6.0–9.0.
Source: Federal Register (1987).

Table 5 Effluent Limitations for Subpart C, Soap by Fatty Acid

Effluent characteristic	Maximum for any 1 day	Average of daily values for 30 consecutive days shall not exceed
(a) BCT	Metric units (kg/1000 kg of anhydrous product)	
BOD_5	0.03	0.01
COD	0.15	0.05
TSS	0.06	0.02
Oil and grease	0.03	0.01
pH	a	a
(b) BAT		
BOD_5	0.02	0.01
COD	0.10	0.05
TSS	0.04	0.02
Oil and grease	0.02	0.01
pH	a	a
(c) NSPS		
BOD_5	0.02	0.01
COS	0.10	0.05
TSS	0.04	0.02
Oil and grease	0.02	0.01
pH	a	a

[a]Within the range 6.0–9.0.
Source: Federal Register (1987).

few manufacturing facilities have employed either ion exchange or the reverse osmosis process.

Flotation or Foam Fractionation

One of the principal applications of vacuum and pressure (air) flotation is in commercial installations with colloidal wastes from soap and detergent factories (Gurnham, 1965). Wastewaters from soap production are collected in traps on skimming tanks, with subsequent recovery floating of fatty acids.

Table 6 Effluent Limitations for Subpart D, Glycerine Concentration

Effluent characteristic	Maximum for any 1 day	Average of daily values for 30 consecutive days shall not exceed
(a) BPT	Metric units (kg/1000 kg of anhydrous product)	
BOD5	4.50	1.50
COD	13.50	4.50
TSS	0.60	0.20
Oil and grease	0.30	0.10
pH	a	a
(b) BAT		
BOD5	0.80	0.40
COD	2.40	1.20
TSS	0.20	0.10
Oil and grease	0.08	0.04
pH	a	a
(c) NSPS		
BOD5	0.80	0.40
COS	2.40	1.20
TSS	0.20	0.10
Oil and grease	0.08	0.04
pH	a	a

[a]Within the range 6.0–9.0.
Source: Federal Register (1987).

Foam separation or fractionation can be used to an extra advantage: Not only do surfactants congregate at the air/liquid interfaces, but other colloidal materials and ionized compounds that form a complex with the surfactants tend to also be concentrated by this method. An incidental, but often important, advantage of air flotation processes is the aerobic condition developed, which tends to stabilize the sludge and skimmings so that they are less likely to turn septic. However, disposal means for the foamate can be a serious problem in the use of this procedure (Ross, 1968). It has been reported that foam separation has been able to remove 70 to 80% of synthetic detergents, at a wide range of costs (Besselievre, 1969). Gibbs (1949) reported the successful use of fine

Table 7 Effluent Limitations for Subpart G, Bar Soaps

Effluent characteristic	Maximum for any 1 day	Average of daily values for 30 consecutive days shall not exceed
(a) BPT	Metric units (kg/1000 kg of anhydrous product)	
BOD5	1.02	0.34
COD	2.55	0.85
TSS	1.74	0.58
Oil and grease	0.12	0.04
pH	a	a
(b) BAT		
BOD5	0.40	0.20
COD	1.20	0.60
TSS	0.68	0.34
Oil and grease	0.06	0.03
pH	a	a
(c) NSPS		
BOD5	0.40	0.20
COS	1.20	0.60
TSS	0.68	0.34
Oil and grease	0.06	0.03
pH	a	a

[a] Within the range 6.0–9.0.
Source: Federal Register (1987).

bubble flotation and 40-min detention in treating soap manufacture wastes, where the skimmed sludge was periodically returned to the soap factory for reprocessing. According to Wang (1990), the dissolved air flotation process is also feasible for the removal of detergents and soaps from water.

Activated Carbon Adsorption

Colloidal and soluble organic materials can be removed from solution through adsorption onto granular or powdered activated carbon, such as the particularly troublesome hard surfactants. Refractory substances resistant to biodegradation, such as ABS, are difficult or impossible to remove by conventional

Table 8 Effluent Limitations for Subpart H, Liquid Soaps

Effluent characteristic	Maximum for any 1 day	Average of daily values for 30 consecutive days shall not exceed
(a) BPT	Metric units (kg/1000 kg of anhydrous product)	
BOD5	0.03	0.01
COD	0.15	0.05
TSS	0.03	0.01
Oil and grease	0.03	0.01
pH	a	a
(b) BAT		
BOD5	0.02	0.01
COD	0.10	0.05
TSS	0.02	0.01
Oil and grease	0.02	0.01
pH	a	a
(c) NSPS		
BOD5	0.02	0.01
COS	0.10	0.05
TSS	0.02	0.01
Oil and grease	0.02	0.01
pH	a	a

[a]Within the range 6.0–9.0.
Source: Federal Register (1987).

biological treatment, and so they are frequently removed by activated carbon adsorption (Eckenfelder, 1989). The activated carbon application is made either in mixed-batch contact tanks with subsequent settling or filtration, or in flow-through GAC columns or contact beds. Obviously, because it is an expensive process, adsorption is being used as a polishing step of pretreated waste effluents. Nevertheless, according to Koziorowski and Kucharski (1972), much better results of surfactant removal have been achieved with adsorption than coagulation/settling.

Table 9 Effluent Limitations for Subpart I, Oleum Sulfonation

Effluent characteristic	Maximum for any 1 day	Average of daily values for 30 consecutive days shall not exceed
(a) BPT	Metric units (kg/1000 kg of anhydrous product)	
BOD5	0.09	0.02
COD	0.40	0.09
TSS	0.15	0.03
Surfactants	0.15	0.03
Oil and grease	0.25	0.07
pH	a	a
(b) BAT		
BOD5	0.07	0.02
COD	0.27	0.09
TSS	0.09	0.03
Surfactants	0.09	0.03
Oil and grease	0.21	0.07
pH	a	a
(c) NSPS		
BOD5	0.03	0.01
COS	0.09	0.03
TSS	0.06	0.02
Surfactants	0.03	0.01
Oil and grease	0.12	0.04
pH	a	a

[a]Within the range 6.0–9.0.
Source: Federal Register (1987).

Coagulation/Flocculation/Settling

As mentioned previously in Sec. 2.4, the coagulation/flocculation process was found to be affected by the presence of surfactants in the raw water or wastewater. Such interference was observed for both alum and ferric sulfate coagulant, but the use of certain organic polymer flocculants was shown to overcome this problem. However, chemical coagulation and flocculation for

Table 10 Effluent Limitations for Subpart P, Liquid Detergents

Effluent characteristic	Maximum for any 1 day	Average of daily values for 30 consecutive days shall not exceed
(a) BPT[a]	Metric units (kg/1000 kg of anhydrous product)	
BOD5	0.60	0.20
COD	1.80	0.60
TSS	0.015	0.005
Surfactants	0.39	0.13
Oil and grease	0.015	0.005
pH	c	c
(b) BPT[b]		
BOD5	0.05	
COD	0.15	
TSS	0.002	
Surfactants	0.04	
Oil and grease	0.002	
pH	c	
(c) BAT[a]		
BOD5	0.10	0.05
COS	0.44	0.22
TSS	0.01	0.005
Surfactants	0.10	0.05
Oil and grease	0.01	0.005
pH	c	c
(d) BAT[b]		
BOD5	0.02	
COD	0.07	
TSS	0.002	
Surfactants	0.02	
Oil and grease	0.002	
pH	c	

[a]For normal liquid detergent operations.
[b]For fast turnaround operation of automated fill lines.
[c]Within the range 6.0–9.0.
Source: Federal Register (1987).

Treatment of Soap and Detergent Industry Wastes

Table 10 (Continued)

Effluent characteristic	Maximum for any 1 day	Average of daily values for 30 consecutive days shall not exceed
(e) NSPS[a]	Metric units (kg/1000 kg of anhydrous product)	
BOD5	0.10	0.05
COD	0.44	0.22
TSS	0.01	0.005
Surfactants	0.10	0.05
Oil and grease	0.01	0.005
pH	c	c
(f) NSPS[b]		
BOD5	0.02	
COD	0.07	
TSS	0.002	
Surfactants	0.02	
Oil and grease	0.002	
pH	c	

[a] For normal liquid detergent operations.
[b] For fast turnaround operation of automated fill lines.
[c] Within the range 6.0–9.0.
Source: Federal Register (1987).

settling may not prove to be very efficient for such wastewaters. Wastes containing emulsified oils can be clarified by coagulation, if the emulsion is broken through the addition of salts such as $CaCl_2$, the coagulant of choice for soap and detergent manufacture wastewaters (Eckenfelder, 1989). Also, lime or other calcium chemicals have been used in the treatment of such wastes whose soapy constituents are precipitated as insoluble calcium soaps of fairly satisfactory flocculating ("hardness" scales) and settling properties. Treatment with $CaCl_2$ can be used to remove practically all grease and suspended solids and a major part of the suspended BOD (Gurnham, 1955). Using carbon dioxide (carbonation) as an auxiliary precipitant reduces the amount of calcium chloride required and improves treatment efficiency. The sludge from $CaCl_2$ treatment can be removed either by sedimentation or by air or vacuum flotation.

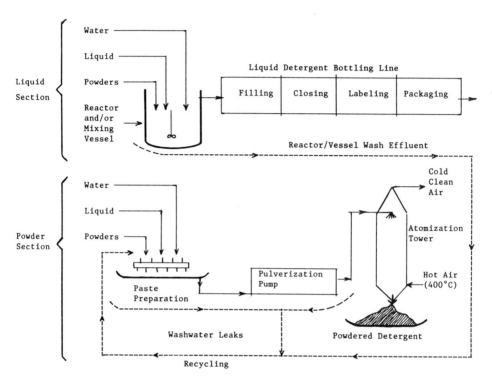

Figure 20 Process modification for wastewater recycling in detergent manufacture. (After Overcash, 1986.)

Table 11 Treatment Methods in the Soap and Detergent Industry

Pollutant and method	Efficiency (percentage of pollutant removed)
Oil and grease	
API-type separation	Up to 90% of free oils and greases. Variable on emulsified oil.
Carbon adsorption	Up to 95% of both free and emulsified oils.
Flotation	Without the addition of solid phase, alum, or iron, 70–80% of both free and emulsified oil. With the addition of chemicals, 90%.
Mixed-media filtration	Up to 95% of free oils. Efficiency in removing emulsified oils unknown
Coagulation/sedimentation with iron, alum, or solid phase (bentonite, etc.)	Up to 95% of free oil. Up to 90% of emulsified oil.
Suspended solids	
Mixed-media filtration	70–80%
Coagulation/sedimentation	50–80%
BOD and COD	
Bioconversions (with final clarifier)	60–95% or more
Carbon adsorption	Up to 90%
Residual suspended solids	
Sand or mixed-media filtration	50–95%
Dissolved solids	
Ion exchange or reverse osmosis	Up to 90%

Source: EPA (1977).

Ion Exchange and Exclusion

The ion-exchange process has been used effectively in the field of waste disposal. The use of continuous ion-exchange and resin regeneration systems has further improved the economic feasibility of the applications over the fixed-bed systems. One of the reported (Abrams and Lewon, 1962) special

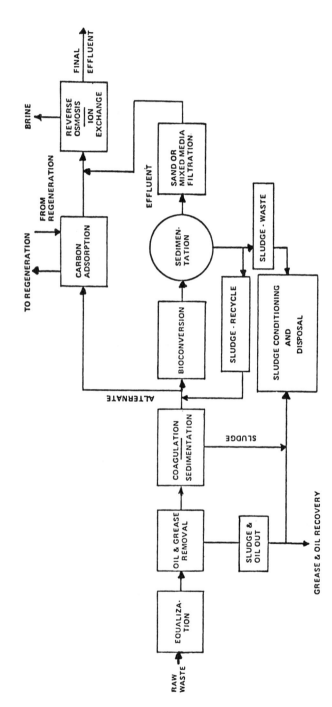

Figure 21 Composite flowsheet of waste treatment in soap and detergent industry. (After EPA, 1977.)

Treatment of Soap and Detergent Industry Wastes

applications of the ion-exchange resins has been the removal of ABS by the use of a Type II porous anion exchanger that is a strong base and depends on a chloride cycle. This resin system is regenerated by removing a great part of the ABS absorbed on the resin beads with the help of a mixture of HCl and acetone. Other organic pollutants can also be removed by ion-exchange resins, and the main problem is whether the organic material can be eluted from the resin using normal regeneration or it is economically advisable to simply discard the used resin. Wang and Wood (1982) successfully used the ion-exchange process for the removal of cationic surfactant from water.

The separation of ionic from nonionic substances can be effected by the use of ion exclusion (Ross, 1968). Ion exchange can be used to purify glycerine for the final product of chemically pure glycerine and reduce losses to waste, but the concentration of dissolved ionizable solids or salts (ash) largely impacts on the overall operating costs. Economically, when the crude or sweet water contains under 1.5% ash, straight ion exchange using a cation–anion mixed bed can be used, whereas for higher percentages of dissolved solids, it is economically feasible to follow the ion exchange with an ion-exclusion system. For instance, waste streams containing 0.2 to 0.5% ash and 3 to 5% glycerine may be economically treated by straight ion exchange, while waste streams containing 5 to 10% ash and 3 to 5% glycerine have to be treated by the combined ion-exchange and ion-exclusion processes.

Biological Treatment

Regarding biological destruction, as mentioned previously, surfactants are known to cause a great deal of trouble due to foaming and toxicity in municipal treatment plants. The behavior of these substances depends on their type (Koziorowski and Kucharski, 1972), i.e., anionic and nonionic detergents increase the amount of activated sludge, whereas cationic detergents reduce it, and also the various compounds decompose to a different degree. The activated sludge process is feasible for the treatment of soap and detergent industry wastes but, in general, not as satisfactory as trickling filters. The turbulence in the aeration tank induces frothing to occur, and also the presence of soaps and detergents reduces the absorption efficiency from air bubbles to liquid aeration by increasing the resistance of the liquid film.

On the other hand, detergent production wastewaters have been treated with appreciable success on fixed-film process units such as trickling filters (Besselievre, 1969). Also, processes such as lagoons, oxidation or stabilization ponds, and aerated lagoons have all been used successfully in treating soap and detergent manufacturing wastewaters. Finally, Vath (1964) demonstrated that both linear anionic and nonionic ethoxylated surfactants underwent degradation, as shown by a loss of surfactant properties, under anaerobic treatment.

7. CASE STUDIES OF TREATMENT FACILITIES

Soap and detergent manufacture and formulation plants are situated in many areas in the United States and other countries. At most, if not all of these locations, the wastewaters from production and cleanup activities are discharged to municipal sewer systems and treated together with domestic, commercial, institutional, and other industrial wastewaters. Following the precipitous reduction in production and use of "hard" syndets such as ABS, no discernible problems in operation and treatment efficiency due to the combined treatment of surfactant manufacture wastes at these municipal sewage treatment plants (most of which employ biological processes) have been reported. In fact, there is a significantly larger portion of surfactants and related compounds being discharged to the municipal facilities from user sources. In most cases, the industrial discharge is simply surcharged due to its high-strength BOD concentration.

7.1 Colgate-Palmolive Plant

Possibly the single largest treatment facility that handles wastewaters from the production of soaps, detergents, glycerines, and personal care products is Colgate-Palmolive Company's plant at Jeffersonville, Indiana (Brownell et al., 1975). The production wastes had received treatment since 1968 (Herin et al., 1970) in a completely mixed activated sludge plant with a 0.6-MGD design flow and consisting of a 0.5-MG mixed equalization and storage basin, aeration basin, and final clarifier. The treated effluent was discharged to the Ohio River, combined with rain drainage and cooling waters. During operation, it was observed that waste overloads to the plant caused a deterioration of effluent quality and that the system recovered very slowly, particularly from surfactant short-term peaks. In addition, the fact that ABS had been eliminated and more LAS and nonionic surfactants were being produced, as well as the changes in product formulation, may have been the reasons for the Colgate treatment plant's generally less than acceptable effluent quality. (Note that 1 MG = 3785 m^3, 1 MGD = 3785 m^3/day.)

Due to the fact that the company considered the treatment efficiency of need of more dependable results, in 1972–73 several chemical pretreatment and biological treatment studies were undertaken in order to modify and improve the existing system. As a result, a modified treatment plant was designed, constructed, and placed in operation. A new 1.5-MG mixed flow and pollutant load equalization basin is provided prior to chemical pretreatment, and a flash mixer with lime addition precedes a flocculator/clarifier unit. Ahead of the preexisting equalization and aeration basins, the capability for pH adjustment and nutrient supplementation was added. Chemical sludge is wasted to two lagoons where thickening and dewatering (normally 15–30% solids) take place.

The intermediate storage basin helps equalize upsets in the chemical pretreatment system, provides neutralization contact time, and allows for storage of pretreated wastewater to supply to the biological treatment unit whenever a prolonged shutdown of the chemical pretreatment occurs. Such shutdowns are planned for part of the weekend and whenever manufacturing stoppage occurs in order to cut down on costs. According to Brownell et al. (1975), waste loads to the pretreatment plant diminish during plant-wide vacations and production shutdowns, and bypassing the chemical pretreatment allows for a more constant loading of the aeration basins at those times. In this way, the previously encountered problems in the start-up of the biological treatment unit after shutdowns were reduced.

The pollutant removal efficiency of this plant is normally quite high, with overall MBAS (methylene blue active substances) removals at 98 to 99% and monthly average overall BOD_5 removals ranging from 88 to 98% (most months averaging about 95%). The reported MBAS removals achieved in the chemical pretreatment units normally averaged 60 to 80%. Occasional high MBAS concentrations in the effluent from the chemical pretreatment system were controlled through the addition of $FeCl_2$ and an organic polymer that supplemented the regular dose of lime and increased suspended solids' capture. Also, high oil and grease concentrations were occasionally observed after spills of fatty acid, mineral oil, olefin, and tallow, and historically this caused problems with the biological system. In the chemical pretreatment units, adequate oil and grease removals were obtained through the addition of $FeCl_2$. Finally, COD removals in the chemical system were quite consistent and averaged about 50% (COD was about twice the BOD_5).

In the biological step of treatment, removal efficiency for BOD_5 was very good, often averaging over 90%. During normal operating periods, the activated sludge system appeared incapable of treating MBAS levels of over 100 lb/day (45.4 kg/day) without significant undesirable foaming. The BOD_5 loading was normally kept at 0.15 to 0.18 gm/day/gm (or lb/day/lb) MLVSS, but it had to be reduced whenever increased foaming occurred. Finally, suspended solids' concentrations in the secondary clarifier effluent were occasionally quite high, although the overflow rate averaged only 510 gal/day/ft^2 and as low as 320 gpd/ft^2 (13–20.8 m^3/day/m^2). The use of polymer flocculants considerably improved the effluent turbidity, reducing it by 50 to 75%, and since higher effluent solids contribute to high effluent BOD$_5$, it was reduced as well. Therefore, although the Colgate-Palmolive waste treatment plant occasionally experiences operating problems, it generally achieves high levels of pollutant removal efficiencies.

7.2 Combined Treatment of Industrial and Municipal Wastes

Most soap and detergent manufacturing facilities, as mentioned previously, discharge their untreated or pretreated wastes into municipal systems. The compositions of these wastewaters vary widely, with some being readily biodegradable and others inhibitory to normal biological treatment processes. In order to allow and surcharge such an effluent to a municipal treatment plant, an evaluation of its treatability is required. Such a detailed assessment of the wastewaters discharged from a factory manufacturing detergents and cleaning materials in the vicinity of Pinxton, England was reported by Shapland (1986). The average weekly effluent discharged from a small collection and equalization tank was 119 m^3/day (21.8 gpm), which contributes about 4% of the flow to the Pinxton sewage treatment plant.

Monitoring of the diurnal variation in wastewater pollutant strengths on different days showed that no regular diurnal pattern exists and the discharged wastewaters are changeable. In particular, the pH value was observed to vary rapidly over a wide range and, therefore, pH correction in the equalization tank would be a minimum required pretreatment prior to discharge into the sewers in such cases. The increase in organic loading contributed to the Pinxton plant by the detergent factory is much higher than the hydraulic loading, representing an average of 32% BOD increase in the raw influent and 60% BOD increase in the primary settled effluent, but it does not present a problem because the plant is biologically and hydraulically underloaded.

The treatability investigation of combined factory and municipal wastewaters involved laboratory-scale activated sludge plants and rolling tubes (fixed-film) units. The influent feed to these units was settled industrial effluent (with its pH adjusted to 10) mixed in various proportions with settled municipal effluent. The variation of hydraulic loading enabled the rotating tubes to be operated at similar biological loadings. In the activated sludge units, the mixed

Treatment of Soap and Detergent Industry Wastes

liquor suspended solids (MLSS) were maintained at about 3000 mg/L, a difficult task since frothing and floc break-up caused solids' loss. The overall results showed that more consistent removals were obtained with the fixed-film system, probably due to the loss of solids from the aeration units.

At 3 and 6% by vol. industrial waste combination, slight to no biological inhibition was caused either to the fixed-film or activated sludge system. The results of sample analysis from the inhibitory runs showed that in two of the three cases, the possible cause of inhibition was the presence of chloroxylenes and brominated compounds. The third case represented only temporary inhibition, since the rolling tubes provided adequate treatment after a period of acclimation. Finally, the general conclusion reached in the investigation was that the detergent factory effluent may be accepted at 3% by vol. equalized flow to the municipal fixed-film treatment plant, i.e., up to 200 m^3/day (36.7 gpm), without any noticeable efficiency reduction.

7.3 Treatability of Oily Wastes from Soap Manufacture

McCarty et al. (1972) addressed the subject of the treatability of animal and vegetable oils and fats in municipal treatment systems. In general, certain reported treatments difficulties in biological systems are attributed to the presence of fats, oils, and other "grease" components in wastewaters. However, as opposed to mineral-type oils, animal and vegetable oils and fats such as those discharged by soap manufacture plants are readily biodegradable and generally nontoxic, although differences exist as to the difficulties caused depending on the form (floatable or emulsified) and type (hydrocarbons, fatty acids, glycerides, sterols, etc.). In general, shorter-chain-length fatty acids, unsaturated acids, and soluble acids are more readily degraded than longer-chain, saturated, and insoluble ones. The more insoluble and larger fatty acid particles have been found to require greater time for degradation than those with opposite characteristics. It has also been reported that animal and vegetable oils, fats, and fatty acids are metabolized quickly in anaerobic systems and generate the major portion of methane in regular anaerobic sludge digestion.

McCarty et al. (1972) also reported on the results of laboratory investigations in the treatability of selected industrial oily wastes from soap manufacturing and food processing by the Procter & Gamble Co. in Cincinnati, Ohio, when combined with municipal sewage or sludge. The grease content of the industrial wastes was high in all cases, ranging from 13 to 32% of the waste COD, and it was about 2.9 g of COD per gram of grease. It was found that it is possible to treat about equal COD mixtures of the industrial wastes with municipal sewage using the activated sludge process and achieve removal efficiencies similar to those for municipal sewage alone.

The grease components of the industrial wastes were readily degraded by anaerobic treatment, with removal efficiencies ranging from 82 to 92%. Sludges from the anaerobic digestion of an industrial/municipal mixture could be dewatered with generally high doses of chemical conditioning ($FeCl_2$), but these stringent requirements seemed due to the hard-to-dewater municipal waste sludge. In conclusion, the Procter & Gamble Co. industrial wastes were readily treated when mixed with municipal sewage without significant adverse impacts, given sufficient plant design capacity to handle the combined wastes hydraulically and biologically. Also, there was no problem with the anaerogic digestion of combined wastes, if adequate mixing facilities are provided to prevent the formation of scum layers.

7.4 Removal of Nonionic Surfactants by Adsorption

Nonionic surfactants, as mentioned previously, have been widely adopted due to their characteristics and properties and, especially, because they do not require the presence of undesirable phosphate or caustic builders in detergent formulation. However, the relative lesser degree of biodegradability is an important disadvantage of the nonionic surfactants compared to the ionic ones. Adsorption on activated carbon and various types of clay particles is, therefore, one of the processes that has been effective in removing heterodisperse nonionic surfactants—those that utilize a polyhydroxyl alcohol as a lipophilic phase—from wastewaters (Carberry et al., 1977). In another study by Carberry and Geyer (1977), the adsorptive capacity kinetics of polydisperse nonionic surfactants'—those that utilize a hydrocarbon species as a lipophilic base—removal by granular activated carbon and clay were investigated. Both clay particulates of different types and various activated carbons were tested and proven efficient in adsorbing nonionic surfactants. Of all the clays and carbons studied, Bentolite-L appeared to be the superior adsorbent (9.95% mol/kg vs. 0.53 mol/kg for Hydrodarco 400), but reaction rate constants for all adsorbents tested appeared to be strikingly similar.

7.5 Removal of Anionic Detergents with Inorganic Gels

Inorganic gels exhibiting ion-exchange and sorption characteristics are more stable than synthetic organic resins, which have also been used for the removal of detergents from wastewaters (Srivastava et al., 1981). The sorption efficiency and number of cycles for which inorganic gels can be used without much loss in sorption capacity would compensate the cost involved in their preparation. Zinc and copper ferrocyanide have been shown to possess promising sorption characteristics for cationic and anionic surfactants. Of the two, copper ferrocyanide is a better scavenger for anionic detergents, which have a relatively small rate and degree of biodegradation and their presence in raw water causes problems in coagulation and sedimentation.

The cation-exchange capacity of the copper ferrocyanide gel used was found to be about 2.60 meq/g and its anion-exchange capacity about 0.21 meq/g. In all cases of various doses of gel used and types of anionic surfactants being removed, the tests indicated that a batch contact time of about 12 hr was sufficient for achieving maximum removals. Trials with various fractions of particle size demonstrated that both uptake and desorption (important in material regeneration) were most convenient and maximized on 170 to 200 BSS mesh size particles. Also, the adsorption of anionic surfactants was found to be maximum at pH 4 and decreased with an increase in pH.

The presence of NaCl and $CaCl_2$ salts (mono and bivalent cations) in solution was shown to increase the adsorption of anionic surfactants in the pH range 4 to 7, whereas the presence of $AlCl_3$ salt (trivalent cation) caused a greater increase in adsorption in the same pH range. However, at salt concentrations greater than about $0.6M$, the adsorption of the studied anionic surfactants started decreasing. On the other hand, almost complete desorption could be obtained by the use of K_2SO_4 or a mixture of H_2SO_4 and alcohol, both of which were found to be equally effective. In conclusion, although in these studies the sorption capacity of the adsorbent gel was not fully exploited, the anionic detergent uptake on copper ferrocyanide was found to be comparable to fly ash and activated carbon.

7.6 Optimum Removal Process for Cationic Surfactants

There are few demonstrated methods for the removal of cationic surfactants from wastewater, as mentioned previously, and ion exchange and ultrafiltration are two of them. Chiang and Etzel (1983) developed a procedure for selecting from these the optimum removal process for cationic surfactants from wastewaters. Preliminary batch-test investigations led to the selection of one resin (Rohm & Haas "Amberlite", Amb-200) with the best characteristics possible (i.e., high exchange capacity with a rapid reaction rate, not very fine mesh resin that would cause an excessive pressure drop and other operational problems, macroporous resin that has advantages over the gel structure resins for the exchange of large organic molecules) to be used in optimizing removal factors in the column studies vs. the performance of ultrafiltration membranes (Sepa-97 CA RO/UF selected). The cyclic operation of the ion-exchange (H+) column consisted of the following steps: backwash, regeneration, rinse, and exhaustion (service).

The ion-exchange tests indicated that the breakthrough capacity or total amount adsorbed by the resin column was greater for low-molecular-weight rather than high-molecular-weight surfactants. Furthermore, the breakthrough capacity for each cationic surfactant was significantly influenced (capacity decreases as the influent concentration increases) by the corresponding relationship of the influent concentration to the surfactant critical micelle concentration (CMC). A NaCl/ethanol/water (10% NaCl plus 50% ethanol) solution was found to be optimum in regenerating the exhausted resin.

In the separation tests with the use of a UF membrane, the rejection efficiency for the C_{16} cationic surfactants was found to be in the range of 90 to 99%, whereas for the C_{12} surfactants it ranged from 72 to 86%, when the feed concentration of each surfactant was greater than its corresponding CMC value. Therefore, UF rejection efficiency seems to be dependent on the respective hydrated micelle diameter and CMC value. In conclusion, the study showed that for cationic surfactants' removal, if the feed concentration of a surfactant is higher than its CMC value, then the UF membrane process is found to be the best. However, if the feed concentration of a surfactant is less than its CMC value, then ion exchange is the best process for its removal.

REFERENCES

Abrams, I. M. and Lewon, S. M. (1962). *J. Amer. Water Wks. Assoc., 54* (5).
Besselievre, E. B. (1969). *The Treatment of Industrial Wastes*. McGraw-Hill, New York.
Brownell, R. P., et al. (1975). Chemical-biological treatment of surfactant wastewater. In *Proceedings of the 30th Industrial Waste Conference,* Purdue Univ., Lafayette, Ind., Vol. 30, p. 1085.
Callely, A. G., et al. (1976). *Treatment of Industrial Effluents*. Halsted Press, New York.
Carberry, J. B. and Geyer, A. T. (1977). Adsorption of non-ionic surfactants by activated carbon and clay. In *Proceedings of the 32nd Industrial Waste Conference*, Purdue Univ., Lafayette, Ind., Vol. 32, p. 867.
Carberry, J. B., et al. (1977). Clay adsorption treatment of non-ionic surfactants in wastewater. *J. Water Poll. Control Fed., 49*, p. 452.
Chambon, M. and Giraud, A. (1960). *Bull. Ac. Nt. Medecine* (France), *144*, 623–628.

Treatment of Soap and Detergent Industry Wastes 291

Chiang, P. C. and Etzel, J. E. (1983). Procedure for selecting the optimum removal process for cationic surfactants. In *Toxic and Hazardous Waste* (LaGrega and Hendrian, eds.). Butterworth, Boston.

Cohen, J. M. (1962). Taste and Odor of ABS. Dept. of Health, Education and Welfare Rept., Cincinnati, Ohio.

Cohen, J. M., et al. (1959). *J. Amer. Water Wks. Assoc., 51*, 1255–1266.

Eckenfelder, W. W. (1989). *Industrial Water Pollution Control*. McGraw-Hill, New York.

Environmental Protection Agency (EPA) (1974). Development document on guidelines for soap and detergent manufacturing. EPA-440/1-74-018a, U.S. Government Printing Office, Washington, D.C., Construction Grants Program.

Environmental Protection Agency (EPA) (1977). Federal Guidelines on State and Local Pretreatment Programs. EPA-430/9-76-017c, U.S. Government Printing Office, Washington, D.C., Construction Grants Program, pp. 8-13-1 to 8-13-25.

Federal Register (1987). Code of Federal Regulations. U.S. Government Printing Office, Washington, D.C., CFR 40, Part 417, pp. 362–412.

Federal Register (1990). Notices, Appendix A, Master Chart of Industrial Categories and Regulations. U.S. Government Printing Office, Washington, D.C., Jan. 2, 55, No. 1, pp. 102, 103.

Gameson, A. L. H., et al. (1955). *J. Inst. Water Engrs.* (U.K.), *9*, 571.

Gibbs, F. S. (1949). The removal of fatty acids and soaps from soap manufacturing wastewaters. In *Proceedings of the 5th Industrial Waste Conference*, Purdue Univ., Lafayette, Ind., Vol. 5, p. 400.

Greek, B. F. (1990). Detergent industry ponders products for new decade. *Chem. & Eng. News*, Jan. 29, 37–60.

Gurnham, C. F. (1955). *Principles of Industrial Waste Treatment*. Wiley, New York.

Gurnham, C. F. (ed.) (1965). *Industrial Wastewater Control*. Academic Press, New York.

Herin, J. L., et al. (1970). Development and operation of an aeration waste treatment plant. In *Proceedings of the 25th Industrial Waste Conference*, Purdue Univ., Lafayette, Ind., Vol. 25, p. 420.

Koziorowski, B. and Kucharski, J. (1972). *Industrial Waste Disposal*. Pergamon Press, Oxford, U.K.

Langelier, W. F., et al. (1952). *Proc. Amer. Soc. Civil Engrs., 78* (118), February.

Lawson, R. (1959). *Sewage and Industrial Wastes, 31*(8), 877–899.

McCarty, P. L., et al. (1972). Treatability of oily wastewaters from food processing and soap manufacture. In *Proceedings of the 27th Industrial Waste Conference*, Purdue Univ., Lafayette, Ind., Vol. 27, p. 867.

McGauhey, P. H. and Klein, S. A. (1959). *Sewage and Industrial Wastes, 31*(8), 877–899.

Nemerow, N. L. (1978). *Industrial Water Pollution*. Addison-Wesley, Reading, Mass.

Nichols, M. S. and Koepp, E. (1961). *J. Amer. Water Wks. Assoc., 53*, 303.

Osburn, Q. W. and Benedict, J. H. (1966). Polyethoxylated alkyl phenols: Relationship of structure to biodegradation mechanism. *J. Amer. Oil Chem. Soc., 43*, 141.

Overcash, M. R. (1986). *Techniques for Industrial Pollution Prevention*. Lewis Publishers, Mich.
OECD (1971). Pollution par les Detergents. Rept. by Expert Group on Biodegradability of Syndets, Paris, France.
Payne, W. J. (1963). Pure culture studies of the degradation of detergent compounds. *Biotech. and Bioeng., 5*, 355.
Prat, J. and Giraud, A. (1964). La Pollution des Eaux par les Detergents. Rept. 16602, Scientific Committee of OECD, Paris, France.
Renn, E. and Barada, M. (1961). *J. Amer. Water Wks. Assoc., 53*, 129–134.
Robeck, G., et al. (1962). *J. Amer. Water Wks. Assoc., 54*, 75.
Ross, R. D. (ed.) (1968). *Industrial Waste Disposal*. Reinhold, New York.
Shapland, K. J. (1986). Industrial effluent treatability. *J. Water Poll. Control* (U.K.), p. 75.
Smith, R. S., et al. (1956). *J. Amer. Water Wks. Assoc., 48*, 55.
Srivastava, S. K., et al. (1981). Use of inorganic gels for the removal of anionic detergents. In *Proceedings of the 36th Industrial Waste Conference*, Purdue Univ., Lafayette, Ind., Vol. 36, Sec. 20, p. 1162.
Swisher, R. D. (1968). Exposure levels and oral toxicity of surfactants, *Arch. Environ. Health, 17*, 232.
Swisher, R. D. (1970). *Surfactant Biodegradation*. Marcel Dekker, New York.
Todd, R. (1954). *Water Sewage Works, 101*, 80.
Vath, C. A. (1964). *Soap and Chem. Specif.*, March.
Wang, L. K. and Wood, G. W. (1982). Water Treatment by Disinfection, Flotation and Ion Exchange Process System, PB82-213349, U.S. Dept. of Commerce, Nat. Tech. Information Serv., Springfield, Va. May.
Wang, L. K. (1990). Treatment of various industrial wastewaters by dissolved air flotation. Presented at N.Y.-N.J. Environmental Expo., Oct., Secaucus, N.J.
Wright, K. A. and Cain, R. B. (1972). Microbial metabolism of pyridinium compounds, *Biochemical J., 128*, p. 543.

6

Treatment of Acid Pickling Wastes of Metals

Veysel Eroglu

Istanbul Technical University, Ayazaga, Istanbul, Turkey

Ferruh Erturk

TUBITAK Mamara Research Center, Gebze, Kocaeli, Turkey

1. INTRODUCTION

Metal industries use substantial quantities of water in processes such as metal finishing and galvanized pipe manufacturing in order to produce corrosion-resistant products. Effluent wastewaters from such processes contain toxic substances, metal acids, alkalis, and other substances that must be treated, such as detergents, oil, and grease. These effluents may cause interference with biological treatment processes at sewage treatment plants. In the case when the effluents are to be discharged directly to a watercourse, treatment requirements will be more stringent and costly.

The pickling process is used to remove the oxide or scale of metals and corrosion products, in which acids or acid mixtures are used. Lime or alkali substances are used to neutralize the waste pickle liquor. In addition, chemical oxygen demand (COD), suspended matter, oil and grease, ammonium nitrogen (NH_4-N), pH, cyanides, fish toxicity, and several relevant metal ions such as cadmium (Cd), iron (Fe), zinc (Zn), nickel (Ni), copper (Cu), and chromate (Cr^{+6}) have to be reduced below the maximum allowable limits.

2. ORIGIN OF PICKLING WASTES

Pickle solutions that are used in the removal metal oxides or scales and corrosion products are acids or acid mixtures. The most common acid used for pickling is sulfuric acid. Other acids such as hydrochloric, phosphoric, hydrofluoric, or nitric acids are also used individually or as mixtures. Sulfuric or hydrochloric acids are used for pickling carbon steels, and phosphoric, nitric, and hydrofluoric acids are used together with sulfuric acid for stainless steel. Water is used in pickling and rinsing. The quantity of water used can vary from less than 100 L/ton to 3000 L/ton, depending on whether once-through or recycle systems are used (Barnes, Forster, and Hrudey, 1987).

Depending on the product being pickled, the process can be batch or continuous. In continuous strip pickling, more water is required for several operations such as the uncoilers, looping pit, and coilers. In the case of pickling hot rolled coils, the coils are transported to the pickling line. In the uncoiler section, the coil is fed through a pit containing water for washing off the surface dirt and then fed through the pickling line.

3. CHARACTERISTICS OF PICKLING WASTES

During the application of the pickling process in the finishing of steel, in which steel sheets are immersed in a heated bath of acid (sulfuric, hydrochloric, phosphoric, etc.), scale (metallic oxides) are chemically removed from the metal surface. The process can be batch or continuous. In these processes, water is used in pickling and rinsing operations. In continuous pickling, wet fume scrubbing systems are also used. The effluent water from the pickling tanks, which is called the waste pickle liquor (WPL), consists of spent acid and iron salts. Waste hydrochloric liquor contains 0.5 to 1% free hydrochloric acid and 10% dissolved iron, and the production of WPL is approximately 1 kg free hydrochloric acid and 10 kg dissolved iron per ton of steel pickled (Barnes, Forster, and Hrudey, 1987). In waste sulfuric acid pickle liquor, the free acid and dissolved iron content are approximately 8% each, resulting in 10 kg each of free sulfuric and dissolved iron per ton of steel pickled. WPL may also contain other metal ions, sulfates, chlorides, lubricants, and hydrocarbons. Rinsewater, which contains smaller concentrations of the above contaminants, ranges in quantities from 200 to 2000 L/ton. Fume scrubber water requirements range from 10 to 200 L/ton (Barnes, Foster, and Hrudey, 1987).

Acid Pickling Wastes of Metals

In hot rolling processes, pickling is used for further processing to obtain the surface finish and proper mechanical properties of a product. In the case of pickling hot rolled coils, the coil is fed through a pit containing water for washing off surface dirt and then fed through the pickling line. In the pickling section, the coil strip comes in contact with the pickle liquor (sulfuric or hydrochloric acid). Wastewater sources are processor water, spent pickling solution (WPL), and rinsewater.

In the case of batch pickling, the product is dipped into a pickling tank and then rinsed in a series of tanks. The quantity of wastewater discharged from a batch process is less than that from continuous operation. The wastewater is usually treated by neutralization and sedimentation.

In sulfuric acid pickling, ferrous sulfate is formed from the reaction of iron oxides with sulfuric acid

$$FeO + H_2SO_4 \rightarrow FeSO_4 + H_2O \tag{1}$$

The ferrous sulfate that is formed in the above reaction is either monohydrate or heptahydrate ($FeSO_4 \cdot 7H_2O$).

4. TREATMENT OF PICKLING WASTES

The wastewater effluents, in general, can be either discharged to a watercourse or a public sewer system. In the former case, the treatment requirements will be more stringent.

The waste pickle liquor, rinsewater discharges, and fume scrubber effluent can be combined in an equalization tank for subsequent treatment. Basically three methods are used to treat the WPL:

1. Neutralization and sedimentation
2. Crystallization of ferric sulfates and regeneration of the acid
3. Deep well disposal

The most commonly used methods are the first two.

4.1 Neutralization and Sedimentation

In old plants, neutralization and sedimentation are applied to the treatment of wastewaters in general, including WPL. A typical treatment system for continuous pickling water is shown in Fig. 1 (Eroglu, Topacik, and Ozturk, 1989).

In an integrated steel mill, a central wastewater treatment system is used to treat wastewater from pickling lines, cold rolling mills, and coating lines.

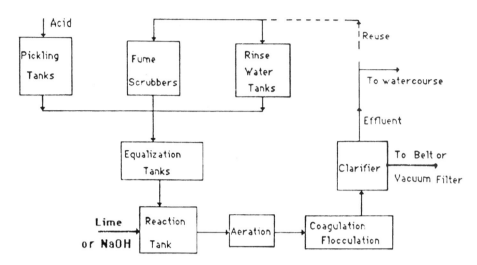

Figure 1 Typical treatment system for pickling. (From Eroglu, Topacik, and Ozturk, 1989.)

The pickler wastewater has a low pH and contains dissolved iron and other metals. The blowdown and dumps from the cold rolling mill solutions, which may contain up to 8% oil, are collected in emulsion-breaking tanks in which the emulsions are broken by heat and acid. The oil is then skimmed, and the water phase containing 200 to 300 mg/L of oils is treated together with the wastewaters from pickler, cold rolling, and coating lines. The combined wastewater flows to a settling and skimming tank where solids and oil are removed. The effluent from the settling/skimming tank is then treated in a series of settling tanks where chemicals (coagulants and/or lime) and air are added to oxidize the remaining iron to Fe^{3+}, to further break the oil emulsions and neutralize the excess acid in the wastewater. The effluent from the mixing tanks then enters a flocculator/clarifier system, the overflow from the clarifier is discharged, and the settled sludge is pumped to a dewatering system consisting of centrifuges, belt, or vacuum filters. The dewatered sludge is disposed and the water phase returned to the clarifier effluent.

4.2 Crystallization and Regeneration

The use of lime or other alkaline substances to neutralize acid is quite costly, especially when large capacities are involved. Also there are potential values in

Acid Pickling Wastes of Metals

the acids and ferrous ion, and therefore, recovery of these substances will not only reduce the pollution load, but their sale or reuse will represent a profit to the industry.

Crystallization is one of the treatment methods for sulfuric acid waste pickle liquor. Thus, it is possible to decrease the pollution load and at the same time recover various hydrates of $FeSO_4$. The crystallization of $FeSO_4$ depends on the characteristics of the water and acid, and solubility of $FeSO_4$. The solubility of ferrous sulfate as a function of temperature and sulfuric acid concentration is shown in Fig. 2 (Abwasser Technik, 1986). In this figure, $FeSO_4 \cdot 7H_2O$ is dominant in region A; $FeSO_4 \cdot 4H_2O$ in region B; and $FeSO_4 \cdot H_2O$ in C.

The crystallization of ferrous sulfate as heptahydrate is commonly used today. The concentration of iron in the acid bath is approximately 80 g/L as Fe^{3+}. The crystallization of $FeSO_4 \cdot 7H_2O$ is achieved by cooling the acid waters in heat exchangers or evaporation under vacuum after pickling. Make-up acid must be added to the bath. During countercurrent cooling, the acid bath waste passes through two to three crystallization tanks and is cooled down to between 0 to 5°C. The crystallized ferric sulfates are recovered by centrifuging. A typical flow diagram for $FeSO_4 \cdot 7H_2O$ crystallization is shown in Fig. 3. The

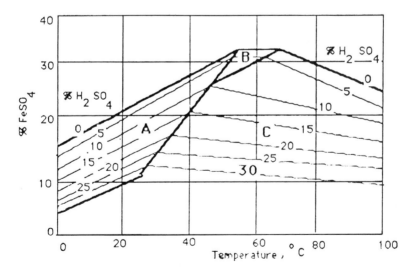

Figure 2 Solubility of $FeSO_4$ as a function of temperature and H_2SO_4 concentration. (From Abwasser Technik, 1986.)

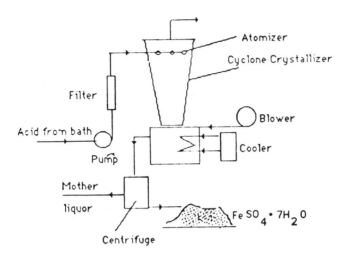

Figure 3 Flow diagram for FeSO$_4$·7H$_2$O crystallization. (From Abwasser Technik, 1986.)

waste pickle liquor is sprayed above a cyclone crystallizer, and air is blown from the bottom countercurrent to the liquid. A packing material is also present in order to increase the area of contact between the air and the liquid. The acid wastewaters are then cooled, and the FeSO$_4$·7H$_2$O crystals are recovered by centrifuging.

In the Ruthner process (Besselievre and Schwartz, 1976), the WPL is first concentrated in an evaporator. The concentrate is then pumped to a reactor where it is combined with hydrochloric acid gas, in which ferrous chloride and sulfuric acid are formed. The sulfuric acid is then separated by centrifuging. The ferrous chloride goes to a roaster in which it is converted to ferric oxide. The gases liberated from the roaster and the acid from the centrifuge go to a degassing chamber, and the sulfuric acid is removed and returned to the pickling process, or can be sold. The remaining gases from the degasser are passed through an absorption system and then reused in the reaction chamber.

In the Lurgi process that was developed in Germany, hydrochloric acid is recovered from the WPL. The acid is regenerated in a fluidized bed. During pickling with HCl, the acid circulates between a pickling tank and a storage tank and the acid reacts with the iron oxide scale from the steel producing ferric chloride, resulting in increasing concentration if dissolved iron and decreasing concentration of acid.

In the Lurgi system, the acid level in the pickling liquor stays constant at about 10%. A continuous-bleed stream is removed from the system at the same rate as it is pickled. The bleed stream, or spent pickle, is fed to a preevaporator and heated with gases from the regeneration reactor. Concentrated liquor from the then enters the lower part of the reactor containing 13% acid and 20% ferrous chloride. The reactor contains a fluidized bed of sand and is fired by oil or gas to maintain an operating temperature of about 800°C. The reaction products leave from the top of the reactor. The ferric oxide is removed by a cyclone, and the hot gases enter the preevaporator. The overhead from the evaporator, which is at a temperature of about 120°C, contains water vapor, HCl, combustion products, and also some HCl that vaporizes directly from the plant liquor which enters the system. The gas mixture from the preevaporator enters the bottom of the adiabatic absorption tower, where HCl is absorbed by another bleed stream of the pickle liquor, and thus the regenerated acid is placed back to the pickle liquor circuit. The regenerated acid contains 12% acid and about 70 g/L of iron. The unabsorbed gases go to a condenser.

5. CASE STUDY: TREATMENT OF WASTEWATER FROM ACID PICKLING TANKS IN A GALVANIZED PIPE MANUFACTURING FACTORY

5.1 General Description

This study was conducted at Cayirova Boru Sanayii A.S. (a galvanized pipe manufacturing factory) in Gebze, Kocaeli, Turkey (Eroglu, Topacik, and Ozturk, 1989). At this plant, batch pickling is applied. During the manufacturing process, the pipes are immersed in an acid bath that contains 25% sulfuric acid at 80°C and then prepared for the galvanization process, by passing through cold water, hot water, and flux baths. The purpose of a cold water bath is to clean the acid from the surface of the pipes following pickling. A hot bath is applied in order to dry and prevent water and acid from entering a flux bath. The purpose of the flux bath, in which ammonium zinc chloride (NH_4ZnCl_3) is used, is to prepare a suitable surface for galvanization and prevent oxidation of the pipe. The flow diagram of the baths is shown in Fig. 4.

Acid bath wastewaters are usually discharged once a week. The average flow rate of these wastewaters is 4 m^3/hr, with maximum of 8 m^3/hr. The hot and cold water baths are discharged once every 15 days. The quantities and

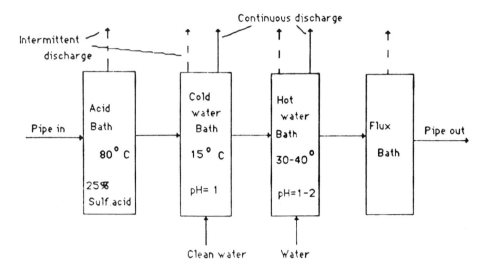

Figure 4 Flow diagram showing sources of wastewaters in galvanized pipe manufacturing process. (From Eroglu, Topacik, and Ozturk, 1989.)

flow rates of these wastewaters are shown in Table 1 (Eroglu, Topacik, and Ozturk, 1989).

5.2 Characteristics of Wastewaters

Wastewater characteristics must be known in order to select a suitable treatment system. For this purpose, the wastewater samples taken from the sources were analyzed to determine various parameters. Also, the quantities of chemicals (NaOH) required for neutralization and settling characteristics were determined. These were made separately for continuous and batch discharges. Since the system is to be designed according to the continuous discharge of wastewaters from the batch system to the treatment plant, "mixed wastewater" was prepared in quantities proportional to the flow rates. The quantity of NaOH required for 1000 mL of mixed wastewater is shown in Table 2 (Eroglu, Topacik, and Ozturk, 1989).

Since the continuous discharge quantities are much larger compared to batch discharges, they were analyzed separately. The wastewaters from continuous discharge were neutralized with $2N$ NaOH. The results are given in Table 3

Acid Pickling Wastes of Metals

Table 1 Types and Quantities of Wastewaters in Acid and Flux Baths

Wastewater source	Average	Maximum
Continuous discharge from hot and cold water baths	4 m^3/hr	8 m^3/hr
Intermittent discharge (once every 7 days)	15 m^3	15 m^3
Cold water bath (once every 15 days)	15 m^3	15 m^3
Hot water bath (every 15 days)	15 m^3	15 m^3
Flux bath	5m^3	5m^3

(Eroglu, Topacik, and Ozturk, 1989). The settling characteristics of the continuous discharge wastewaters are shown in Table 4 (Eroglu, Topacik, and Ozturk, 1989). The experimental results form the "mixed wastewaters," the quantities of which were shown in Table 2, are given in Table 5 (Eroglu, Topacik, and Ozturk, 1989). The settling characteristics of the mixed wastewaters are shown in Table 6 (Eroglu, Topacik, and Ozturk, 1989).

Neutralization can also be carried out by a combination of NaOH and lime. Experiments were conducted in order to determine the optimum combination of NaOH and lime. For this purpose, various quantities of lime were added to 1 L of mixed wastewater, and then the amount of NaOH required was

Table 2 Quantities of Wastewater Required for 1000 mL "Mixed Wastewater"

Units	Flow rate (m^3/2 months)	Quantity of NaOH required for 1000 mL
Continuous discharge	8640	971
Acid bath	129	14.5
Cold water bath	60	6.8
Hot water bath	60	6.8
Flux bath	5	0.56
Total	8894	1000

Table 3 Experimental Results for Continuous Discharge

Parameter	Unit	Original sample	After neutralization and separation
Total iron	mg/L	5980	350
Chromate	mg/L	0	0
Lead	mg/L	0	0
COD	mg/L	350	20
Zinc	mg/L	0	0
pH	—	1.6	8.0
Color	—	—	Greenish

Table 4 Settling Characteristics of Continuous Discharge Wastewaters

Time	Volume of clear phase, h (mL/L)
15 min	20
30 min	50
1.0 hr	90
2.5 hr	240
3.5 hr	400
4.5 hr	460
5.5 hr	500
20 hr	720

Table 5 Experimental Results for Mixed Wastewater Samples

Parameter	Original sample	After neutralization and separation in clear phase
Total iron, mg/L	6,100	300
Sulfate, mg/L	19,000	16,000
Chromate, mg/L	0	0
Lead, mg/L	0	0
Zinc, mg/L	15	0
COD, mg/L	360	15
pH	0.7	8.5

Table 6 Settling Characteristics of Mixed Wastewaters

Time	Volume of clear phase, (mL/L)
30 min	40
1 hr	100
2.5 hr	220
3.5 hr	350
4.5 hr	410
5.5 hr	460
20 hr	700

determined to obtain a pH of 8.5. The results are shown in Table 7 (Eroglu, Topacik, and Ozturk, 1989).

As can be seen from Table 7, the required dosage of NaOH does not increase significantly when the lime dosage is more than 20 g/1000 mL. The mixed wastewater, which was treated with the dosages of lime and NaOH shown in Table 7, was then aerated for 15 min after pH reached 8.5. After aeration it was allowed to settle for a period of 30 to 120 min. An analysis of the clear phase after settling is shown in Table 8 (Eroglu, Topacik, and Ozturk, 1989).

The wastewater was treated with 15 g/L of lime and NaOH to attain a pH of 8.5, aerated for 1 hr, mixed for 23 hr, and 1 additional hr was allowed for

Table 7 Quantities of NaOH Required for Different Quantities of Lime to Obtain a pH of 8.5

I 10 g Lime		II 20 g Lime		III 26 g Lime		IV 32 g Lime	
pH	NaOH added (mL)	pH	NaOH added (mL)	pH	NaOH added (mL)	pH	NaOH added (mL)
3.4	0	6.8	0	7.1	0	7.4	0
6.5	20	7.4	8	7.6	4	7.6	4
7.4	28	8.4	16	7.9	12	8.6	12
8.0	30	8.5	17.2	8.6	15.2		
8.5	32						

Table 8 Analysis of Mixed Wastewater after Neutralization, Aeration, and Settling

Parameter (mg/L)	Settling time (min)	Lime +NaOH (g)			
		10	20	26	32
Iron	30	125	30	5	0
	120	0	0	0	0
Sulfate	30	5750	5759	5000	3000
	120	5750	5750	5000	2750
Settlable matter	30	120	280	320	440
	120	400	520	410	480

settling. The analysis of the clear effluent is shown in Table 9 (Eroglu, Topacik, and Ozturk, 1989).

5.3 Treatment Methods

As was indicated in the previous section, the concentration of iron in the mixed wastewaters ranged from 5980 to 6100 mg/L; its pH was 0.7 and zinc concentration 15 µg/L. Since these wastewaters come only from acid baths and not from other processes of the plant, parameters like cadmium and fluoride are not encountered. The discharge standards for the metal industry effluents set by the Turkish Water Pollution Control Regulation (Official Gazette, Table 15.7, Sept. 4, 1988) are shown in Table 10 (Eroglu, Topacik, and Ozturk, 1989).

Table 9 Analysis of Wastewater After Neutralization, Aeration, and Settling

Parameter	Concentration (mg/L)
COD	0
Total iron	0
Zinc	0
Sulfate	2100

Acid Pickling Wastes of Metals

Table 10 Effluent Standards for Metal Industry Wastewaters

Parameter	2-h Composite sample (mg/L)
COD	200
Suspended solids	125
Oil and grease	20
Ammonium nitrogen	400
Cd	0.1
Fe	3
Flouride	50
Zn	5
Fish toxicity	10
pH	6–9

6. SUMMARY

As stated in the previous sections, the basic unit operations required for the treatment system are

1. Neutralization with NaOH and/or lime to increase the pH
2. Physicochemical methods to remove COD and iron

The iron present in the wastewater appears in the form of ferrous ion (Fe^{2+}), which is soluble in water. Ferrous ion can be removed either by oxidation to ferric (Fe^{3+}) or by crystallization that was mentioned in the previous section.

The experiments conducted on wastewaters, the results of which were shown in the previous section, indicated that neutralization/aeration/settling gave satisfactory results. The sludge formed must be disposed after dewatering either in a filterpress, horizontal belt filter, or centrifuge. An equalization tank is required in order to compensate the effects of intermittent discharges. The treated wastewater can then be recycled to be used in the process or discharged to the river. The flow diagram of the selected system is shown in Fig. 5.

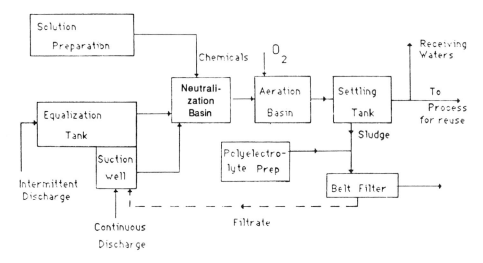

Figure 5 Flow diagram of selected treatment system. (From Eroglu, Topacik, and Ozturk, 1989.)

REFERENCES

Abwasser Technik, ATV (1986). *Behandling von Beizereiabwassern und Abbeizen bei der Halbzeugfertigung*, 4–7, 149–154.

Barnes, Forster, and Hrudey (1987). Surveys in industrial wastewater treatment. In *Manufacturing and Chemical Industries, The Treatment of Wastewaters from Steel Plants*, Vol. 3. Longman Scientific & Technical, Essex, England.

Eroglu, V., Topacik, D., and Ozturk, I. (1989). Proj. rep. of Wastewater Treatment Plant for Cayirova Pipe Factory, Environmental Eng. Dept., Istanbul Technical Univ. (in Turkish).

7
Treatment of Textile Wastes

Sanjoy K. Bhattacharya
Tulane University, New Orleans, Louisiana

1. ORIGIN OF WATER POLLUTION WITHIN THE TEXTILE COMPANY

The textile industry actually represents a range of industries with operations and processes as diverse as its products. It is almost impossible to describe a "typical" textile effluent because of such diversity. After fabrics are manufactured, they are subjected to several wet processes collectively known as "finishing," and it is in these finishing operations that the major waste effluents are produced (Abo-Elela et al., 1988). These finishing processes are complex and ever-changing. This is a fact of life that is reflected in the variety of chemicals that find their way into textile finishing wastewaters.

It is important to know the various operations involved in finishing before the problems of treatment of wastewaters from these operations can be appreciated. It is beyond the scope of this chapter to discuss all these operations in detail. The finishing operations for cotton, wool, and synthetic fiber processing are discussed very briefly in the following sections.

1.1 Cotton

Slashing is the first process in which liquid treatment is involved. In this process, the wrap yarns are coated with "sizing" in order to give them textile strength to withstand the process exerted on them during the weaving operation (Jones, 1973). The common sizes are starch, polyvinyl alcohol, carboxy methyl cellulose, polyacrylic acid, and polyester size. Starch is commonly used especially in combination with other sizes such as polyvinyl alcohol. Starch is very high in BOD. Most other sizes exert very little BOD even though they may contribute to COD. Sizes represent the single largest group of chemicals used in the industry.

Desizing removes the substance applied to the yarns in the slashing operation, by hydrolyzing the size into a soluble form. Acid desizing and enzyme desizing are the two common methods. The method of desizing is dependent on the actual size used.

Scouring follows desizing. Noncellulosic components of the cotton are removed by soap solution or hot alkaline detergents. Solvent processes, on the other hand, will use little water. Even though the projected solvent recovery is between 90 and 95%, a significant amount of solvent may reach the atmosphere and the waste stream (Porter, 1970).

Bleaching is performed to remove the yellowish color of the fiber and render it white. Sodium hypochlorite, hydrogen peroxide, and sodium chlorite are the three bleaches most commonly used for cotton (Jones, 1973).

Mercerization uses sodium hydroxide, water, and an acid wash. This process increases the dye affinity, tensile strength, and surface luster.

Dyeing in most cases is performed in dye houses or color shops when the goods are sent there after mercerizing. A few large plants may do the dyeing in the same facility where the other finishing operations are performed. Depending on the type of dye used, the waste will vary. The general process is to use rolling machines that contain a paste of dye, thickener, hygroscopic substances, dyeing assistants, water, and other chemicals.

1.2 Wool

In the processing of woolen fibers, five sources of pollution load exists: scouring, dyeing, washing after fulling, neutralizing, and bleaching (Jones, 1973). Scouring removes impurities from the woolen fibers. Wool scouring produces one of the strongest industrial wastes in terms of BOD by contributing 55 to 75% of the total BOD load in the wool finishing (Porter, 1970). The BOD load contributed by dyeing may represent a small percent of the mill's

Treatment of Textile Wastes

total BOD load, but again, this percentage may vary depending on the plant. Washing after fulling may be the second largest source of BOD in wool finishing wastewaters. Neutralization after carbonizing the fabric does not significantly contribute to the BOD load. The amount of wool fabric bleached is very small because bleaching is required for white and lightly shaded fabrics only. Hence, bleaching (like neutralization) does not significantly contribute to effluent BOD.

1.3 Synthetic Fibers

The demand for synthetics has experienced phenomenal growth in the past several decades. Synthetic textile fibers may be cellulosic or noncellulosic. Rayon and cellulosic acetate are examples of cellulosic fibers. The major noncellulosic fibers are acrylics, nylon, polyester, etc. Different aqueous processes are involved in the production of these fibers that lead to varying pollution loads (Porter, 1970). Desizing, scouring, and dyeing are the chief sources of water pollution from finishing processes for synthetic fibers (Jones, 1973).

1.4 Dyes

In this section, a discussion of dyes has been included since color is a perplexing problem in treating textile mill wastewaters. Dyeing of fabrics is a common process for all fabrics, as discussed in the earlier sections.

It is important to note that all colored substances are not necessarily dyes. A dye is a colored substance that can be applied in solution or dispersion to a substrate (e.g., a textile fiber, paper, or foodstuff), thus giving it a colored appearance (Allen, 1971). The substrate has a natural affinity for appropriate dyes and readily absorbs them from solutions or aqueous dispersion. Concentrations of dyes, temperature, and pH are important factors in controlling the dyeing process. The presence of auxiliary substances may be necessary to control the rate of dyeing.

Dyes are classified according to either their chemical structure or method of application. The Color Index (CI) includes a comprehensive classification of known commercial dyes according to usage and chemical constitution. There are over 7000 dye structures listed in the CI, and the major application classes are well known. The dyes are not readily classified under conventional chemical headings because of structural variety.

Dyes consists of a chromogen and an auxochrome (Sherve, 1967). The chromogen contains the chromophore (color giver) and is represented by chemical radicals such as azo group, nitroso group, nitro group, ethylene group, carbonyl group, etc. The auxochromes normally contain -NH_2, -OH, -COOH, or -SO_3H.

On the basis of their chromophoric system, the dyes may be classified as nitroso, nitro, azo, azoic, anthraquinone, indigoid, etc. According to their application, the dyes may be classified as acid, direct disperse, vat, reactive, etc. Abrahart (1977) has described these classifications in detail.

Shaul et al. (1986) reported that azo dyes constitute a significant portion of submissions to the Premanufacture Notification (PMN) process under the Toxic Substances Control Act (TSCA). Azo dyes are characterized by the presence of one or more azo groups (-N=N-). Two reactions form the basis of azo dye chemistry. These reactions are diazotization and coupling (Allen, 1971). Figure 1 shows the structure of some azo dyes. Some of the azo dyes, dye precursors, or their degradation products such as aromatic amines have been shown to be, or are suspected to be, carcinogenic (Helemes, 1984).

2. CHARACTERISTICS OF TEXTILE WASTEWATERS

From the preceding sections, it is apparent that the characteristics of the wastewaters from a textile plant will depend on the specific operations in the plant. It is misleading to speak of a typical textile effluent. The type of fiber involved and machinery employed are the main factors determining the type and quantities of chemicals present in the textile wastewaters.

Finishing processes discussed earlier may be either batch or continuous. For batch processes, the discharge is intermittent, with the interval between discharged depending on the operations. All the wastewaters from a batch process are likely to come from the same operation, the first being the most heavily contaminated and the last rinse the most dilute. For continuous processes, a steady flow of effluents with moderate concentrations is expected.

Developing a database from sampling the textile effluent discharges at frequent intervals should lead to establishing reliable average values. In addition to frequent sampling, the other approach to determining the characteristics of textile wastewaters is to study the process, its waste components, and volumes. For example, the amount of organic matter that is removed from a fabric in the course of normal textile processing can be visualized when one considers that about 10% of the gross weight of a cotton fabric consists of natural impurities that may be removed in processing. For a firm that manufacturers 20 tons of fabric per week, about 2 tons per week of impurities are

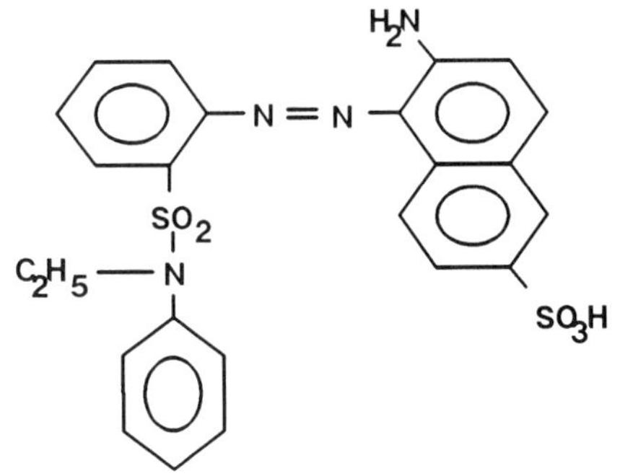

Acid Red 37 (Monoazo)

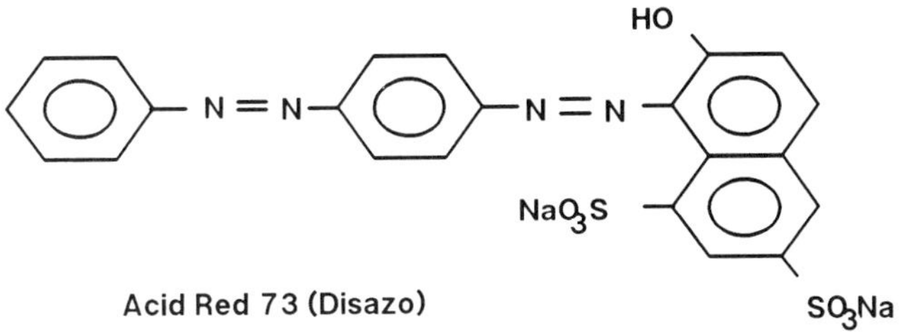

Acid Red 73 (Disazo)

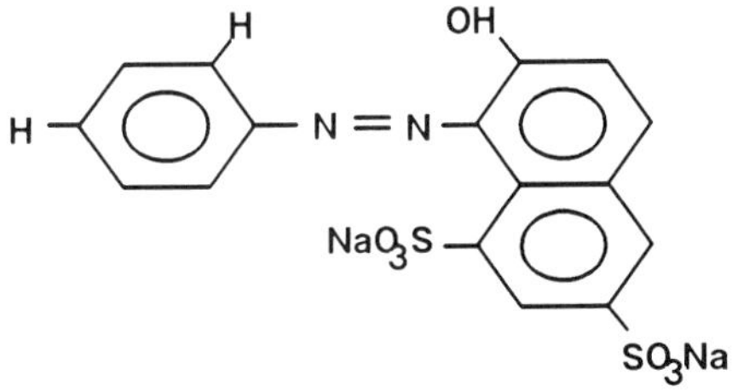

Acid Orange 10 (Monoazo)

Figure 1 Structures of some common azo dyes.

discharged to the sewer (Junkins, 1982). With known discharge volumes, the concentrations of the impurities in the effluent may be estimated.

Kremer et al. (1982) reviewed the pollutants generated by the textile industry. They divided the pollutants into four general groups, i.e., sizes, detergents, dyes, and priority pollutants. They reported that most of the priority pollutants contained in textile industry effluents are aromatics, halogenated hydrocarbons, and heavy metals.

Cotton and linen contribute organic matter from the noncellulosic materials that are present in the natural fibers, whereas wool contains sand and grease that are removed during scouring. Synthetic fibers may contain spinning oils and antistatic dressings. Textile wastes are generally colored, highly alkaline, and high in BOD, suspended solids, and temperature (Junkins, 1982). The raw wastewater (pH=9) of a bleachery had 660 mg/L of BOD, 2080 mg/L of COD, 34 mg/L of oil, and 2700 mg/L of TDS. Randall and King (1980) reported that wastewaters from textile dyeing and finishing operations may be characterized as high in organic matter, both biodegradable and nonbiodegradable, total solids and color, and very variable in pH. In addition, they tend to be high in surfactants and contain potentially significant concentrations of oils, phenols, and heavy metals such as chromium, zinc, and copper. Randall and King (1980) reported the characteristics of raw wastewaters from three plants. The BOD_5 varied between 260 and 560 mg/L. The color (APHA units) varied between 1000 and 1335. The hue was brown, red to black, or yellow to black.

Kertell and Hill (1982) reported the characteristics of the wastewater from a dye and finishing company. The average BOD was 371 mg/L and the average color was 113 (Pt-Co units). About 70 mg/L of oil and grease were present. The total solids concentration was about 480 mg/L. Troxler and Hopkins (1981) reported that the introduction of continuous dyeing machines had significantly increased the strength of carpet finishing wastewater. This is due to the use of natural bean gum thickeners as a viscosity modifier for the dye solutions. The average BOD_5 and COD from "beck dyeing" wastewater were 232 and 943 mg/L, respectively. The average BOD_5 and COD from continuous dyeing wastewater were 930 and 2912 mg/L, respectively.

Davis et al. (1981) reported that the BOD_5 of a dyehouse wastewater varied between 20 and 1250 mg/L with an ave rage of 634 mg/L. The color (APHA units) varied between 7700 and 13,100 mg/L. The average flow of this plant was 1.58 mgd.

A wide variety of methods have been used for reporting results related to color-removal processes (Stahr et al., 1980). These include the use of American Public Health Association (APHA) color units, transmittance, hue, intensity, etc. The current EPA standards are, however, based on a color analysis procedure developed by the American Dye Manufacturer's Institute (1973).

3. TREATMENT OF TEXTILE WASTEWATERS

To solve the problems of treatment of wastewaters from a textile plant, several alternatives should be included. The alternatives are the following:

1. In-plant control for waste reduction
2. Treatment to "reuse standard" on an external (end-of-line) basis or by closed-loop recycle systems
3. Direct discharge to municipal waste treatment systems (i.e., POTWs)
4. "On-site treatment" of textile wastewaters at POTWs before combining with municipal wastewaters
5. "Pretreatment" of textile wastewaters at the plant before discharging to sewer

3.1 In-Plant Control for Waste Reduction

Jones (1973) had reviewed the various ways to reduce textile wastes by in-plant control. In addition to in-plant measures to reduce wastewater volume, the importance of reduction of process chemicals, recovery and reuse of process chemicals, process modifications, and substitution of chemicals were discussed.

3.2 Treatment to Reuse

Groves and Buckley (1980) evaluated membrane separation technology for the reuse of textile effluents. They studied two pilot-plant applications: (1) high-temperature *ultrafiltration* of desizing effluent for polymer size recovery and water reuse and (2) hyperfiltration (reverse osmosis) of mixed cotton/synthetic fiber dyehouse effluents for water reuse. The membrane separation processes may offer potential for the recovery of various chemicals like sizing agents. Groves and Buckley (1980) concluded that the use of closed-loop recycle systems is technically feasible for textile wastewaters. They also discussed the "fouling" problems and requirements of cleanings for the restoration of the design flux. Membrane processes are expected to be expensive even though no cost information was reported by these authors.

Davis et al. (1981) reported that ultrafiltration has several applications for the recovery of textile sizes. Also, latex recovery can be accomplished using ultrafiltration membranes (Hoffman and Zakha, 1978). Tinghul et al. (1983) showed how the science of reverse osmosis offers a basis for the choice of membrane materials for use in reverse osmosis applications involving the separation of dyes in aqueous solutions.

The reuse of a textile effluent may be economic only if the plant faces an acute shortage of water. Complete reuse will probably be unrealistic under any circumstances for many more years.

3.3 Direct Discharge to POTWs

For many textile plants, direct discharge to POTWs may be the best alternative. A mill's wastewater may be clean enough or low enough in volume to be treated by the POTW at little or no extra cost. Even if preliminary or primary treatment is required, the cost to the mill may be much less than if a complete treatment facility were required (Slade, 1968). Jones (1973) listed three advantages of combined treatment, i.e., the direct discharge of textile waters to POTWs. Potential economy of operation is the first advantage. Textile wastewaters may not contain enough nutrients (nitrogens and phosphorus) required for biological treatment. Hence, combining such wastewaters with nutrient-rich domestic wastewaters appears to be another advantage. Finally, in combined treatment, the dilution of highly concentrated textile wastes can be achieved, which prevents shock loads of toxic materials from killing the bacteria in the treatment plant (Jones, 1973).

Newlin (1969) studied the economic feasibility of treating textile wastes in municipal treatment plants. The general conclusion, based on 26 municipalities serving some 100 textile mills, was that problems and conditions were so diverse that each case must be given individual attention. Three of the six textile mills covered by the study would have saved money by direct discharges, as compared with the costs of treating their own wastes. However, three plants were saving money by treating their own wastewaters. In general, the mills with small amounts of wastewaters were better off paying their service charges to the POTWs (Newlin, 1969).

Jones (1973) concluded that the findings regarding the cost of wastewater treatment in relation to total textile manufacturing costs in the southeastern United States raise doubts about the significance of wastewater treatment requirements as an important factor in a competitive strategy as long as the effluent standards do not change drastically. On the other hand, some old plants may face treatment costs that are high enough to create financial problems for them.

3.4 On-Site Treatment of Textile Wastewaters

Figure 2 shows schematic diagrams for alternatives 4 and 5. It is important to note how "on-site treatment" and "pretreatment" are defined *for this discussion*.

Treatment of Textile Wastes

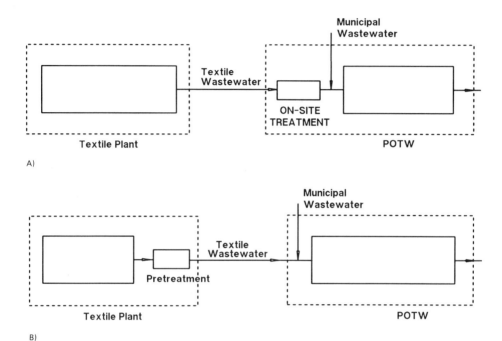

Figure 2 Systemic flow diagrams for (A) on-site treatment and (B) pretreatment of textile wastewaters.

On-site treatment refers to any additional treatment at the POTW before combining the textile effluents with the municipal wastewaters for the subsequent treatment. Such additional treatment of the textile wastewaters at the POTW may be physico-chemical or biological. On-site treatment appears to be feasible when several small, closely located textile plants discharge the wastewaters to the same POTW.

3.5 Pretreatment of Textile Wastewater

Pretreatment refers to the treatment of wastewaters by the textile plants before discharging to the sewer (Fig. 2). Again, such treatments may be physico-chemical or biological. A combination of physico-chemical and biological (both anaerobic and aerobic) processes may also be feasible. Such pretreatment appears to be feasible for large textile plants.

3.6 Physical/Chemical Treatment

Textile wastewaters may be treated using physical/chemical processes either at the POTW (on-site) or at the plant (pretreatment). Experience has demonstrated that chemical processes remove biodegradable organic matter (Randall and King, 1980). Some of the physical/chemical processes are coagulation/clarification, multimedia filtration, granular carbon adoption, dissolved air flotation, and ozonation.

Coagulation/clarification is an effective process for textile wastewater treatment. This method may be especially effective for color removal. Typical coagulants are alum, ferrous sulfate, ferric sulfate, and ferric chloride with lime or sulfuric acid for pH control (Brower and Reed, 1986). Other widely used coagulants are cationic, anionic, and nonionic organic polymers.

For effective coagulation, the experimental determination of the optimum dosage is required. Some wastewaters may require very high coagulant dosage. Chemical dosages in the range of 500 to 1000 mg/L are not uncommon for textile wastewaters (Stahr et al., 1980). The addition of large amounts of chemicals result in the production of significant quantities of waste solids. The ultimate disposal of these wastes may be very expensive. Hence, the cost calculations of coagulation/clarification should include the additional costs for ultimate disposal. Stahr et al. (1980) reported that the most economical approach to color removal by coagulation appears to involve the use of a cationic polymer coagulant aid with alum as the primary coagulant. The resulting sludges from this process are reported to be more easily dewatered and conditioned than the sludges produced through the use of alum alone. Stahr et al. (1980) also reported that the critical parameters defining the optimum polymer dose for color removal include the waste solution pH, the concentration and types of dyes present, and the charge density of the polymer being added.

Abo-Elela et al. (1988) reported that coagulation using a lime-ferrous sulfate combination was effective in removing the organic contaminants of textile wastewaters. Davis et al. (1981) reported that the optimum ferric chloride coagulant dose for a composite dyehouse wastewater was 400 mg/L. Brower and Reed (1986) studied the treatment of textile dye wastes with sodium hydroxide and ferric chloride. Sodium hydroxide was used to minimize additional sludge production at the design pH of 7. Adding sodium hydroxide before coagulation produced a weak floc with little color removed, whereas adding sodium hydroxide after coagulation removed more than 85% of the color. More information on color removal by coagulation is available in the report by Beszedits et al. (1980).

Multimedia filtration performed well in most cases for total suspended solids (TSS) removal. If the TSS value of a wastewater was approximately 100 mg/L or less, multimedia filtration can be expected to be effective as an initial process for TSS reduction. Multimedia filtration was also an effective process for reducing TSS after coagulation/clarification (Storey and Schroeder, 1980).

Granular carbon adsorption worked well for some textile wastewaters, whereas for others it was found that a portion of the organic removal occurred from physical filtering rather than an adsorption mechanism (Storey and Schroeder, 1980). McKay (1984) evaluated a model to explain the adsorption of selected dyes on activated carbon. The feasibility of activated carbon treatment of dyehouse wastewaters frequently depends on costs associated with the regeneration of spent carbon. Thermal regeneration has been the primary means of regenerating granular activated carbon. Posey and Kim (1987) studied the feasibility of solvent regeneration of exhausted activated carbon using methanol as the organic solvent. They found that for the three dye compounds tested, solvent regeneration was not cost-competitive with thermal regeneration because of the large amounts of methanol required.

Dissolved air flotation is an effective pretreatment process for textile mill effluent. The U.S. Environmental Protection Agency (EPA) has conducted a survey and compiled removal data on dissolved air flotation (Wang and Wang, 1990). Table 1 indicates the removal data of dissolved air flotation for treatment of a textile mill effluent. The treated wastewater was from a woven fabric operation with a flow rate of 1730 m^3/day. Cationic polymer was dosed as the flotation aid. It is important to know that BOD removal was over 50%. Suspended solids and phenol removals were 85 and 72%, respectively. The removals of lead, bis(2-ethylexyl)phtalate, di-n-butyl phthalate, and naphthalene were all over 92%.

Ozonation is a very effective oxidation process for color removal. Beszedits et al. (1980) reviewed several studies on ozonation for color removal. A combination of ozonation with other processes like ultraviolet light or fluidized bed systems was found to be more efficient than the individual processes.

Other physical/chemical processes especially applicable to color removal include chlorination, photochemical degradation, treatment by irradiation, ion exchange, liquid–liquid extraction, etc. These processes may show special advantages for specific, small-volume, dye wastewaters.

Table 1 Treatment of Textile Mills Effluent by Dissolved Air Flotation.

Pollutant/parameter	Concentration		Percent removal
	Influent	Effluent	
Classical pollutants (mg/L)			
BOD (5)	400	<200	>50
COD	1000	720	28
TSS	200	32	84
Total phenol	0.092	0.026	72
Toxic pollutants (mg/L)			
Copper	320	81	75
Lead	14	ND	>99
Bis(2-ethylexyl) phthalate	570	45	92
Di-n-butyl phthalate	13	ND	>99
Pentachlorophenol	37	30	19
Phenol	94	26	72
Benzene	18	12	33
Ethylbenzene	460	160	65
Toluene	320	130	59
Naphthalene	250	ND	>99

ND = not detected.

3.7 Biological Treatment

Dyes have to be more and more resistant to ozone, nitric oxides, light, hydrolysis, and other degradative environments to be successful in the commercial market. It is not surprising that most studies on the biological degradation of dyestuffs yield negative results when dyes are designed to resist this type of treatment (Porter, 1970). Of those dyes that are known to undergo biodegradation, the azo dyes are perhaps the most commonly studied, although they tend not to be readily biodegradable in sewage treatment works. Shaul et al. (1986) reported that some azo dyes (e.g., acid yellow 151 and acid red 337) are moderately adsorbed but not biodegraded by the activated sludge process. Azo dyes like acid blue 113 appeared to be strongly adsorbed and possibly biodegraded; acid orange 7 and acid red 88 were moderately adsorbed and significantly biodegraded (Shaul et al., 1986). They also reported that of the various dye removal mechanisms in an activated sludge process, adsorption and/or biodegradation appeared to be the only two removal mechanisms.

In most cases, the biodegradation of dyes does not occur spontaneously in aerobic conditions. Lengthy adaptation periods are generally necessary. It is assumed that despite the presence of oxygen, the first step of degradation is a reduction of the azo bridge to the corresponding amines. In long sewers and primary settling tanks anaerobic conditions occur, and hence, azo reduction is likely (Richardson, 1983). The obligate anaerobic organisms are assumed to be generally capable of reducing azo compounds, but this assumption is yet to be systematically verified for the common azo dyes. It is, however, fairly well established that most of the azo dyes are not toxic to the biological systems. Grady (1972) reported that anaerobic processes are effective in decolorizing many dyes, suggesting that they could be used in the treatment of specific dye streams. The mechanism of color removal was not determined and could have been due to surface adsorption, reduction reactions, microbial activity, or combinations of the three. Kremer (1987) reported that anaerobic digestion is an effective means for the removal of selected monoazo dyes (acid red 88 and acid orange 7), up to 10 mg/L in the presence of a supplemental carbon source. The metabolites of acid red 88 were naphthionic acid, 2-naphthol, 1,2-naphthoquinone, and isoquinoline.

An aerobic process following the anaerobic process should be able to biodegrade the intermediates formed by the anaerobic degradation of the dyes. Such a combination of anaerobic/aerobic processes appears to be a very promising concept that has yet to be proven. At the time of writing, the author was serving as the principal investigator of an EPA project seeking to verify this concept for the treatment of acid orange 10 and acid yellow 151.

The fibers in the textile wastewaters caused operational problems in several of the municipal plants studied by Troxler and Hopkins (1981). These fibers tend to float and form dense mats. Problems caused include fouling of mechanical aerators, plugging of centrifugal pumps, and the formation of dense blankets of floating fibers in an anaerobic digester. A good pretreatment program to remove these fibers at the factory and routine maintenance at the POTWs can reduce these problems.

From the preceding discussion on the physical/chemical and biological treatments of textile wastewaters, it appears that a combination of these two processes should be more efficient than either of the individual processes. Interestingly, several researchers have studied the advantages and disadvantages of such combinations. Randall and King (1980) studied three full-scale treatment plants treating relatively similar dyeing and finishing wastewaters. The first plant consisted of only an extended aeration activated sludge system plus sand dyeing beds. The other two plants combined biological and chemical (lime or aluminum chloride was the primary coagulant) processes in an attempt to attain complete treatment of the textile wastewaters. One design placed the chemical system ahead of the biological unit, whereas the other design positioned the chemical/physical processes after the biological system. Randall and King (1980) concluded that both lime and aluminum chloride are satisfactory primary coagulants and anionic polymers are best as secondary coagulants for the removal of color from textile dyeing and finishing wastewaters. They also concluded that combined physical/chemical and biological processes are required for high removals of both BOD, color, and suspended solids. Randall and King (1980) additionally reported that biological systems preceding chemical processes should be designed to minimize biological solids carry-over in an effort to minimize operating problems.

3.8 Land Treatment of Textile Wastewaters

Overcash and Kendall (1981) studied the feasibility of land treatment of textile wastewaters. They concluded that from an investment perspective, wool scouring with dyeing and finishing, and woven fabric dyeing and finishing require the greatest expenditure for land treatment. At the other extreme, knit fabric dyeing and finishing involve the smallest investment. The same authors' general conclusion was that relative to 1983 industry pollution standards, textile mills which can utilize land treatment will enjoy a considerable economic advantage.

3.9 Summary

The information included in this chapter on the treatment of textile wastewaters encompasses only a fragment of the research in this area. Due to the variability of textile wastewater characteristics, treatability studies should be conducted on a case-by-case basis in order to identify and confirm the required design parameters. For textile wastewater treatment, physical/chemical, biological, or a combination of both processes may be suitable. A combined aerobic/anaerobic treatment may be shown to be a feasible process for the treatment of dye wastewaters. Land treatment may be suitable for some dye wastes. After identifying the technically feasible processes, the final selection should be based on the economics of the processes.

REFERENCES

Abo-Elela, S. I., El-Gohary, F. A., Ali, H. I., and Wahaab, R. S. A. (1988). Treatability studies of textile wastewater. *Environ. Technol. Lett., 9*, 101–108.

Abrahart, E. N. (1977). *Dyes and Their Intermediates*, 2nd ed. Chemical Publishing, New York.

Allen, R. L. M. (1971). *Colour Chemistry*. Appleton-Century-Crofts, Meredith Corp., New York.

American Dye Manufacturers Institute (1973). *Dyes and the Environment, Reports on Selected Dyes and Their Effects*s, Vol. 1.

Beszedits, S., Lugowski, A., and Miyamoto, H. K. (1980). Color Removal from Textile Mill Effluents. B & L Inform. Services, Toronto, Canada.

Brower, G. R. and Reed, G. D. (1986). Economical pretreatment for color removal from textile dye wastes. In *Proceedings of 41st Industrial Waste Conference*, 612.

Davis, C. H., Smith, M. J., Turschmid, K. H., and Main, C. T. (1981). Innovative technology in textile wastewater treatment. In *The Textile Industry and the Environment*, Environ. Symp. Amer. Assoc. Text. Chemists and Colorists, Washington, D.C.

Grady, C. P. L., Jr. (1972). Effects of Dyes on the Anaerobic Digestion of Primary Sewage Sludge. Final rep., Amer. Dye Manufacturers Inst.

Groves, G. R. and Buckley, C. A. (1980). Treatment of textile effluents by membrane separation processes. In *Proceedings of 7th International Symposium on Fresh Water from the Sea*, Athens, Greece, Vol. 2, p. 249.

Helemes, C. T. (1984). A study of azo and nitro dyes for the selection of candidates for carcinogen bioassay. *J. Environ. Sci. and Health, A19*, 2, 97.

Hoffman, C. R. and Zakha, J. (1978). Ultrafiltration of latex emulsions. Presented at Latex Technol., and Appl. Seminar, Clemson Univ., Clemson, S.C., Aug.

Jones, H. R. (1973). *Pollution Control in the Textile Industry*. Noyes Data Corp., Park Ridge, N.J.

Junkins, R. (1982). Case history: Pretreatment of textile wastewater. In *Proceedings of 37th Industrial Waste Conference*, Purdue Univ., Lafayette, Ind., Vol. 37, p. 139.

Kertell, C. R. and Hill, G. F. (1982). Textile dyehouse wastewater treatment: A case history. In *Proceedings of 37th Industrial Waste Conference*, Purdue Univ., Lafayette, Ind., Vol. 37, p. 147.

Kremer, F. (1987). Anaerobic degradation of monoazo dyes. Ph.D. diss., Civil and Environ. Eng. Dept., Univ. of Cincinnati, Ohio.

Kremer, F., Brromfield, B., and Fradlin, L. (1982). Energy and materials recovery options for the textile industry. In *Proceedings of 37th Industrial Waste Conference*, Purdue Univ., Lafayette, Ind., Vol. 37, p. 157.

McKay, G. (1984). The adsorption of dyestuffs from aqueous solutions using the activated carbon adsorption model to determine breakthrough curves. *Chem. Eng. J., 28*, 95.

Newlin, K. D. (1969). An Economic analysis of treatment of textile wastes in municipal plants. M.S. thesis, Clemson Univ., Clemson, S.C., May.

Overcash, M. R. and Kendall, R. L. (1981). Assessment of land treatment for the textile mill industry. In *Proceedings of 36th Industrial Waste Conference*, Purdue Univ., Lafayette, Ind., Vol. 36, p. 766.
Porter, J. J. (1970). The changing nature of textile processing and waste treatment technology. *Text. Chemist and Colorists, 2*, 336.
Posey, R. J. and Kim, B. R. (1987). Solvent regeneration of dye-laden activated carbon. *J. Water Poll. Control Fed., 59*, 1, 47.
Randall, C. W. and King, P. H. (1980). Full scale physical-chemical-biological treatment of textile wastewaters. *Prog. Wat. Tech., 12*, 3, 231.
Richardson, M. L. (1983). Dyes—The aquatic environment and the mess made by metabolites. *JSDC, 19*.
Shaul, G. M., Dempsey, C. R., Dostal, K. A., and Lieberman, R. J. (1986). Fate of azo dyes in the activated sludge process. In *Proceedings of 41st Industrial Waste Conference*, Purdue Univ., Lafayette, Ind., Vol. 41, p. 603.
Sherve, R. N. (1967). *Chemical Process Industries*, 3rd ed., p. 824.
Slade, F. H. (1968). Process water and textile effluent problems, part 2. *Text. Manufacturer, 94*, 3, 89.
Stahr, R. W., Boepple, C. P., and Knocke, W. R. (1980). Textile waste treatment: Color removal and solids handling characteristics. In *Proceedings of 35th Industrial Waste Conference*, Purdue Univ., Lafayette, Ind., Vol. 35, p. 186.
Storey, W. A. and Schroeder, E. J. (1980). Pilot plant evaluation of the 1974 BATEA guidelines for the textile industry. In *Proceedings of 35th Industrial Waste Conference*, Purdue Univ., Lafayette, Ind., Vol. 35, p. 160.
Tinghul, L., Matsuura, T., and Sourirajan, S. (1983). Effect of membrane materials and average pore sizes on reverse osmosis separation of dyes. *Ind. Eng. Chem. Prod. Res. Dev., 22*, 77.
Troxler, R. W. and Hopkins, K. S. (1981). Case histories: Carpet manufacturing wastewater treatment in municipal plants. In *Proceedings of 36th Industrial Waste Conference*, Purdue Univ., Lafayette, Ind., Vol. 36, p. 755.
Wang, L. K. and Wang, M. H. S. (1990). Decontamination of groundwater and hazardous industrial effluents by high-rate air flotation processes. presented at Great Lakes '90 Conf., Hazardous Materials Control Res. Inst., Silver Springs, Md., Sept.

8

Treatment of Phosphate Industry Wastes

Constantine Yapijakis

The Cooper Union, New York, New York

1. INTRODUCTION

The phosphate manufacturing and phosphate fertilizer industry includes the production of elemental phosphorus, various phosphorus-derived chemicals, phosphate fertilizer chemicals, and other nonfertilizer phosphate chemicals. Chemicals that are derived from phosphorus include phosphoric acid (dry process), phosphorus pentoxide, phosphorus pentasulfide, phosphorus trichloride, phosphorus oxychloride, sodium tripolyphosphate, and calcium phosphates (EPA, 1977). The nonfertilizer phosphate production part of the industry includes defluorinated phosphate rock, defluorinated phosphoric acid, and sodium phosphate salts. The phosphate fertilizer segment of the industry produces the primary phosphorus nutrient source for the agricultural industry and for other applications of chemical fertilization. Many of these fertilizer products are toxic to aquatic life at certain levels of concentration, and many are also hazardous to human life and health when contact is made in a concentrated form.

1.1 Sources of Raw Materials

The basic raw materials used by the phosphorus chemicals, phosphates, and phosphate fertilizer manufacturing industry are mined phosphate rock and phosphoric acid produced by the wet process.

Ten to 15 millions of years ago, many species of marine life withdrew minute forms of phosphorus dissolved in the oceans, combined with such substances as calcium, limestone, and quartz sand, in order to construct their shells and bodies (Specht, 1960). When these multitudes of marine organisms died, their shells and bodies (along with sea-life excretions and inorganic precipitates) settled to the ocean bottom where thick layers of such deposits—containing phosphorus among other things—were eventually formed. Land areas that formerly were at the ocean bottom millions of years ago and where such large deposits have been discovered are now being commercially mined for phosphate rock. About 70% of the world supply of phosphate rock comes from such an area around Bartow in central Florida, which was part of the Atlantic Ocean 10 million years ago (Anonymous, 1951). Other significant phosphate rock mining and processing operations can be found in Jordan, Algeria, and Morocco (Shahalam et al., 1985).

1.2 Characteristics of Phosphate Rock Deposits

According to a literature survey conducted by Shahalam et al. (1985), the contents of various chemicals found in the natural mined phosphate rocks vary widely, depending on location, as is shown in Table 1. For instance, the mineralogical and chemical analyses of low-grade hard phosphate from the different mined beds of phosphate rock in the Rusaifa area of Jordan indicated that the phosphates are of three main types: carbonate, siliceous, and silicate-carbonate. Phosphate deposits in this area exist in four distinct layers, of which the two deepest—first and second—ones (the thickness of bed is about 3 and 3.5 m, respectively, and depth varies from about 20 to 30 m) appear to be suitable for a currently cost-effective mining operation. A summary of the data from chemical analyses of the ores is shown in Table 2 (Shahalam et al., 1985).

Screen tests of the size fraction obtained from rocks mined from these beds, which were crushed through normal crushers of the phosphate processing plant in the area, indicated that the best recovery of phosphate in the first (deepest) bed is obtained from phosphate gains recovered at grain sizes of mesh 10 to 20 (standard). The high dust (particles of less than 200 mesh) portion of 11.60% by wt. of the ores remains as a potential air pollution source; however, the

Table 1 Range of Concentrations of Various Chemicals in Phosphate Ores

Chemical	Range
Fluorine	2.8–5.6%[a]
Sulphur (SO_3)	0.8–7.52%[a]
Carbon (CO_2)	2.07–10.7%[a]
Strontium	180–1683 ppm[b]
Manganese	0.001–0.004%[a]
Barium	0.044–0.40%[a]
Chlorine	0.20–1.42%[a]
Zinc	59–765 ppm[b]
Nickel	7–244 ppm[b]
Cobalt	31–34 ppm[b]
Chromium	12–895 ppm[b]
Copper	18–46 ppm[b]
Vanadium	0.03–0.08%[a]
Cadmium	0.038–1.5 ppm[b]
Uranium	4–8 ppm[b]
P_2O_5	40–55%[c]
Silica	3–34%[c]
Carbon (C)	14–48%[c]

[a]% by wt.
[b]Parts per million.
[c]Kusaifa Rocks only (% by wt.).
Source: Shahalam et al. (1985).

chemical analyses of these ores showed that crushing to smaller grain sizes tends to increase phosphate recovery. The highest percentage of phosphate from the second bed (next deepest) is also recovered from grain sizes of 10 to 20 mesh; however, substantial amounts of phosphate are also found in sizes of 40 to 100 mesh. Currently, the crushing operation usually maintains a maximum grain size between 15 to 30 mesh.

The phosphate rock deposits in the Florida region are in the form of small pebbles embedded in a matrix of phosphatic sands and clays (Wakefield, 1952). These deposits are overlain with lime rock and nonphosphate sands and can be found at varying depths from a few feet to hundreds of feet, although the current economical mining operations seldom reach beyond 18.3 m (60 ft) of depth.

Table 2 Chemical Analysis of Different Size Fractions of Phosphate in Mining Beds at Rusaifa

Size fractions in mesh	First bed, average chemical composition			Second bed, average chemical composition			Fourth bed, average chemical composition		
	$P_2O_5\%$	$CaCO_3\%$	Insoluble %	$P_2O_5\%$	$CaCO_3\%$	Insoluble %	$P_2O_5\%$	$CaCO_3\%$	Insoluble %
<10	21.35	10.01	36.46	15.73	35.64	23.35	17.99	47.95	2.08
10 + 20	21.03	10.67	36.94	18.04	32.60	21.26	21.33	36.58	2.25
20 + 30	21.03	10.46	37.30	19.63	29.23	19.82	26.95	30.81	2.48
30 + 40	21.82	10.56	36.21	21.80	25.24	17.72	29.38	28.51	2.00
40 + 60	22.04	10.65	33.70	25.96	22.40	14.78	32.26	20.66	2.49
60 + 100	24.12	11.27	29.64	26.65	18.13	11.75	30.88	26.76	1.23
100 + 150	24.32	10.90	28.64	26.76	21.71	13.17	26.55	33.51	1.41
150 + 200	25.92	11.95	24.30	24.46	23.18	14.45	23.18	40.25	2.52
>200	25.28	12.50	23.42	23.37	28.06	12.99	21.29	40.25	4.24
Average total sample	Chemical					25.19			
Mineralogical	47.0	31.0[a]	18.0[b]	27.0[a]	50.0	15.0[b]	55.0	35.0[a]	3.5[b]

[a]Carbonaceous materials.
[b]Silica.
Source: Shahalam et al. (1985).

1.3 Mining and Phosphate Rock Processing

Mechanized open-cut mining is used to first strip off the overburden and then to excavate in strips the exposed phosphate rockbed matrix. In the Rusaifa area of Jordan, the stripping ratio of overburden to phosphate rock is about 7 to 1 by wt. (Shahalam et al., 1985). Following crushing and screening of the mined rocks in which the dust (less than 200 mesh) is rejected, they go through "beneficiation" processing. The unit processes involved in this wet treatment of the crushed rocks for the purpose of removing the mud and sand from the phosphate grains include slurrification, wet screening, agitation and hydrocycloning in a two-stage operation, followed by rotating filtration and thickening, with a final step of drying the phosphate rocks and separating the dusts. The beneficiation plant makes use of about 85% of the total volume of process water used in phosphate rock production.

Phosphate rocks from crushing and screening, which contain about 60% tricalcium phosphate, are fed into the beneficiation plant for upgrading by rejection of the larger than 4 mm over-size particles. Two stages of agitation follow the hydrocycloning, the underflow of which (over 270 mesh particles) is fed to rotary filters from which phosphatic cakes result (with 16–18% moisture). The hydrocyclone overflow contains undesirable slimes of silica carbonates and clay materials and is fed to gravity thickeners. The thickener underflow consisting of wastewater and slimes is directly discharged, along with wastewater from dust-removing cyclones in the drying operation, into the nearby river.

In a typical mining operation in Florida, the excavated phosphate rockbed matrix is dumped into a pit where it is slurrified by mixing it with water and subsequently carried to a washer plant (Wakefield, 1952). In this operation, the larger particles are separated by the use of screens, shaker tables, and size-separation hydrocyclone units. The next step involves recovery of all particles larger than what is considered dust, i.e., 200 mesh, through the use of both clarifiers for hydraulic sizing and a flotation process in which selective coating (using materials such as caustic soda, fuel oil, and a mixture of fatty acids and resins from the manufacture of chemical wood pulp known as tall oil, or resin oil from the flotators) of phosphate particles takes place after pH adjustment with NaOH.

The phosphate concentration in the tailings is upgraded to a level adequate for commercial exploitation through removal of the nonphosphate sand particles by flotation, in which the silica solids are selectively coated with an amine and floated off following a slurry dewatering and sulfuric acid treatment step. The commercial quality, kiln-dried phosphate rock product is sold directly as fertilizer, processed to normal superphosphate or triple superphosphate, or burned in electric furnaces to produce elemental phosphorus or phosphoric acid, as described in Sec. 2.

2. INDUSTRIAL OPERATION AND WASTEWATER

The phosphate manufacturing and phosphate fertilizer industry is a basic chemical manufacturing industry, in which essentially both the mixing and chemical reactions of raw materials are involved in production. Also, short- and long-term chemical storage and warehousing, as well as loading/unloading and transportation of chemicals, are involved in the operation. In the case of fertilizer production, only the manufacturing of phosphate fertilizers and mixed and blend fertilizers containing phosphate along with nitrogen and/or potassium is presented here.

Regarding wastewater generation, volumes resulting from the production of phosphorus are several orders of magnitude greater than the wastewaters generated in any of the other product categories. Elemental phosphorus is an important wastewater contaminant common to all segments of the phosphate manufacturing industry, if the phossy water (water containing colloidal phosphorus) is not recycled to the phosphorus production facility for reuse.

2.1 Categorization in Phosphate Production

As previously mentioned, the phosphate manufacturing industry is broadly subdivided into two main categories: phosphorus-derived chemicals and other nonfertilizer phosphate chemicals. For the purposes of raw waste characterization and delineation of pretreatment information, the industry is further subdivided into six subcategories. The following categorization system (Table 3) of the various main production streams and their descriptions are taken from the federal guidelines (EPA, 1977) pertaining to state and local industrial pretreatment programs. It will be used in the following discussion to identify process flows and characterize the resulting raw waste. Figure 1 shows a flow diagram for the production streams of the entire phosphate manufacturing industry.

The manufacture of phosphorus-derived chemicals is almost entirely based on the production of elemental phosphorus from mined phosphate rock. Ferrophosphorus, widely used in the metallurgical industries, is a direct by-product of the phosphorus production process. In the United States, over 85% of

Table 3 Categorization System in Phosphorus-Derived and Nonfertilizer Phosphate Chemicals Production

Main category	Subcategory	Code
1. Phosphorus-derived chemicals	Phosphorus production	(A)
	Phosphorus-consuming	(B)
	Phosphate	(C)
2. Other nonfertilizer phosphate chemicals	Defluorinated phosphate rock	(D)
	Defluorinated phosphoric acid	(E)
	Sodium phosphates	(F)

Source: EPA (1977).

elemental phosphorus production is used to manufacture high-grade phosphoric acid by the furnace or dry process as opposed to the wet process that converts phosphate rock directly into low-grade phosphoric acid. The remainder of the elemental phosphorus is either marketed directly or converted into phosphorus chemicals. The furnace-grade phosphoric acid is marketed directly, mostly to the food and fertilizer industries. Finally, phosphoric acid is employed to manufacture sodium tripolyphosphate, which is used in detergents and for water treatment, and calcium phosphate, which is used in foods and animal feeds.

On the other hand, defluorinated phosphate rock is utilized as an animal feed ingredient. Defluorinated phosphoric acid is mainly used in the production of animal foodstuffs and liquid fertilizers. Finally, sodium phosphates, produced from wet process acid as the raw material, are used as intermediates in the production of cleaning compounds.

2.2 Phosphorus and Phosphate Compounds

Phosphorus Production

Phosphorus is manufactured by the reduction of commercial-quality phosphate rock by coke in an electric furnace, with silica used as a flux. Slag, ferrophosphorus (from iron contained in the phosphate rock), and carbon monoxide are reaction by-products. The standard process, as shown in Fig. 2, consists of three basic parts: phosphate rock preparation, smelting in an electric furnace, and recovery of the resulting phosphorus. Phosphate rock ores are first blended

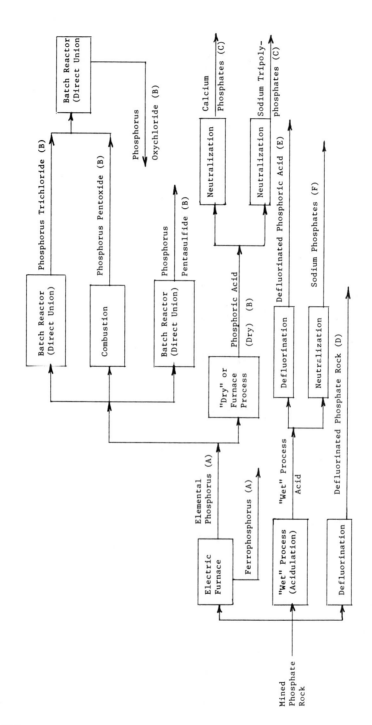

Figure 1 Phosphate manufacturing industry flow diagram. (After EPA, 1977.)

330

Treatment of Phosphate Industry Wastes

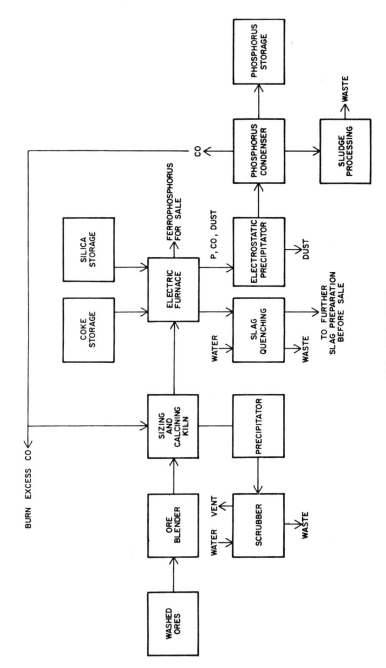

Figure 2 Standard phosphorus process flow diagram. (After EPA, 1977.)

so that the furnace feed is of uniform composition and then pretreated by heat drying, sizing or agglomerating the particles, and heat treatment.

The burden of treated rock, coke, and sand is fed to the furnace (which is extensively water-cooled) by incrementally adding weighed quantities of each material to a common conveyor belt. Slag and ferrophosphorus are tapped periodically, whereas the hot furnace gases (90% CO and 10% phosphorus) pass through an electrostatic precipitator that removes the dust before phosphorus condensation. The phosphorus is condensed by direct impingement of a hot water spray, sometimes enhanced by heat transfer through water-cooled condenser walls. Liquid phosphorus drains into a water sump, where the water maintains a seal from the atmosphere. Liquid phosphorus is stored in steam-heated tanks under a water blanket and transferred into tank cars by pumping or hot water displacement. The tank cars have protective blankets of water and are equipped with steam coils for remelting at the destination.

There are numerous sources of fumes from the furnace operation, such as dust from the raw materials feeding and fumes emitted from electrode penetrations and tapping. These fumes that consist of dust, phosphorus vapor (immediately oxidized to phosphorus pentoxide), and carbon monoxide are collected and scrubbed. Principal wastewater streams consist of calciner scrubber liquor, phosphorus condenser and other phossy water, and slag-quenching water.

Phosphorus Consuming

This subcategory involves phosphoric acid (dry process), phosphorus pentoxide, phosphorus pentasulfide, phosphorus trichloride, and phosphorus oxychloride. In the standard dry process for phosphoric acid production, liquid phosphorus is burned in the air, the resulting gaseous phosphorus pentoxide is absorbed and hydrated in a water spray, and the mist is collected with an electrostatic precipitator. Regardless of the process variation, phosphoric acid is made with the consumption of water and no aqueous wastes are generated by the process.

Solid anhydrous phosphorus pentoxide is manufactured by burning liquid phosphorus in an excess of dried air in a combustion chamber and condensing the vapor in a roomlike structure. Condensed phosphorus pentoxide is mechanically scraped from the walls using moving chains and is discharged from the bottom of the barn with a screw conveyor. Phosphorus pentasulfide is manufactured by directly reacting phosphorus and sulfur, both in liquid form, in a highly exothermic batch operation. Since the reactants and products are highly flammable at the reaction temperature, the reactor is continuously purged with nitrogen and a water seal is used in the vent line.

Phosphorus trichloride is manufactured by loading liquid phosphorus into a jacketed batch reactor. Chlorine is bubbled through the liquid, and phosphorus trichloride is refluxed until all the phosphorus is consumed. Cooling water is used in the reactor jacket and care is taken to avoid an excess of chlorine and the resulting formation of phosphorus pentachloride. Phosphorus oxychloride is manufactured by the reaction of phosphorus trichloride, chlorine, and solid phosphorus pentoxide in a batch operation. Liquid phosphorus trichloride is loaded to the reactor, solid phosphorus pentoxide added, and chlorine bubbled through the mixture. Steam is supplied to the reactor jacket, water to the reflux condenser is shut off, and the product is distilled over and collected.

Because phosphorus is transported and stored under a water blanket, phossy water is a raw waste material at phosphorus-consuming plants. Another source of phossy wastewater results when reactor contents (containing phosphorus) are dumped into a sewer line due to operator error, emergency conditions, or inadvertent leaks and spills.

Phosphate

This subcategory involves sodium tripolyphosphate and calcium phosphates. Sodium tripolyphosphate is manufactured by the neutralization of phosphoric acid by soda ash or caustic soda and soda ash, with the subsequent calcining of the dried mono- and di-sodium phosphate crystals. This product is then slowly cooled or tempered to produce the condensed form of the phosphates.

The nonfertilizer calcium phosphates are manufactured by the neutralization of phosphoric acid with lime. The processes for different calcium phosphates differ substantially in the amount and type of lime and amount of process water used. Relatively pure, food-grade monocalcium phosphate (MCP), dicalcium phosphate (DCP), and tricalcium phosphate (TCP) are manufactured in a stirred batch reactor from furnace-grade acid and lime slurry, as shown in the process flow diagram of Fig. 3. Dicalcium phosphate is also manufactured for livestock feed supplement use, with much lower specifications on product purity.

Sodium tripolyphosphate manufacture generates no process wastes. Wastewaters from the manufacture of calcium phosphates are generated from a dewatering of the phosphate slurry and wet scrubbing of the airborne solids during product operations.

Defluorinated Phosphate Rock

The primary raw material for the defluorination process is fluorapatite phosphate rock. Other raw materials used in much smaller amounts, but critical to

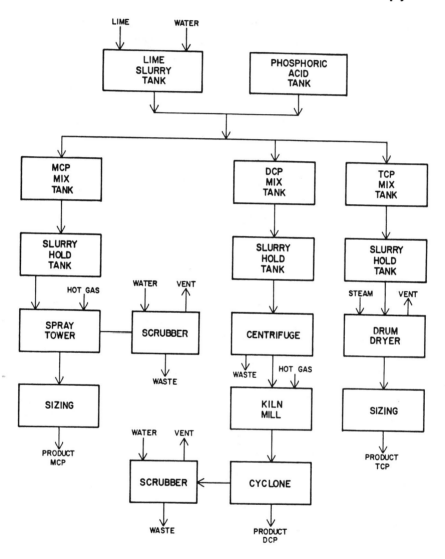

Figure 3 Standard process for food-grade calcium phosphates. (After EPA, 1977.)

Treatment of Phosphate Industry Wastes

the process, are sodium-containing reagents, wet process phosphoric acid, and silica. These are fed into either a rotary kiln or a fluidized bed reactor that requires a modular and predried charge. Reaction temperatures are maintained in the 1205 to 1366°C range, whereas the retention time varies from 30 to 90 min. From the kiln or fluidized bed reactor, the defluorinated product is quickly quenched with air or water, followed by crushing and sizing for storage and shipment. A typical flow diagram for the fluidized bed process is shown in Fig. 4.

Wastewaters are generated in the process of scrubbing contaminants from gaseous effluent streams. This water requirement is of significant volume and process conditions normally permit the use of recirculated contaminated water for this service, thereby effectively reducing the discharged wastewater volume. Leaks and spills are routinely collected as part of process efficiency and housekeeping and, in any case, their quantity is minor and normally periodic.

Defluorinated Phosphoric Acid

One method used in order to defluorinate wet process phosphoric acid is vacuum evaporation. The concentration of 54% P_2O_5 acid to a 68 to 72% P_2O_5 strength is performed in vessels that use high-pressure (30.6–37.4 at or 450–550 psig) steam or an externally heated Dowtherm solution as the heat energy source for evaporation of water from the acid. Fluorine removal from the acid occurs concurrently with the water vapor loss. A typical process flow diagram for vacuum-type evaporation is shown in Fig. 5.

A second method of phosphoric acid defluorination entails the direct contact of hot combustion gases (from fuel oil or gas burners) with the acid by bubbling them through the acid. Evaporated and defluorinated product acid is sent to an acid cooler, while the gaseous effluents from the evaporation chamber flow to a series of gas scrubbing and absorption units. Finally, aeration can also be used for defluorinating phosphoric acid. In this process, diatomaceous silica or spray-dried silica gel is mixed with commercial 54% P_2O_5 phosphoric acid. Hydrogen fluoride in the impure phosphoric acid is converted to fluosilicic acid, which in turn breaks down to SiF_4 and is stripped from the heated mixture by simple aeration.

The major wastewater source in the defluorination processes is the wet scrubbing of contaminants from the gaseous effluent streams. However, process conditions normally permit the use of recirculated contaminated water for this service, thereby effectively reducing the discharged wastewater volume.

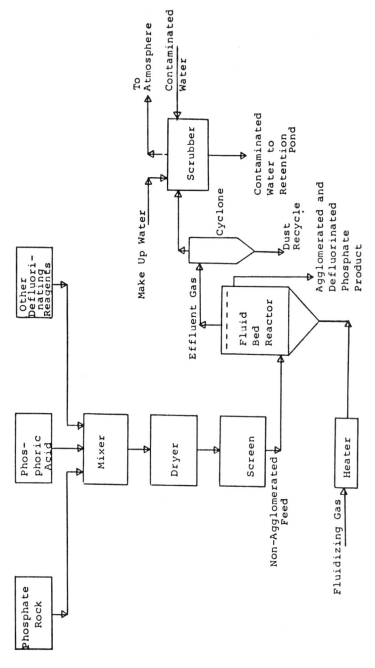

Figure 4 Defluorinated phosphate rock fluid bed process. (After EPA, 1977.)

Treatment of Phosphate Industry Wastes

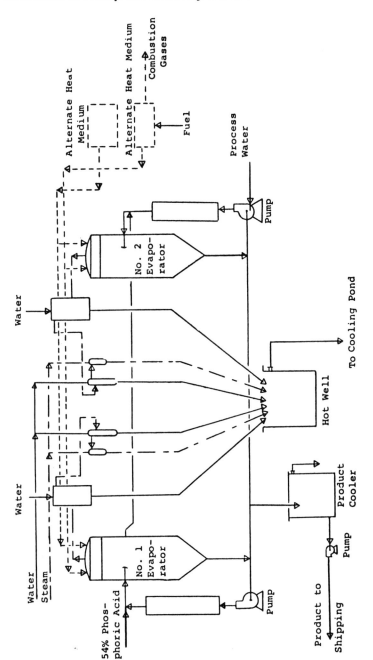

Figure 5 Defluorinated phosphoric acid vacuum process. (After EPA, 1977.)

Sodium Phosphates

In the manufacture of sodium phosphates, the removal of contaminants from the wet process acid takes place in a series of separate neutralization steps. The first step involves the removal of fluosiliccates with recycled sodium phosphate liquor. The next step precipitates the minor quantities of arsenic present by adding sodium sulfide to the solution, while barium carbonate is added to remove the excess sulfate. The partially neutralized acid still contains iron and aluminum phosphates, and some residual fluorine.

A second neutralization is carried out with soda ash to a pH level of about 4. Special heating, agitation, and retention are next employed to adequately condition the slurry so that filtration separation of the contaminants can be accomplished. The remaining solution is sufficiently pure for the production of monosodium phosphate that can be further converted into other compounds such as sodium metaphosphate, disodium phosphate, and trisodium phosphate. A typical process flow diagram is shown in Fig. 6. Wastewater effluents from these processes originate from leaks and spills, filtration backwashes, and gas scrubber wastewaters.

2.3 Categorization in Phosphate Fertilizer Production

The fertilizer industry comprises nitrogen-based, phosphate-based, and potassium-based fertilizer manufacturing, as well as combinations of these nutrients in mixed and blend fertilizer formulations. Only the phosphate-based fertilizer industry is discussed here and, therefore, the categorization mainly involves two broad divisions: (1) the phosphate fertilizer industry (A) and (2) the mixed and blend fertilizer industry (G) in which one of the components is a phosphate compound. The following categorization system of the various separate processes and their production streams and descriptions is taken from the federal guidelines (EPA, 1977) pertaining to state and local industrial pretreatment programs. It will be used in the discussion that ensues to identify process flows and characterize the resulting raw waste. Figure 7 shows a flow diagram for the production streams of the entire phosphate and nitrogen fertilizer manufacturing industry.

2.4 Phosphate and Mixed and Blend Fertilizer Manufacture

Phosphate Fertilizer (A)

The phosphate fertilizer industry is defined as eight separate processes: phosphate rock grinding, wet process phosphoric acid, phosphoric acid

Treatment of Phosphate Industry Wastes

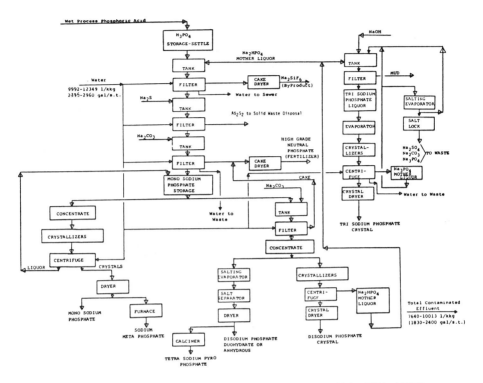

Figure 6 Sodium phosphate process from wet process. (After EPA, 1977.)

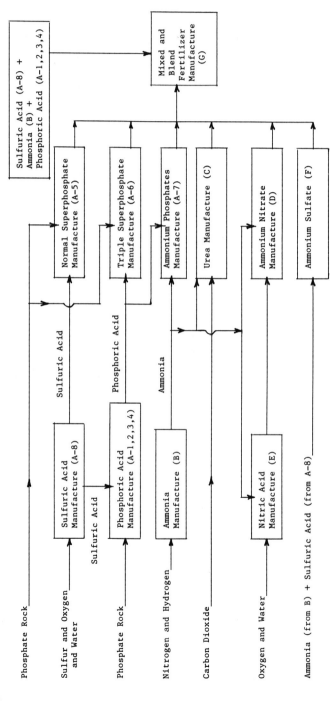

Figure 7 Flow diagram of fertilizer products manufacturing. (After EPA, 1977.)

concentration, phosphoric acid clarification, normal superphosphate, triple superphosphate, ammonium phosphate, and sulfuric acid. Practically all phosphate manufacturers combine the various effluents into a large recycle water system. It is only when the quantity of recycle water increases beyond the capacity to contain it that effluent treatment is necessary.

Phosphate Rock Grinding. Phosphate rock is mined and mechanically ground to provide the optimum particle size required for phosphoric acid production. There are no liquid waste effluents.

Wet Process Phosphoric Acid. A production process flow diagram is shown in Fig. 8. Insoluble phosphate rock is changed to water-soluble phosphoric acid by solubilizing the phosphate rock with an acid, generally sulfuric or nitric. The phosphoric acid produced from the nitric acid process is blended with other ingredients to produce a fertilizer, whereas the phosphoric acid produced from the sulfuric acid process must be concentrated before further use. Minor quantities of fluorine, iron, aluminum, silica, and uranium are usually the most serious waste effluent problems.

Phosphoric Acid Concentration. Phosphoric acid produced with sulfuric acid cannot be used for processing due to its very low concentration. It is, therefore, concentrated to 50 to 54% by evaporation. Waste streams contain fluorine and phosphoric acid.

Phosphoric Acid Clarification. When the phosphoric acid has been concentrated, iron and aluminum phosphates, gypsum and fluorosilicates become insoluble and can pose problems during acid storage. They are, therefore, removed by clarification and/or centrifugation.

Normal Superphosphate. Normal superphosphate is produced by the reaction between ground phosphate rock and sulfuric acid, followed by three to eight weeks of curing time. Obnoxious gases are generated by this process.

Triple Superphosphate (TSP). Triple superphosphate is produced by the reaction between ground phosphate rock and phosphoric acid by either of two processes. One utilizes concentrated phosphoric acid and generates obnoxious gases. The dilute phosphoric acid process permits the ready collection of dusts and obnoxious gases generated.

Ammonium Phosphate. Ammonium phosphate, a concentrated water-soluble plant food, is produced by reacting ammonia and phosphoric acid. The resultant slurry is dried, stored, and shipped to marketing.

Sulfuric Acid. Essentially, all sulfuric acid manufactured in this industry is produced by the "contact" process, in which SO_2 and oxygen contact each other on the surface of a catalyst (vanadium pentoxide) to form SO_3 gas. Sulfur

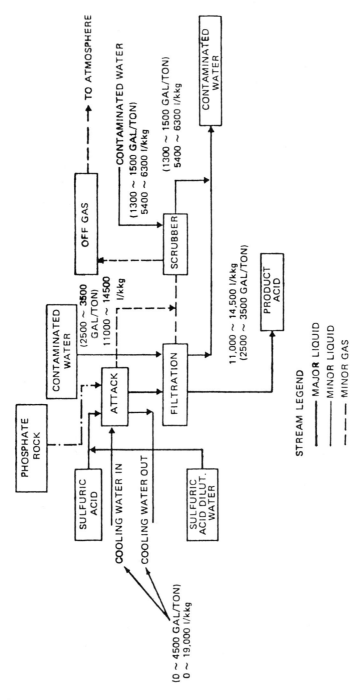

Figure 8 Wet process phosphoric acid-H_2SO_4 acidulation. (After EPA, 1977.)

Treatment of Phosphate Industry Wastes

trioxide gas is added to water to form sulfuric acid. The sulfur dioxide used in the process is produced by burning elemental sulfur in a furnace.

In addition, the process is designed to capture a high percentage of the energy released by the exothermic reactions occurring in the oxidation of sulfur to sulfur trioxide. This energy is used to produce steam that is then utilized for other plant unit operations or converted to electrical energy. It is the raw water treatment necessary to condition water for this steam production that generates essentially all the wastewater effluent from this process.

Mixed and Blend Fertilizer (G)

Mixed Fertilizer. The raw materials used to produce mixed fertilizers include inorganic acids, solutions, double nutrient fertilizers, and all types of straight fertilizers. The choice of raw materials depends on the specific nitrogen, phosphate, potassium (N-P-K) formulation to be produced and on the cost of the different materials from which they can be made.

The mixed fertilizer process involves the controlled addition of both dry and liquid raw materials to a granulator, which is normally a rotary drum, but pug mills are also used. Raw materials, plus some recycled product material, are mixed to form an essential homogeneous granular product. Wet granules from the granulator are discharged into a rotary drier where the excess water is evaporated and dried granules from the drier are then sized on vibrating screens. Over- and under-sized granules are separated for use as recycle material in the granulator. Commercial-product-size granules are cooled and then conveyed to storage or shipping.

Blend Fertilizer. Raw materials used to produce blend fertilizers are a combination of granular dry straight and mixed fertilizer materials with an essentially identical particle size. Although many materials can be utilized, the five most commonly used in this process are ammonium nitrate, urea, triple superphosphate, diammonium phosphate, and potash. These raw materials are stored in a multicompartmented bin and withdrawn in the precise quantities needed to produce the nitrogen-phosphorus-potassium (N-P-K) formulation desired. Raw material addition is normally done by batch weighing, and the combination of batch-weighed and granular raw materials is then conveyed to a mechanical blender for mixing. From the blender, the product is conveyed to storage or shipping.

2.5 Wastewater Characteristics and Sources

Wastewaters from the manufacturing, processing, and formulation of inorganic chemicals such as phosphorus compounds, phosphates, and phosphate fertilizers cannot be exactly characterized. The wastewater streams are usually expected to contain trace or large concentrations of all raw materials used in the plant; all intermediate compounds produced during manufacture; all final products, co-products, and by-products; and the auxiliary or processing chemicals employed. It is desirable from the viewpoint of economics that these substances not be lost, but some losses and spills appear unavoidable and some intentional dumping does take place during housecleaning, vessel emptying, and preparation operations.

The federal guidelines (EPA, 1977) for state and local pretreatment programs reported the raw wastewater characteristics (Table 4) in mg/L concentration, and flows and quality parameters (Table 5) based on the production of 1 ton of the product manufactured, for each of the six subcategories of the phosphate manufacturing industry. Few fertilizer plants discharge wastewaters to municipal treatment systems. Most use ponds for the collection and storage of wastewaters, pH control, chemical treatment, and settling of suspended solids. Whenever available retention pond capacities in the phosphate fertilizer industry are exceeded, the wastewater overflows are treated and discharged to nearby surface water bodies. The federal guidelines (EPA, 1977) reported the range of wastewater characteristics (Table 6) in mg/L concentrations for typical retention ponds used by the phosphate fertilizer industry.

The specific types of wastewater sources in the phosphate fertilizer industry are (1) water treatment plant wastes from raw water filtration, clarification, softening and deionization, which principally consist of only the impurities removed from the raw water (such as carbonates, hydroxides, bicarbonates, and silica) plus minor quantities of treatment chemicals; (2) closed-loop cooling tower blowdown, the quality of which varies with the makeup of water impurities and inhibitor chemicals used [the only cooling water contamination from process liquids is through mechanical leaks in heat exchanger equipment, and Table 7 shows the normal range of contaminants that may be found in cooling water blowdown systems (Robasky and Koraido, 1973)]; (3) boiler blowdown, which is similar to cooling tower blowdown but the quality differs as shown in Table 8 (Robasky and Koraido, 1973); (4) contaminated water or gypsum pond water, which is the impounded and reused water that accumulates sizable concentrations of many cations and anions, but mainly fluorine and phosphorus [concentrations of 8500 mg/L F and in excess of 5000 mg/L P are not unusual; concentrations of radium 226 in recycled gypsum pond water are 60 to 100 picocuries/L, and its acidity reaches extremely high levels (pH 1–2)]; (5) wastewater from spills and leaks that, when possible, is reintroduced

Table 4 Phosphate Manufacturing Industry Raw Waste Characteristics

Parameter (mg/L)	Phosphorus production A	Phosphorus consuming B	Phosphate C	Defluorinated phosphate rock D	Defluorinated phosphoric acid E	Sodium phosphate F
Flow type	C	B	B	B	B	B
BOD5				3	15	31
SS	100		24,000–54,000	16	30	460
TDS			1900–7000[a]	2250[a]	28,780[a]	1640[a]
COD				48	306	55
pH				1.65[a]	1.29[a]	7.8
Phosphorus	21					
PO4	59		7000[a]			
SO4	260			350	4770	240
F	126			1930	967	15
HCl		0–800				
H2SO3		0–34				
H3PO3 + H3PO4		17–500				
HF, H2SiF6, H2SiO3			1900[a]			
Chloride				101	65	
Calcium				40	1700[a]	90
Magnesium				12	106	95
Aluminum				58	260	
Iron				8[a]	180[a]	
Arsenic				0.38[a]	0.83[a]	
Zinc				5.2[a]	5.3[a]	
Total acidity	128					
Total phosphorus				600	5590[a]	250

[a] In high levels, these parameters may be inhibitory to biological systems.
B = Batch process.
C = Continuous process.
Source: EPA (1977).

Table 5 Phosphate Manufacturing Industry Raw Waste Characteristics Based on Production

Parameter (kg/kkg)	Phosphorus production A	Phosphorus consuming B	Phosphate C	Defluorinated phosphate rock D	Defluorinated phosphoric acid E	Sodium phosphate F
Flow range (L/kkg)	425,000	38,000	10,920	45,890	18,020–70,510	7640–10,020
Flow type	C	B	B	B	B	B
BOD_5						0.2–0.3
SS	42.5		22.5–50	0.73	0.27–1.06	3.5–4.6
TDS			4.0–14.6	103	0.54–2.11	12.5–16.40
COD				2.2	519–2031	0.4–0.52
pH				1.65	5.5–21.5	7.8
Phosphorus	9				1.29	
PO_4	25		15			
SO_4	111			16	86–336	1.8–2.36
F	53.5			88	17.4–68.1	0.1–0.13
HCl		0–3				
H_2SO_3		0–1.0				
$H_3PO_3 + H_3PO_4$		0.5–2.5				
HF, H_2SiF_6, H_2SiO_3			12			
Chloride				4.6	1.17–4.58	0.68–0.90
Calcium				1.8	30.6–120	0.72–0.94
Magnesium				0.6	1.9–7.43	
Aluminum				2.7	4.7–18.39	
Iron				0.37	3.2–12.52	
Arsenic				0.02	0.02–0.08	
Zinc				0.24	0.09–0.35	
Total acidity	54.5					
Total phosphorus				27.5	101–395	1.91–2.51

Source: EPA (1977).

Treatment of Phosphate Industry Wastes

Table 6 Raw Wastewater Characteristics of Phosphate Fertilizer Industry Retention Ponds

Quality parameter	Phosphate (A)
Suspended solids (mg/L)	800–1200
pH	1–2
Ammonia (mg/L)	450–500
Sulfate (mg/L)	4000
Chloride (mg/L)	58
Total phosphate (mg/L)	3–5M
Fluoride (mg/L)	6–8.5M
Aluminum (mg/L)	110
Iron (mg/L)	85
Radium 226 (picocuries/L)	60–100

M = Thousand.
Source: EPA (1977).

directly to the process or into the contaminated water system; and (6) nonpoint source discharges that originate from the dry fertilizer dust covering the general plant area and then dissolving in rainwater and snowmelt that become contaminated.

In the specific case of wastewater generated from the condenser water bleedoff in the production of elemental phosphorus from phosphate rock in an

Table 7 Range of Concentrations of Contaminants in Cooling Water

Cooling water contaminant	Concentration (mg/L)
Chromate	0–250
Sulfate	500–3000
Chloride	35–160
Phosphate	10–50
Zinc	0–30
TDS	500–10,000
TSS	0–50
Biocides	0–100

Source: Robasky and Koraido (1973).

Table 8 Range of Concentrations of Contaminants in Boiler Blowdown Waste

Boiler blowdown contaminant	Concentration (mg/L)
Phosphate	5–50
Sulfite	0–100
TDS	500–3500
Zinc	0–10
Alkalinity	50–700
Hardness	50–500
Silica (SiO_2)	25–80

Source: Robasky and Koraido (1973).

electric furnace, Horton et al. (1956) reported that the flow varies from 10 to 100 gpm (2.3–23 m^3/hr), depending on the particular installation. The most important contaminants in this waste are elemental phosphorus, which is colloidally dispersed and may ignite if allowed to dry out, and fluorine that is also present in the furnace gases. The general characteristics of this type of wastewater (if no soda ash or ammonia were added to the condenser water) are given in Table 9.

Table 9 Range of Concentrations of Contaminants in Condenser Waste from Electric Furnace Production of Phosphorus

Quality parameter	Concentration or value
pH	1.5–2.0
Temperature	120–150°F
Elemental phosphorus	400–2500 ppm
Total suspended solids	1000–5000 ppm
Fluorine	500–2000 ppm
Silica	300–700 ppm
P_2O_5	600–900 ppm
Reducing substances as (I_2)	40–50 ppm
Ionic charge of particles	Predominantly positive (+)

Source: Horton et al. (1956).

Treatment of Phosphate Industry Wastes

As previously mentioned, fertilizer manufacturing may create problems within all environmental media, i.e., air pollution, water pollution, and solid wastes disposal difficulties. In particular, the liquid waste effluents generated from phosphate and mixed and blend fertilizer production streams originate from a variety of sources and may be summarized (Kiff, 1987, and Search et al., 1979) as follows: (1) ammonia-bearing wastes from ammonia production; (2) ammonium salts such as ammonium phosphate; (3) phosphates and fluoride wastes from phosphate and superphosphate production; (4) acidic spillages from sulfuric acid and phosphoric acid production; (5) spent solutions from the regeneration of ion-exchange units; (6) phosphate, chromate, copper sulfate, and zinc wastes from cooling tower blowdown; (7) salts of metals such as iron, copper, manganese, molybdenum, and cobalt; (8) sludge discharged from clarifiers and backwash water from sand filters; and (9) scrubber wastes from gas purification processes.

Considerable variation, therefore, is observed in quantities and wastewater characteristics at different plants. According to a UNIDO report (1974), the most important factors that contribute to excessive in-plant materials losses and, therefore, probable subsequent pollution are the age of the facilities (low efficiency, poor process control), the state of maintenance and repair (especially of control equipment), variations in feedstock and difficulties in adjusting processes to cope, and an operational management philosophy such as consideration for pollution control and prevention of materials loss. Because of process cooling requirements, fertilizer manufacturing facilities may have an overall large water demand, with the wastewater effluent discharge largely dependent on the extent of in-plant recirculation (Kiff, 1987). Facilities designed on a once-through process cooling flowstream generally discharge from 1000 to over 10,000 m^3/hr wastewater effluents that are primarily cooling water.

3. IMPACTS OF PHOSPHATE INDUSTRY POLLUTION

The possibility of the phosphate industry adversely affecting streams did not arise until 1927, when the flotation process was perfected for increasing the recovery of fine-grain pebble phosphate (Fuller, 1949). A modern phosphate mining and processing facility typically has a 30,000 gpm (1892 L/sec) water supply demand and requires large areas for clearwater reservoirs, slime settling basins, and tailings sand storage. With the help of such facilities, the discharge of wastes into nearby surface water bodies is largely prevented, unless heavy rainfall inputs generate volumes that exceed available storage capacity.

According to research results reported by Fuller (1949), the removal of semicolloidal matter in settling areas or ponds seems to be one of the primary problems concerning water pollution control. The results of DO and BOD surveys indicated that receiving streams were actually improved in this respect by the effluents from phosphate operations. On the other hand, no detrimental effects on fish were found, but there is the possibility of destruction of fish food (aquatic microorganisms and plankton) under certain conditions.

The wastewater characteristics vary from one production facility to the next, and even the particular flow magnitude and location of discharge will significantly influence its aquatic environmental impact. The degree to which a receiving surface water body dilutes a wastewater effluent at the point of discharge is important, as are the minor contaminants that may occasionally have significant impacts. Fertilizer manufacturing wastes, in general, affect water quality primarily through the contribution of nitrogen and phosphorus, whose impacts have been extensively documented in the literature. Significant levels of phosphates assist in inducing eutrophication, and in many receiving waters they may be more important (growth-limiting agent) than nitrogenous compounds. Under such circumstances, programs to control eutrophication have generally attempted to reduce phosphate concentrations in order to prevent excessive algal and macrophyte growth (Griffith, 1978).

In addition to the above major contaminants, pollution from the discharge of fertilizer manufacturing wastes may be caused by such secondary pollutants as oil and grease, hexavalent chromium, arsenic, and fluoride. As reported by Beg et al. (1982), in certain cases, the presence of one or more of these pollutants may have adverse impacts on the quality of a receiving water, due primarily to toxic properties, or can be inhibitory to the nitrification process. Finally, oil and grease concentrations may have a significant detrimental effect on the oxygen transfer characteristics of the receiving surface water body.

The manufacture of phosphate fertilizers also generates great volumes of solid wastes known as phosphogypsum, which creates serious difficulties especially in large production facilities (Koziorowski and Kucharski, 1972). The disposal of phosphogypsum wastes requires large areas impervious to the infiltration of effluents, because they usually contain fluorine and phosphorus compounds that would have a harmful impact on the quality of a receiving water. Dumping of phosphogypsum in the sea would be acceptable only at coastal areas of deep oceans with strong currents that guarantee thorough mixing and high dilution.

4. U.S. CODE OF FEDERAL REGULATIONS

The information presented here has been taken from the U.S. Code of Federal Regulations, 40 CFR, containing documents related to the protection of the environment (Federal Register, 1987). In particular, the regulations contained in Part 418, Fertilizer Manufacturing Point Source Category (Subpart A, Phosphate Subcategory, and Subpart G, Mixed and Blend Fertilizer Production Subcategory), and Part 422, Phosphate Manufacturing Point Source Category, pertain to effluent limitations guidelines and pretreatment or performance standards for each of the six subcategories shown in Table 3.

4.1 Phosphate Fertilizer Manufacture

The effluent guideline regulations and standards of 40 CFR, Part 418 were promulgated on July 29, 1987. According to the most recent notice in the Federal Register (Federal Register, 1990) regarding industrial categories and regulations, no review is under way or planned and no revision proposed for the fertilizer manufacturing industry. The effluent guidelines and standards applicable to this industrial category are (1) the best practicable control technology currently available (BPT); (2) the best available technology economically achievable (BAT); (3) the best conventional pollutant control technology (BCT); (4) standards of performance for new sources (NSPS); and (5) pretreatment standards for new sources (PSNS).

The provisions of 40 CFR, Part 418, Subpart A, Phosphate Subcategory, are applicable to discharges resulting from the manufacture of sulfuric acid by sulfur burning, wet process phosphoric acid, normal superphosphate, triple superphosphate, and ammonium phosphate. The limitations applied to process wastewater, which establish the quantity of pollutants or pollutant properties that may be discharged by a point source into a surface water body after the application of various types of control technologies, are shown in Table 10. The total suspended solids limitation is waived for process wastewater from a calcium sulfate (phosphogypsum) storage pile runoff facility, operated separately or in combination with a water recirculation system, which is chemically treated and then clarified or settled to meet the other pollutant limitations. The concentrations of pollutants discharged in contaminated nonprocess wastewater, i.e., any water including precipitation runoff that comes into incidental contact with any raw material, intermediate or finished product, by-product, or waste product by means of precipitation, accidental spills, or leaks and other nonprocess discharges, should not exceed the values given in Table 10(e).

Table 10 Effluent Limitations (mg/L) for Subpart A, Phosphate Fertilizer

Effluent characteristic	Maximum for any 1 day	Average of daily values for 30 consecutive days shall not exceed
(a) BPT		
Total phosphorus (as P)	105	35
Fluoride	75	25
TSS	150	50
(b) BAT		
Total phosphorus (as P)	105	35
Fluoride	75	25
(c) BCT		
TSS	150	50
(d) NSPS		
Total phosphorus (as P)	105	35
Fluoride	75	25
TSS	150	50
(e) Contaminated nonprocess wastewater		
Total phosphorus (as P)	105	35
Fluoride	75	25

Source: Federal Register (1987).

The provisions of Subpart G, Mixed and Blend Fertilizer Production Subcategory, are applicable to discharges resulting from the production of mixed fertilizer and blend fertilizer (or compound fertilizers), such as nitrogen/phosphorus (NP) or nitrogen/phosphorus/potassium (NPK) balanced fertilizers of a range of formulations. The plant processes involved in fertilizer compounding comprise mainly blending and granulation plants, with in-built flexibility to produce NPK grades in varying proportions (Markham, 1983). According to Subpart G, "mixed fertilizer" means a mixture of wet and/or dry straight fertilizer materials, mixed fertilizer materials, fillers, and additives prepared through chemical reaction to a given formulation, whereas "blend fertilizer" means a mixture of dry, straight, and mixed fertilizer materials. The effluent limitations guidelines for BPT, BCT, and BAT, and the standards of performance for new sources, allow no discharge of process wastewater pollutants to navigable waters. Finally, the pretreatment standards establishing the quantity of pollutants that may be discharged to publicly owned treatment works (POTW) by a new source are given in Table 11.

4.2 Phosphate Manufacturing

The effluent guideline regulations and standards of 40 CFR, Part 422 were promulgated on July 9, 1986. According to the most recent notice in the Federal Register (Federal Register, 1990) regarding industrial categories and regulations, no review is underway or planned and no revision proposed for the phosphate manufacturing industry. The effluent guidelines and standards applicable to this industrial category are (1) the best practicable control technology currently available (BPT); (2) the best conventional pollutant control technology (BCT); (3) the best available technology economically achievable (BAT); and (4) standards of performance for new sources (NSPS).

The provisions of 40 CFR, Part 422, Phosphate Manufacturing, are applicable to discharges of pollutants resulting from the production of the chemicals described by the six subcategories shown in Table 3. The effluent limitations guidelines for Subpart D, Defluorinated Phosphate Rock Subcategory, are shown in Table 12, and the limitations for contaminated nonprocess wastewater do not include a value for TSS. Tables 13 and 14 show the effluent limitations guidelines for Subpart E, Defluorinated Phosphoric Acid, and

Table 11 Effluent Limitations (mg/L) for Subpart G, Mixed and Blend Fertilizer

Effluent characteristic	Average of daily values for 30 consecutive days shall not exceed
BOD_5	—
TSS	—
pH	—
NH_3	30
NO_3	30
Total P	35

Source: Federal Register (1987).

Table 12 Effluent Limitations (mg/L) for Subpart D, Defluorinated Phosphate Rock

Effluent characteristic	Maximum for any 1 day	Average of daily values for 30 consecutive days shall not exceed
(a) BPT and NSPS		
Total phosphorus (as P)	105	35
Fluoride (as F)	75	25
TSS	150	50
pH	a	a
(b) BPT and BAT for nonprocess wastewater, and BAT for process		
Total phosphorus (as P)	105	35
Fluoride (as F)	75	25
pH	a	a
(c) BCT		
TSS	150	50
pH	a	a

[a]Within the range 6.0–9.5.
Source: Federal Register (1987).

Subpart F, Sodium Phosphate Subcategories, respectively, and again the limitations for contaminated nonprocess wastewater do not include a value for TSS. As can be seen, only for Subpart F are the effluent limitations given as kilograms of pollutant per ton of product (or lb/1000 lb).

4.3 Effluent Standards in Other Countries

The control of wastewater discharges from the phosphate and phosphate fertilizer industry in various countries differs significantly, as is the case with the effluents from other industries. The discharges may be regulated on the basis of the receiving medium, i.e., whether the disposal is to land, municipal sewer system, inland surface water bodies, or coastal areas. Consideration may be given to environmental, socio-economic, and water-quality requirements and objectives, as well as to an assessment of the nature and impacts of the specific industrial effluents, which leads to an approach of either specific industry

Table 13 Effluent Limitations (mg/L) for Subpart E, Defluorinated Phosphoric Acid

Effluent characteristic	Maximum for any 1 day	Average of daily values for 30 consecutive days shall not exceed
(a) BPT and NSPS		
Total phosphorus (as P)	105	35
Fluoride (as F)	75	25
TSS	150	50
pH	a	a
(b) BAT for process and nonprocess wastewater, and BPT for nonprocess		
Total phosphorus (as P)	105	35
Fluoride (as F)	75	25
(c) BCT		
TSS	150	50
pH	a	a

[a] Within the range 6.0–9.5.
Source: Federal Register (1987).

subcategories or classification of waters, or on a case-by-case basis. To a more limited extent than in the United States, the Indian Central Board for Prevention and Control of Water Pollution established a fertilizer industry subcommittee that adopted suitable effluent standards, proposed effective pollution control measures, and established subcategories for the fertilizer industry (Fertilization Association of India, 1979).

Pollution control legislation and standards in many countries are based on the adoption of systems of water classification. This approach of environmental management can make use of either a broad system of classifications with a limited number of subcategories or, as in Japan, a detailed system of subcategories such as river groups for various uses, lakes, and coastal waters. Within such a framework, specific cases of discharge standards could also be considered under circumstances of serious localized environmental impacts. Other countries, i.e., the U.K. and Finland, have a more flexible approach and review discharge standards for fertilizer plants on a case-by-case basis, with no

Table 14 Effluent Limitations (mg/L) for Subpart F, Sodium Phosphates

Effluent characteristic	Maximum for any 1 day	Average of daily values for 30 consecutive days shall not exceed
(a) BPT		
TSS	0.50	0.25
Total phosphorus (as P)	0.80	0.40
Fluoride (as F)	0.30	0.15
pH	a	a
(b) BAT		
Total phosphorus (as P)	0.56	0.28
Fluoride (as F)	0.21	0.11
(c) NSPS		
TSS	0.35	0.18
Total phosphorus (as P)	0.56	0.28
Fluoride (as F)	0.21	0.11
pH	a	a
(d) BCT		
TSS	0.35	0.18
pH	a	a

[a]Within the range 6.0–9.5.
Source: Federal Register (1987).

established uniform guidelines (Kiff, 1987). The assessment of each case is based on the nature and volume of the discharge, the characteristics of receiving waters, and the available pollution control technology.

5. WASTEWATER CONTROL AND TREATMENT

The sources and characteristics of wastewater streams from the various subcategories in phosphate and phosphate fertilizer manufacturing, as well as some of the possibilities for recycling and treatment, were discussed in Sec. 2. The pollution control and treatment methods and unit processes used are discussed in more detail in the following. The details of the process design criteria for these unit treatment processes can be found in any design handbook.

Treatment of Phosphate Industry Wastes

5.1 In-Plant Control, Recycle, and Process Modification

The primary consideration for in-plant control of pollutants that enter waste streams through random accidental occurrences, such as leaks, spills, and process upsets, is establishing loss prevention and recovery systems. In the case of fertilizer manufacture, a significant portion of contaminants may be separated at the source from process wastes by dedicated recovery systems, improved plant operations, retention of spilled liquids, and the installation of localized interceptors of leaks such as oil drip trays for pumps and compressors (Kiff, 1987). Also, certain treatment systems installed (i.e., ion-exchange, oil recovery, and hydrolyzer-stripper systems) may, in effect, be recovery systems for direct or indirect reuse of effluent constituents. Finally, the use of effluent gas scrubbers to improve in-plant operations by preventing gaseous product losses may also prevent the airborne deposition of various pollutants within the general plant area, from where they end up as surface drainage runoff contaminants.

Cooling Water

Cooling water constitutes a major portion of the total in-plant wastes in fertilizer manufacturing and it includes water coming into direct contact with the gases processed (largest percentage) and water that has no such contact. The latter stream can be readily used in a closed-cycle system, but sometimes the direct contact cooling water is also recycled (after treatment to remove dissolved gases and other contaminants, and clarification). By recycling, the amount of these wastewaters can be reduced by 80 to 90%, with a corresponding reduction in gas content and suspended solids in the wastes discharged to sewers or surface water (Koziorowski and Kucharski, 1972).

Phosphate Manufacturing

Significant in-plant control of both waste quantity and quality is possible for most subcategories of the phosphate manufacturing industry. Important control measures include stringent in-process abatement, good housekeeping practices, containment provisions, and segregation practices (EPA, 1977). In the phosphorus chemicals industry (subcategories A, B, and C in Table 3), plant effluent can be segregated into noncontact cooling water, process water, and auxiliary streams comprising ion-exchange regenerants, cooling tower blowdowns, boiler blowdowns, leaks, and washings. Many plants have accomplished the desired segregation of these streams, often by a painstaking rerouting of the sewer lines. The use of once-through scrubber waste should be discouraged; however, there are plants that recycle the scrubber water from a sump, thus satisfying the scrubber water flow rate demands on the basis of mass transfer considerations while retaining control of water usage.

The containment of phossy water from phosphorus transfer and storage operations is an important control measure in the phosphorus-consuming subcategory B. Although displaced phossy water is normally shipped back to the phosphorus-producing facility, the usual practice in phosphorus storage tanks is to maintain a water blanket over the phosphorus for safety reasons. This practice is undesirable because the addition of makeup water often results in the discharge of phossy water, unless an auxiliary tank collects phossy water overflows from the storage tanks, thus ensuring zero discharge. A closed-loop system is then possible if the phossy water from the auxiliary tank is reused as makeup for the main phosphorus tank.

Another special problem in phosphorus-consuming subcategory B is the inadvertent spills of elemental phosphorus into the plant sewer pipes. Provision should be made for collecting, segregating, and bypassing such spills, and a recommended control measure is the installation of a trap of sufficient volume just downstream of reaction vessels. In the phosphates subcategory C, an important area of concern is the pickup by stormwater of dust originating from the handling, storing, conveying, sizing, packaging, and shipping of finely divided solid products. Airborne dust can be minimized through air pollution abatement practices, and stormwater pickup could be further controlled through strict dust cleanup programs.

In the defluorinated phosphate rock (D) and defluorinated phosphoric acid (E) subcategories, water used in scrubbing contaminants from gaseous effluent streams constitutes a significant part of the process water requirements. In both subcategories, process conditions do permit the use of contaminated water for this service. Some special precautions are essential at a plant producing sodium phosphates (F), where all meta-, tetra-, pyro-, and poly-phosphate wastewater in spills should be diverted to the reuse pond. These phosphates do not precipitate satisfactorily in the lime treatment process and interfere with the removal of fluoride and suspended solids. Since unlined ponds are the most common treatment facility in the phosphate manufacturing industry, the prevention of pond failure is vitally important. Failures of these ponds sometimes occur because they are unlined and may be improperly designed for containment in times of heavy rainfall. Design criteria for ponds and dikes should be based on anticipated maximum rainfall and drainage requirements. Failure to put in toe drainage in dikes is a major problem, and massive contamination from dike failure is a major concern for industries utilizing ponds.

Treatment of Phosphate Industry Wastes

Process Modifications

The following are possible process modifications and plant arrangements (EPA, 1974) that could help reduce wastewater volumes, contaminant quantities, and treatment costs. (1) In ammonium phosphate production and mixed and blend fertilizer (G) manufacturing, one possibility is the integration of an ammonia process condensate steam stripping column into the condensate-boiler feedwater systems of an ammonia plant, with or without further stripper bottoms treatment depending on the boiler quality makeup needed. (2) Contaminated wastewater collection systems designed so that common contaminant streams can be segregated and treated in minor quantities for improved efficiencies and reduced treatment costs. (3) In ammonium phosphate and mixed and blend fertilizer (G) production, another possibility is to design for a lower-pressure steam level (i.e., 42–62 atm) in the ammonia plant to make process condensate recovery easier and less costly. (4) When possible, the installation of air-cooled vapor condensers and heat exchangers would minimize cooling water circulation and subsequent blowdown.

In a recent document (Overcash, 1986) presenting techniques adopted by the French for pollution prevention, a new process modification for steam segregation and recycle in phosphoric acid production is described. As shown in Fig. 9, raw water from the sludge/fluorine separation system is recycled to the heat-exchange system of the sulfuric acid dilution unit and the wastewater used in plaster manufacture. Furthermore, decanted supernatant from the phosphogypsum deposit pond is recycled for treatment in the water filtration unit. The claim was that this process modification permits an important reduction in pollution by fluorine, and that it makes the treatment of effluents easier and in some cases allows specific recycling. Finally, the new process produced a small reduction in water consumption, either by recycle or discharging a small volume of polluted process water downstream, and required no particular equipment and very few alterations in the mainstream lines of the old process.

5.2 Wastewater Treatment Methods

Phosphate Manufacturing

Nemerow (1978) summarized the major characteristics of wastes from phosphate and phosphorus compounds production (i.e., clays, slimes and tall oils, low pH, high suspended solids, phosphorus, silica, and fluoride) and suggested the major treatment and disposal methods such as lagooning, mechanical clarification, coagulation, and settling of refined wastewaters. The various

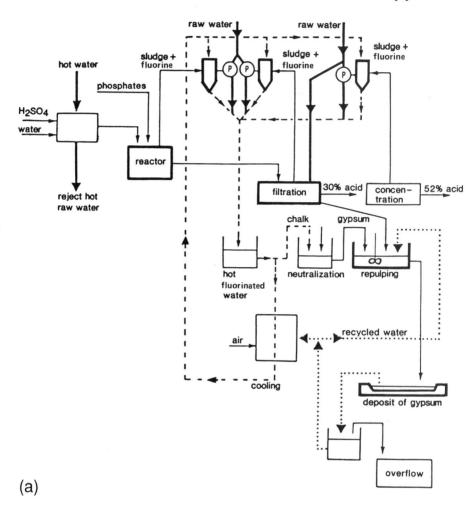

(a)

Figure 9 (a) The old process flow diagram and (b) the new process modification for steam segregation and recycle in phosphoric acid production. (From Overcash, 1986.)

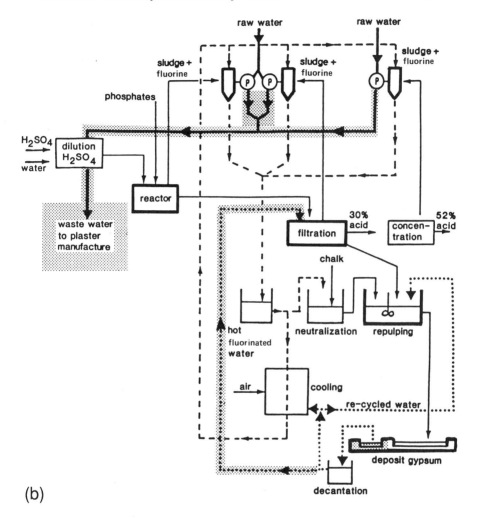

Figure 9 (Continued)

wastewater treatment practices for each of the six subcategories (Table 3) of the phosphate manufacturing industry were summarized by the EPA (1977) as shown in Table 15. The percent removal efficiencies indicated in this table pertain to the raw waste loads of process effluents from each of these six subcategories. As can be seen from Table 15, the predominant method of removal of primary pollutants such as TDS, TSS, total phosphate, phosphorus, fluoride, sulfate, and for pH adjustment or neutralization is lime treatment followed by sedimentation.

Phosphate Fertilizer Production

Contaminated water from the phosphate fertilizer subcategory A is collected in gypsum ponds and treated for pH adjustment and control of phosphorus and fluorides. Treatment is achieved by "double liming" or a two-stage neutralization procedure, in which phosphates and fluorides precipitate out (EPA, 1974). The first treatment stage provides sufficient neutralization to raise the pH from 1 to 2 to a pH level of at least 8. The resultant effectiveness of the treatment depends on the point of mixing of lime addition and on the constancy of pH control. Fluosilisic acid reacts with lime and precipitates calcium fluoride in this step of the treatment.

The wastewater is again treated with a second lime addition to raise the pH level from 8 to at least 9 (where phosphate removal rates of 95% may be achieved), although two-stage dosing to pH 11 may be employed. Concentrations of phosphorus and fluoride with a magnitude of 6500 and 9000 mg/L, respectively, can be reduced to 5 to 500 mg/L P and 30 to 60 mg/L F. Soluble orthophosphate and lime react to form an insoluble precipitate, calcium hydroxy apatite (Kiff, 1987). Sludges formed by lime addition to phosphate wastes from phosphate manufacturing or fertilizer production are generally compact and possess good settling and dewatering characteristics, and removal rates of 80 to 90% for both phosphate and fluoride may be readily achieved (Ghokas, 1983).

The seepage collection of contaminated water from phosphogypsum ponds and reimpoundment is accomplished by the construction of a seepage collection ditch around the perimeter of the diked storage area and the erection of a secondary dike surrounding the first (EPA, 1977). The base of these dikes is usually natural soil from the immediate area, and these combined earth/gypsum dikes tend to have continuous seepage through them (see Fig. 10). The seepage collection ditch between the two dikes needs to be of sufficient depth and size to not only collect contaminated water seepage, but also to permit collection of seeping surface runoff from the immediate outer perimeter of the seepage

Treatment of Phosphate Industry Wastes

Table 15 Phosphate Manufacturing Industry Wastewater Treatment Practices and Unit Removal Efficiencies (%)

Pollutant and method	Phosphorus production A	Phosphorus-consuming B	Phosphate C	Defluorinated phosphate rock D	Defluorinated phosphoric acid E	Sodium phosphate F
TDS						
Lime treatment and sedimentation[a]	99		99			
TSS						
Lime treatment and sedimentation[a]	99		99			
Flocculation, clarification, and dewatering		92				
Total Phosphate						
Lime treatment and sedimentation[a]	97	73–97	97			
Phosphorus						
Lime treatment and sedimentation[a]				90	99	88
Flocculation, clarification, and dewatering		92				
Sulfate						
Lime treatment and sedimentation[a]	98		98			
Fluoride						
Lime treatment and sedimentation[a]	99		99	98	96	0
pH (effluent level)						
Lime treatment and sedimentation[a] (neutralization)				6–8	6–8	6–8

[a]Preceded by recycle of phossy water and evaporation of some process water in subcategories A, B and C.
Source: EPA (1977).

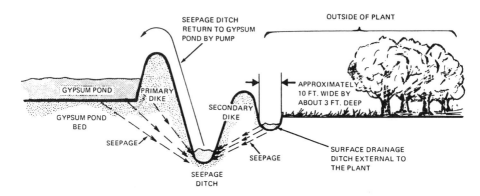

Figure 10 Phosphogypsum pond water seepage control. (After EPA, 1974.)

ditch. This is accomplished by the erection of the small secondary dike, which also serves as a backup or reserve dike in the event of a failure of the primary major dike.

The sulfuric acid plant has boiler blowdown and cooling tower blowdown waste streams, which are uncontaminated. However, accidental spills of acid can and do occur, and when they do, the spills contaminate the blowdown streams. Therefore, neutralization facilities should be supplied for the blowdown waste streams (see Table 15), which involves the installation of a reliable

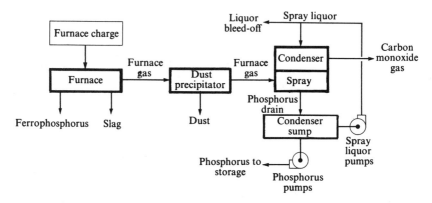

Figure 11 Flow diagram of electric furnace process for phosphorus production. (After Horton et al., 1956.)

Treatment of Phosphate Industry Wastes

pH or conductivity continuous-monitoring unit on the plant effluent stream. The second part of the system is a retaining area through which noncontaminated effluent normally flows. The detection and alarm system, when activated, causes a plant shutdown that allows location of the failure and initiation of necessary repairs. Such a system, therefore, provides the continuous protection of natural drainage waters, as well as the means to correct a process disruption.

Mixed fertilizer (subcategory G) treatment technology consists of a closed-loop contaminated water system, which includes a retention pond to settle suspended solids. The water is then recycled back to the system. There are no liquid waste streams associated with the blend fertilizer (subcategory G) process, except when liquid air scrubbers are used to prevent air pollution. Dry removals of air pollutants prevent a wastewater stream from being formed.

Phosphate and Fluoride Removal

Phosphates may be removed from wastewaters by the use of chemical precipitation as insoluble calcium phosphate, aluminum phosphate, and iron phosphate (EPA, 1971). The liming process has been discussed previously, lime being typically added as a slurry, and the system used is designed as either a single- or two-stage one. Polyelectrolytes have been employed in some plants to improve overall settling, and clariflocculators or sludge-blanket clarifiers are used in a number of facilities (Fertilizer Association of India, 1979). Alternatively, the dissolved air flotation process is also feasible for phosphate and fluoride removal (Wang and Wang, 1990).

A number of aluminum compounds, such as alum and sodium aluminate, have also been used as phosphate precipitants at an optimum pH range of 5.5 to 6.5 (Layer and Wang, 1984), as have iron compounds such as ferrous sulfate, ferric sulfate, ferric chloride, and spent pickle liquor (Kiff, 1987). The optimum pH range for the ferric salts is 4.5 to 5, and for the ferrous salts it is 7 to 8, although both aluminum and iron salts have a tendency to form hydroxyl and phosphate complexes. As reported by Ghokas (1983), sludge solids produced by aluminum and iron salts precipitation of phosphates are generally less settleable and more voluminous than those produced by lime treatment.

Removal of Other Contaminants

Compressor houses, tank farm areas, cooling water, and loading or unloading bays may be sources of oil in wastewater discharges from phosphate and fertilizer manufacturing. Oil concentrations may range from 100 to 900 mg/L and can be removed by such units as coke filters or recovered by the use of separators (usually operating in series to recover high oil levels). Chromates and dichromates are present in cooling tower blowdown, at levels of about 10 mg/L, because they are used in cooling water for corrosion inhibition. They may be removed (over 90%) from cooling tower discharges through chemical reduction (i.e., the use of sulfuric acid for lowering pH to less than 4 and addition of a reducing agent) and precipitation as the hydroxide by the addition of lime or NaOH. Chromates may also be recovered and reused from cooling tower blowdown by the use of ion exchange, at recovery levels in excess of 99%, employing a special weak base anion resin that is regenerated with caustic soda.

Phosphoric Acid Production

The use of the electric furnace process (see Fig. 11) and acidulation of phosphate bearing rock is made commercially to produce phosphoric acid. In the first method, elemental phosphorus is first produced from phosphate ore, coke, and silica in an electric furnace, and then the phosphorus is burned with air to form P_2O_5 that is cooled and reacted with water to form orthophosphoric acid (see Sec. 2.2). Extremely high acid mist loadings from the acid plant are common, and there are five types of mist-collection equipment generally used: packed towers, electrostatic precipitators, venturi scrubbers, fiber mist eliminators, and wire mesh contactors (Lund, 1971). Choosing one of these control equipments depends on the required contaminant removal efficiency, the required materials of construction, the pressure loss allowed through the device, and capital and operating costs of the installation (with very high removal efficiencies being the primary factor). The venturi scrubber is widely used for mist collection and is particularly applicable to acid plants burning sludge. The sludge burned is an emulsion of phosphorus, water, and solids carried out in the gas stream from the phosphorus electric furnace as dust or volatilized materials. Impurities vary from 15 to 20% and the venturi scrubber can efficiently collect the acid mist and fine dust discharged in the exhaust from the hydrator.

Wet process phosphoric acid is made by reacting pulverized, beneficiated phosphate ore with sulfuric acid to form calcium sulfate (gypsum) and dilute phosphoric acid (see Sec. 2.4). The insoluble calcium sulfate and other solids are removed by filtration, and the weak (32%) phosphoric is then concentrated in evaporators to acid containing about 55% P_2O_5. Mist and gaseous emissions from the gypsum filter, the phosphoric acid concentrator, and the acidulation off-gas are controlled with scrubbers or other equipment. The preparation of the phosphate ore generates dust from drying and grinding operations, and this is generally controlled with a combination of dry cyclones and wet scrubbers (Lund, 1971). The material collected by the cyclones is recycled, and the scrubber water discharged to the waste phosphogypsum ponds. Most frequently, simple towers and wet cyclonic scrubbers are used, but at some plants the dry cyclone is followed by an electrostatic precipitator.

6. CASE STUDIES OF TREATMENT FACILITIES

Phosphate production and phosphate fertilizer manufacturing facilities are situated in many areas in the United States (primarily in Florida and California) and in other countries such as Algeria, Jordan, and Morocco, as previously mentioned. The wastewaters from production and cleanup activities and surface runoff in most of these locations are stored, treated, and recycled and the excess overflows are discharged into natural water systems. In those facilities where wastewater from production and cleanup activities and drainage are discharged into municipal water systems and treated together with domestic, commercial, institutional, and other industrial wastewaters, a degree of pretreatment is required to meet federal guidelines or local ordinances such as those presented in Sec. 4. For instance, according to the EPA (1973), the pretreatment unit operations required for the phosphate fertilizer industry comprise solids separation and neutralization, and it may be achieved by either a suspended biological process, a fixed-film biological process, or an independent physico-chemical system.

6.1 Pebble Phosphate Mining Industry

In one of the earlier reports on the phosphate mining and manufacturing industry in Florida and its water pollution control efforts, Wakefield (1952) gave the following generalized account. Because of the huge volumes of water being used for washing, hydraulic sizing, flotation, and concentration of phosphate ores (i.e., one of the main mines of a larger company requires about 60 MGD or 2.63 m^3/sec), and since make-up water is not readily available and excess wastewater constitutes a major disposal problem, the recovery and reuse of water have always been of great importance. Waste products from the mining and processing operation consist of large quantities of nonphosphating sands and clays, together with unrecovered phosphatic materials less than 300-mesh in size, and they are pumped into huge lagoons. Easily settled sands fill the near-end, leaving the rest to be gradually filled with "slimes" (a semicolloidal water suspension), while a thin layer of virtually clear water at the surface of the lagoon flows over spillways and is returned to the washers for reuse.

The above ideal wastewater management, however, is infeasible during wet weather (rains of 3 in./day or 7.6 cm/day are frequent in the tropical climate of Florida), because more water goes into the lagoons than can be used by the washers, and the excess volume must be discharged into nearby surface waters. Especially in small streams, this results in highly turbid waters due to the fact that the larger of slowly consolidating slimes is near the surface of the lagoons in most cases. Furthermore, there are no core walls (for cost reasons) in the large earthen dams forming the settling lagoons; rather, it is usual to depend on the slimes to seal their inner face and prevent excessive seepage. The entire operation, therefore, involves a delicate balance of slime input and weir discharge to accomplish the objective of a maximum of water reuse with a minimum of danger of dam failure and a minimum of turbid discharge to the stream. Such dams have failed very often and the pollution effect of the volume of, e.g., a 100-acre (0.4 km^2) pond with 25 to 30 ft (7.6–9.1 m) of consolidated slime being discharged into a stream with a mean flow of possibly 300 cfs (8.5 m^3/sec) has been devastating.

There are, however, other much more frequent situations when effluents of higher or lower turbidity are discharged to streams to protect dam structures or as excess flow when it rains heavily. These discharges of turbid wastewater volumes may be due to underdesigning the settling lagoons or because the wastewater slimes are of a more completely colloidal nature and do not clarify too well. They usually continue over extended periods and cause noticeable stream turbidities, although nothing approaching those encountered after a lagoon dam failure. When streams contain appreciable turbidity due to phosphate industry effluents, local sports fishermen claim that it ruins fishing (but mostly they just prefer fishing in clear waters), while industry managers have demonstrated that fish are not affected by stocking mined-out pits and settling ponds with bass and other species. Finally, a positive effect of discharging moderate levels of turbidity noted in a water treatment plant located downstream is a decrease in chemical costs, undoubtedly due to greater turbidity in the raw water that aids coagulation of color and other impurities.

6.2 Phosphate Industry Waste Disposal

In one of the earliest extensive studies and reports on the disposal of wastes from the phosphate mining and processing industry in Florida, Specht (1960) reviewed the waste treatment and disposal practices in the various phases of phosphate and phosphorus manufacturing. Regarding waste disposal from mining and beneficiation operations, he reported the use of specially constructed settling areas for the clay and quartz sand separated in the washing and flotation processes and also for the clarification of water to be reused in the process or discharged into streams. As mining processes, the mined-out areas are then used as supplementary settling lagoons, with the wastewater circulating through them using specially made cuts, similar to the slow movement through settling areas that are frequently divided into compartments. During the dry season, as mentioned previously, very little (if any) water is wasted to the streams, but sometimes an estimated maximum of 10% of the total amount of water used is wasted at some facilities during the rainy season. This may represent a significant volume, given the large quantities of water needed in phosphate mining (2000–8000 gpm or 7.6–30 m^3/min) and at the recovery plants (4000–50,000 gpm or 15–190 m^3/min), depending on the size of the plant and its method of treatment.

Occasionally, the phosphate "slime" is difficult to settle in the lagoons because of its true colloidal nature, and the use of calcium sulfate or other electrolytes can promote coagulation, agglomeration, and settling of the particles. Usually an addition of calcium sulfate is unnecessary, because it is present in the wastewater from the sand-flotation process. Generally, it has been shown (Specht, 1950) that the clear effluent from the phosphate mining and beneficiation operation is not deleterious to fish life, but the occurrence of a dam break may result in adverse effects (Lanquist, 1955).

In superphosphate production, fluoride vapors are removed from the mixing vessel, den or barn, and elevators under negative pressure and passed through water sprays or suitable scrubbers. A multiple-step scrubber is required to remove all the fluorides from the gases and vapors, and the scrubbing water containing the recovered fluorosilisic acid and insoluble silicon hydroxide is recycled to concentrate the acid to 18 to 25%. The hydrated silica is removed from the acid by filtration and is washed with fresh water and then deposited in settling areas or dumps. In triple superphosphate (also known as double, treble, multiple, or concentrated superphosphate) manufacturing, the calcium sulfate cake from the phosphoric acid production is transferred into settling areas after being washed, where the solid material is retained. The clarified water that contains dissolved calcium sulfate, dilute phosphoric acid, and some fluorosilisic acid is either recycled for use in the plant or treated in a two-step process to remove the soluble fluorides, as described in Sec. 5.2. Water from plant washing and the evaporators may also be added to the wastes sent to the calcium sulfate settling area.

According to Sprecht (1960), in the two-step process to remove fluorides and phosphoric acid, water entering the first step may contain about 1700 mg/L F and 5000 mg/L P_2O_5, and it is treated with lime slurry or ground limestone to a pH of 3.2 to 3.8. Insoluble calcium fluorides settle out and the fluoride concentration is lowered to about 50 mg/L F, whereas the P_2O_5 content is reduced only slightly. The clarified supernatant is transferred to another collection area where lime slurry is added to bring the solution to pH 7, and the resultant precipitate of P is removed by settling. The final clear water, which contains only 3 to 5 mg/L F and practically no P_2O_5, is either returned to the plant for reuse or discharged to surface waters. The two-step process is required to reduce fluorides in the water below 25 mg/L F, because a single-step treatment to pH 7 lowers the fluoride content only to 25 to 40 mg/L F. In the process where the triple phosphate is to be granulated or nodulized, the material is transferred directly from the reaction mixer to a rotary dryer, and the fluorides in the dryer gases are scrubbed with water.

In making defluorinated phosphate by heating phosphate rock, one method of fluoride recovery consists of absorption in a tower of lump limestone at temperatures above the dewpoint of the stack gas, where the reaction product separates from the limestone lumps in the form of fines. A second method of recovery consists of passing the gases through a series of water sprays in three separate spray chambers, of which the first one is used primarily as a cooling chamber for the hot exit gases of the furnace. In the second chamber, the acidic water is recycled to bring its concentration to about 5% equivalence of hydrofluoric acid in the effluent, by withdrawing acid and adding fresh water to the system. In the final chamber, scrubbing is supplemented by adding finely ground limestone blown into the chamber with the entering gases. Hydrochloric acid is sometimes formed as a by-product from the fluoride recovery in the spray chambers and this is neutralized with NaOH and lime slurry before being transferred to settling areas.

6.3 Ammonium Phosphate Fertilizer and Phosphoric Acid Plant

The fertilizer industry is plagued with a tremendous problem concerning waste disposal and dust because of the very nature of production that involves large volumes of dusty material. Jones and Olmsted (1962) described the waste disposal problems and pollution control efforts at such a plant, Northwest Cooperative Mills, in St. Paul, Minnesota. Two types of problems are associated with waste from the manufacture of ammonium phosphate: wastes from combining ammonia and phosphoric acid and the subsequent drying and cooling of the products, and wastes from the handling of the finished product arising primarily from the bagging of the product prior to shipping. Because the ammoniation process has to be "forced" by introducing excess amounts of ammonia than the phosphoric acid is capable of absorbing, there is high ammonia content in the exhaust air stream from the ammoniator. Since it is neither economically sound nor environmentally acceptable to exhaust this to the atmosphere, an acid scrubber is employed to recover the ammonia without condensing with it the steam that nearly saturates the exhausted air stream.

Drying and cooling the products of ammonium phosphate production are conventionally achieved in a rotary drum, and a means must be provided to remove the dust particles from the air streams to be exhausted to the atmosphere. At the Minnesota plant, a high-efficiency dry cyclone recovery system followed by a wet scrubber was designed. In this way, material recovered from the dry collector (and recycled to the process) pays for the dry system and minimizes the load and disposal problem in the wet scrubber, because it eliminates the need for a system to recover the wet waste material that is discharged to the gypsum disposal pond for settling.

The remaining problem of removing dust from discharges to the ambient air originating from the bagging and shipping operations is the one most neglected in the fertilizer industry, causing complaints from neighbors. Jones and Olmsted (1962) reported the installation of an elaborate, relatively expensive, system of suction pickups at each transfer point of the products in the entire bagging system and shipping platform areas. The collected dust streams are passed over a positive cloth media collector before discharge to the atmosphere, and the system recovers sufficient products to cover only operating expenses.

The filtered gypsum cake from phosphoric acid production, slurried with water to about 30%, is pumped to the settling lagoons, from where the clarified water is recycled to process. To provide a startup area, approximately two acres (or 8100 m^2) of the disposal area were black-topped to seal the soil surface against seepage, and the gypsum collected in this area was worked outward to provide a seal for enlarging the settling area. A dike-separated section of the disposal area was designated as a collection basin for all drainage waters at the plant site. From this basin, after the settling of suspended solid impurities, these waters are discharged to surface waters under supervision from a continuous monitoring and alarm system that guards against accidental contamination from any other source.

Air streams from the digestion system, vacuum cooler, concentrator, and other areas where fluorine is evolved are connected to a highly-efficient absorption system, providing extremely high volumes of water relative to the stream. The effluent from this absorption system forms part of the recycled water and is eventually discharged as part of the product used for fertilizer manufacture. The Minnesota plant requires a constant recirculating water load in excess of 3000 gpm (11.4 m^3/min), but multiple use and recycle reduce makeup requirements to less than 400 gpm (1.5 m^3/min) or a mere 13% of total water use.

6.4 Rusaifa Phosphate Mining and Processing Plant

As reported by Shahalam et al. (1985), Jordan stands third in the region following Morocco and Algeria with respect to the mining of phosphate rock and production of phosphates, and the Jordanian phosphate industry is bound to grow, with time creating additional environmental problems. The paper presented the results of a study assessing the phosphate deposits and pollution resulting from a phosphate processing operation in Rusaifa, Jordan. The beneficiation plant uses about 85% of the total process water, and the overflow from the hydrocyclones (rejected as slimes of silica carbonates and clay materials) is taken to a gravity thickener, the underflow (about 0.93 m^3/min or 245 gpm) from which is directly discharged to a nearby holding pond as wastewater and slimes (sludge containing about 25% solids by wt.). As shown in Table 16, the sludge contains significant amounts of fluorine, sulfate, P_2O_5, and organics, and it is unsuitable for direct or indirect discharge into natural waters, in accordance with U.S. 40 CFR (see Sec. 4). The water portion of this thick sludge partially evaporates and the remainder percolates into the ground through the bottom of the holding pond.

The second major wastewater discharge (about 1.2 m^3/min or 318 gpm) is from the bottom of the scrubbers used after a dry dust collector cyclone to reduce the dust concentration in the effluent air stream from the phosphate dryer. The underflow of the scrubber contains a high concentration of dust and mud, laden with tiny phosphate organic and inorganic silica and clay particles, and is disposed of in a nearby stream. The main pollutant in the flow to the stream is P_2O_5 at a concentration of nearly 1200 mg/L and it remains mostly as suspended particles.

Finally, most of the P_2O_5 content in the solid wastes generated from Rusaifa Mining is mainly from overburdens and exists as solid particles. Since it does not dissolve in water readily, the secondary pollution potential with respect to

Table 16 Chemical Analysis of the Characteristics of Underflow Sludge from the Gravity Thickener

Parameter	Contents
pH	7.75
Solids	About 10% by sludge volume
P_2O_5	24.4% of dry solid
Total orthophosphate	0.30 mg/L
SiO_2 (inorganic)	14.16% of dry solid
Total SiO_2	16.83% of dry solid
Organics	1.00% of dry solid
Total chlorine	0.09% of dry solid
CaO	40.43% of dry solid
Fluorine	2.96% of dry solid
$CaSO_4$	0.32% of dry solid

Source: Shahalam et al. (1985).

P_2O_5 is practically nonexistent. However, loose overburdens in piles, when carried off by rainwater, may create problems for nearby stream(s) due to the high suspended solids concentration resulting in stormwater runoff. Also, loose overburdens blown by strong winds may cause airborne dust problems in neighboring areas; therefore, a planned land reclamation of the mined-out areas is the best approach to minimizing potential pollution from the solid wastes.

6.5 Furnace Wastes from Phosphorus Manufacture

The electric furnace process (Fig. 11) for the conversion of phosphate rock into phosphorus was described by Horton et al. (1956) in a paper that also presented the results of a pilot plant study of treating the wastes produced. The process, as well as the handling of the various waste streams for pollution control, are discussed in Sec. 5.2. In processing the phosphate, the major source of wastewater is the condenser water bleed off from the reduction furnace, the flow of which varies from 10 to 100 gpm (2.3–22.7 m^3/hr) and its quality characteristics are presented in Table 7.

Phossy water, a waste product in the production of elemental phosphorus by the electric furnace process, contains from 1000 to 5000 mg/L suspended solids that include 400 to 2500 mg/L of elemental phosphorus, distributed as liquid colloidal particles. These particles are usually positively charged, although this

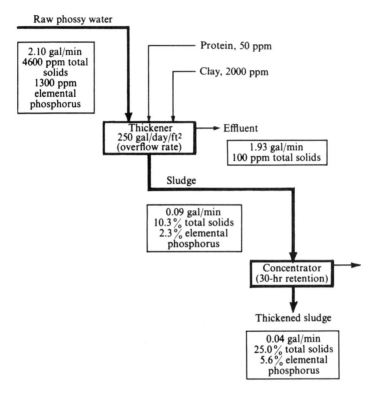

Figure 12 Summary of materials balance in a pilot plant for recovery of phosphorus from phossy water. (After Horton et al., 1956.)

varies depending on the operation of the electrostatic precipitators. Furthermore, the chemical equilibrium between the fluoride and fluosilicate ions introduces an important source of variation in suspended solids that is a pH function. Commonly used coagulants such as alum or ferric chloride were unsatisfactory for wastewater clarification because of the positively charged particles, as were inorganic polyelectrolytes despite their improved performance. However, high-molecular-weight protein molecules at a suitable pH level (which varies for each protein) produced excellent coagulation and were highly successful in clarifying phossy water.

Horton et al. (1956) investigated or attempted such potential treatment and disposal methods as lagooning, oxidizing, settling (with or without prior chemical coagulation), filtering, and centrifuging and concluded that the best

Treatment of Phosphate Industry Wastes

solution appears to be coagulation and settling. The pilot units installed to evaluate this optimum system are shown in Fig. 12, together with a summary of the experimental results. The proper pH for optimum coagulation with proteins alone was obscured at higher pH levels by the formation of silica, which tended to encrust the pipe lines. It was found that the addition of clay, as a weighting agent with the coagulant, eliminated the scale problem without decreasing the settling rates. Finally, it was concluded that in the pilot plant it was possible to obtain a 40-fold concentration of suspended solids (or 25% solids) by a simple coagulation and sedimentation process.

6.6 Phosphate Fertilizer Industry in Eastern Europe

Koziorowski and Kucharski (1972) presented a survey of fertilizer industry experience in Eastern European countries and compared it with the U.S. and Western European equivalents. For instance, they stated that HF and silicofluoric acid are evolved during the process of dissolving phosphorite in the manufacture of normal superphosphate. These constituents are removed from the acidic and highly toxic gases by washing with water or brine in closed condenser equipment, and the wastes from this process are usually slightly acidic, clear, and colorless.

The production of superphosphate is often combined with the manufacture of sodium fluorosilicate and then the amount of wastes is larger.

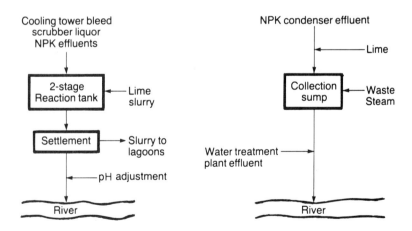

Figure 13 Phosphoric acid and N-P-K fertilizer production waste treatment. (After Kiff, 1987.)

Czechoslovakian experiments have shown that for every 1000 tons of superphosphate (20% P_2O_5 content) produced, 133 m^3 (4700 ft^3) of postcrystallization liquor from the crystallization of sodium fluorosilicate and 67 m^3 (2370 ft^3) of washings are discharged. The liquor contains 20 to 25 g/L NaCl, 25 to 35 g/L HCl, 10 to 15 g/L silisic acid, and 8 to 10 g/L sodium silicofluoride. For this waste, the most significant aspect of treatment is recovering the sodium silicofluoride from the brine used to absorb fluorine compounds from the gaseous waste streams, and this is relatively easy to accomplish since it settles nearly ten times as fast as silicic acid gel and, therefore, it is separated by sedimentation. The silicofluoride recovered is a valuable by-product that, following filtration, washing with water, and drying, is used as a flux in enamel shops, glass works, and other applications.

In Czechoslovakian phosphate fertilizer plants, the superphosphate production wastewaters are further treated by neutralization on crushed limestone beds contained in special tanks that are followed by settling tanks for clarification of the wastewater. The beds have from three to five layers (with a minimum bed height of 0.35–1.60 m), treat a range of acidity of wastes from 438 to 890 meq/L, and are designed for a hydraulic load ranging from 0.13 to 0.52 $cm^3/cm^2 \cdot sec$ (1.9 to 7.7 gpm/ft^2) at operating temperatures of 20 to 28°C. This experience agrees with results reported from Polish plants, the limestone used contains 56% CaO, and it was found in practice that coarse particles of 3 to 5 mm give better results because less material is carried away. In the USSR, superphosphate wastes are neutralized with powdered limestone or milk of lime.

The neutralized wastes leaving the settling tank contain primarily dissolved sodium and calcium chlorides. As previously mentioned, the manufacture of phosphate fertilizers also yields large quantities of phosphogypsum, which often contains significant amounts of fluorine and phosphorus compounds and requires large areas for dumping. It has been estimated that for each ton of phosphorite processed, a wet-process phosphoric acid plant yields 1.4 to 1.6 tons of gypsum containing about 30% water and 66% calcium sulfate. This waste material has been used for the production of building materials such as plasterboard, ammonium sulfate, but primarily sulfuric acid and cement.

6.7 Phosphoric Acid and N-P-K Fertilizer Plant

According to Kiff (1987), the heterogeneous nature of fertilizer production plants precludes the possibility of presenting a "typical" case study of such a

Treatment of Phosphate Industry Wastes

facility (see also Beg et al., 1980). Nevertheless, he presented information regarding wastewater flows and characteristics as well as the treatment systems for a phosphoric acid and N-P-K fertilizer plant, which was part of a large fertilizer manufacturing facility. The full facility additionally included an ammonia plant, a urea plant, a sulfuric acid plant, and a nitric acid plant. The typical effluent flows were 183 m^3/hr (806 gpm) from the phosphoric plant and 4.4 m^3/hr (20 gpm) from the water treatment plant associated with it, whereas in the N-P-K plant they were 420 m^3/hr (1850 gpm) from the barometric condenser and 108 m^3/hr (476 gpm) from other effluent sources.

These wastewater effluents had quality characteristics that could be described as follows. (1) In the phosphoric acid plant, the contributing sources of effluent are the cooling tower bleed off and the scrubber liquor solution that contains concentrations ranging for phosphate from 160 to 200 mg/L and for fluoride from 225 to 7000 mg/L. (2) In the water treatment plant, the wastewater effluent is slightly acidic in nature. (3) In the N-P-K plant, the barometric condenser effluent has a pH range of 5.5 to 8, and concentrations of ammonia-nitrogen about 250 mg/L, fluoride about 10 mg/L, and trace levels of phosphate. (4) The N-P-K plant other effluents contain concentrations of ammonia-nitrogen about 2000 mg/L, fluoride about 350 mg/L, and phosphate about 3000 mg/L.

The wastewater treatment systems utilized for the phosphoric acid and N-P-K plant effluents are shown in Fig. 13. As can be seen, the cooling tower bleed off and scrubber liquor from the phosphoric acid plant are treated together with N-P-K plant effluents by a two-stage lime slurry addition to precipitate out the phosphates and fluorides, reducing them to levels of less than 10 mg/L. The treated effluent pH is adjusted to 5.5 to 7 using sulfuric acid, and it is discharged to a river, while the precipitated slurry containing the phosphates and fluorides is disposed of in lagoons. As can be seen in Fig. 13(B), the effluent of the barometric condenser has its pH adjusted to 11 by adding lime to remove residual ammonia-nitrogen, subsequently waste steam is introduced to remove free ammonia, and the final effluent is mixed with the water treatment plant effluent prior to discharge in a river.

6.8 Environmentally Balanced Industrial Complexes

Unlike common industrial parks where factories are selected simply on the basis of their willingness to share the real estate, environmentally balanced industrial complexes (EBIC) are a selective collection of compatible industrial plants located together in a complex so as to minimize environmental impacts

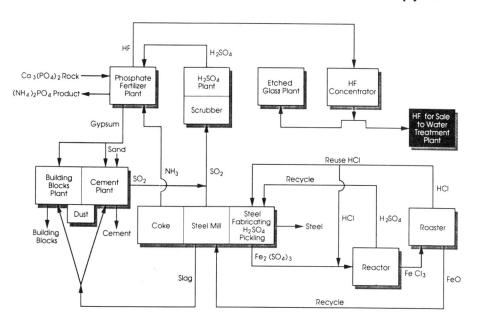

Figure 14 Example of environmentally balanced industrial complex centered about a steel mill plant. (After Nemerow and Dasgupta, 1981.)

and industrial production costs (Nemerow and Dasgupta, 1981). These objectives are accomplished by utilizing the waste materials of one plant as the raw materials for another with a minimum of transportation, storage, and raw materials preparation costs. It is obvious that when an industry neither needs to treat its wastes, nor is required to import, store, and pretreat its raw materials, its overall production costs must be reduced significantly. Additionally, any material reuse costs in an EBIC will be difficult to identify and more easily absorbed into reasonable production costs.

Such EBICs are especially appropriate for large, water-consuming, and waste-producing industries whose wastes are usually detrimental to the environment, if discharged, but they are also amenable to reuse by close association with satellite industrial plants using wastes from and producing raw materials for others within the complex. Examples of such major industries that can serve as the focus industry of an EBIC are fertilizer plants, steel mills, pulp and paper mills, and tanneries. Nemerow and Dasgupta (1981) presented the example of a steel mill complex with a phosphate fertilizer and a building

Treatment of Phosphate Industry Wastes

materials plant as the likely candidates for auxiliary or satellite industries (see Fig. 14).

A second example presented was an EBIC centered about a phosphate fertilizer plant, with a cement production plant, a sulfuric acid plant, and a municipal solid wastes composting plant (its product to be mixed with phosphate fertilizer and sold as a combined product to the agricultural industry) as the satellite industries (see Fig. 15). As previously mentioned, in the usual starting process of producing phosphoric acid and ammonium phosphate fertilizer by dissolving the phosphate rocks with sulfuric acid, a gypsumlike sludge is generated as a by-product and some sulfur dioxide and fluorine are in the waste gases emitted at the high reaction temperatures. The large, relatively impure, quantities of phosphogypsum (5 vol. to 1 P_2O_5 fertilizer produced) are difficult to treat, and the fluorine present in the gas as hydrofluoric acid (concentrations from 1–10%, which is very low for commercial use) requires further costly and extensive treatment. Using such a fertilizer production

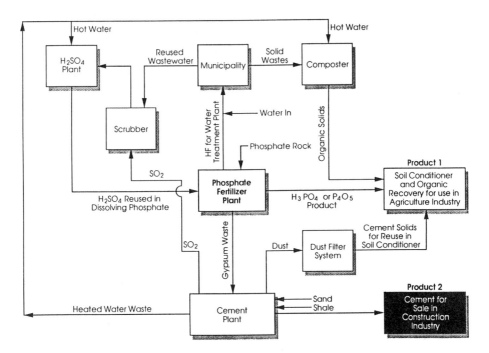

Figure 15 Example of an environmentally balanced industrial complex with a phosphate fertilizer plant as the focus industry. (After Nemerow and Dasgupta, 1981.)

facility as the focus industry of an EBIC would be a feasible solution to the environmental problems if combined, i.e., with such satellite industries as (1) a sulfuric acid plant to feed its products to the phosphoric acid plant and to use some of the hot water effluent from the cement plant and the effluent from the SO_2 scrubber of the phosphate fertilizer plant; (2) a municipal solid wastes composting plant utilizing hot water from the cement plant, serving as disposal facility for the garbage of a city, and producing composted organic solids to be used as fertilizer; (3) a cement and plasterboard production plant utilizing the phosphogypsum waste sludge in the manufacture of products for the construction industry, producing hot water effluent to be used as mentioned above, and waste dust collected by a dust filter and used as a filler for the soil fertilizer produced by mixing composted garbage and phosphate fertilizer.

6.9 Fluoride and Phosphorus Removal from a Fertilizer Complex Wastewater

A laboratory-scale treatability study was conducted for the Mississippi Chemical Corporation to develop a physico-chemical wastewater treatment process for a fertilizer complex wastewater to control nitrogen, phosphorus, and fluoride and to recover ammonia (Arnold and Wolfram, 1975). The removal technique investigated consisted of precipitation of fluorides, phosphorus, and silica by lime addition, a second stage required for the precipitation of ammonia by the use of phosphoric acid and magnesium, and a third stage for further polishing of the wastewater necessary to remove residual phosphate. The wastewater quality parameters included the following concentrations (mg/L): fluoride 2000, ammonia 600, phosphorus as P_2O_5 (P) 145 (63), and an acidic pH level of 3.5. Ammonia removals of 96% were achieved and the insoluble struvite complex produced by the ammonia removal stage is a potentially commercial-grade fertilizer product, whereas the fluoride and phosphorus in the effluent fell below 25 and 2 mg/L, respectively.

The multistage treatment approach was to first remove the fluorides by lime precipitation, with the optimum removal (over 99%) occurring with a two-step pH adjustment to about 10.4 (removal was more a function of lime dosage, rather than a pH solubility controlled phenomenon). Each step was followed by clarification, and the required lime equivalent dosage was 180% of the calcium required stoichiometrically. The effluent from the first stage had an average fluoride content of 135 mg/L (93% removal) and a phosphorus content of less than 5 mg/L, whereas the hydraulic design parameters were a 15 min per stage reaction time and a 990 gpd/ft^2 (40.3 m^3/m^2/day) clarifier overflow rate. The resulting precipitated solids underflow concentration was 7.7% by weight.

The second-stage (ammonia removal) effluent contained unacceptable levels of F and P and had to be subjected to third-stage lime treatment. It raises the pH from 8.5 to 11.4 and produces an effluent with concentrations of F and P equal to 25 and 2 mg/L, respectively. The hydraulic design parameters were a 15-min reaction time and a 265 gpd/ft^2 (10.8 m^3/m^2/d) clarifier overflow rate. The resulting precipitated solids underflow concentration was 0.6% by wt. In all three stages, an anionic polymer was used to aid coagulation, solids settling, and effluent clarification.

The solids resulting from the first- and third-stage treatment consisted of calcium fluoride, calcium phosphate, and fluorapatite-type compounds. Typically, in the fertilizer industry, such sludges are disposed of by lagooning and subsequent landfilling. Other studies have investigated the recovery of the fluoride compounds, such as hydrofluoric and fluosilisic acids, for use in the glass industry and in the fluoridation of drinking water supplies.

REFERENCES

Anonymous (1951). Phosphate, the servant of mankind. *Oil Power, 26* (3).

Arnold, D. W. and Wolfram, W. E. (1975). Ammonia removal and recovery from fertilizer complex wastewaters. In *Proceedings of 30th Industrial Waste Conference*, Purdue Univ., Lafayette, Ind., Vol. 30, pp. 760–767.

Beg, S. A., et al. (1980). Effect of toxicants on biological nitrification for treatment of fertilizer industry wastewater. In *Proceedings of 35th Industrial Waste Conference*, Purdue Univ., Lafayette, Ind., Vol. 35, pp. 826–834.

Beg, S. A., et al. (1982). Inhibition of nitrification by arsenic, chromium and fluoride. *J. Water Poll. Control Fed., 54*, 482–488.

Environmental Protection Agency (EPA) (1971). Process Design Manual for Phosphorus Removal. Office of Technology Transfer, Washington, D.C.

EPA (1973). Pretreatment of Pollutants Introduced into Publicly Owned Treatment Works. Federal Guidelines, Office of Water Program Operations, Washington, D.C., Oct.

EPA (1974). Basic Fertilizer Chemicals. EPA-440/1-74-011a, Effluent Guidelines Div., Washington, D.C.

EPA (1977). Federal Guidelines on State and Local Pretreatment Programs. EPA-430/9-76-017c, Construction Grants Program, Washington, D.C., Jan.

Federal Register (1987). Code of Federal Regulations. CFR 40, U.S. Government Printing Office, Washington, D.C., pp. 412–430 and 729–739.

Federal Register (1990). Notices, Appendix A, Master Chart of Industrial Categories and Regulations. U.S. Government Printing Office, Washington, D.C., Vol. 55, No. 1, Jan. 2, pp. 102, 103.

Fertilizer Association of India (1979). Liquid effluents. In Pollution Control in Fertilizer Industry, Tech. Rep. 4, Part I.

Fuller, R. B. (1949). The position of the pebble phosphate industry in stream sanitation. *Sewage Wks. J., 21* Sept., (5), 944.

Ghokas, S. I. (1983). Treatment of effluents from a fertilizer complex. M.S. thesis, Univ. of Manchester, U.K.

Griffith, E. J. (1978). Modern mankind's influence on the natural cycles of phosphorus. In *Phosphorus and the Environment*, Ciba Foundation, New Series, 57.

Horton, J. P., et al. (1956). Processing of phosphorus furnace wastes. *J. Water Poll. Control Fed., 28* (1), 70–77.

Jones, W. E. and Olmsted, R. L. (1962). Waste disposal at a phosphoric acid and ammonium phosphate fertilizer plant. In *Proceedings of the 17th Industrial Waste Conference*, Purdue Univ., Lafayette, Ind., Vol. 17, pp. 198–202.

Kiff, R. J. (1987). Water pollution control in the fertilizer manufacturing industry. In *Manufacturing and Chemical Industries* (D. Barnes, et al., eds.). Longman Scientific & Technical, Essex, U.K.

Koziorowski, B. and Kucharski, J. (1972). *Industrial Waste Disposal*. Pergamon Press, Oxford, U.K., pp. 142–151.

Lanquist, E. (1955). Peace and Alafia River Stream Sanitation Studies. Florida State Board of Health, June, Suppl. II to Vol. II.

Layer, W. and Wang, L. K. (1984). Water Purification and Wastewater Treatment with Sodium Aluminate. Rept. PB 85-214-492/AS, U.S. NTIS, Springfield, Va.

Lund, H. F. (ed.) (1971). *Industrial Pollution Control Handbook*. McGraw-Hill, New York, pp. 14-6 to 14-8.

Markham, J. H. (1983). Effluent Control on a Fertilizer Manufacturing Site. Fertilizer Soc. of London, April, Proc. 213.

Nemerow, N. L. (1978). *Industrial Water Pollution*. Addison-Wesley, Reading, Mass., pp. 583–588.

Nemerow, N. L. and Dasgupta, A. (1981). Environmentally balanced industrial complexes. In *Proceedings of the 36th Industrial Waste Conference, Purdue Univ., Lafayette, Ind., Vol. 36, pp. 982–989.*

Overcash, M. R. (1986). *Techniques for Industrial Pollution Prevention*. Lewis Publishers, Michigan, pp. 87–89.

Robasky, J. G. and Koraido, D. L. (1973). Gauging and sampling industrial wastewater. *J. Chem. Engrs., 80* (1), 111–120.

Search, W. J., et al. (1979). Source Assessment, Nitrogen Fertilizer Industry Water Effluents. Rept. PB 292 937, Department of Commerce, U.S. NTIS, Springfield, Va.

Shahalam, A. B. M., et al. (1985). Wastes from processing of phosphate industry. In *Proceedings of 40th Industrial Waste Conference*, Purdue Univ., Lafayette, Ind., Vol. 40, pp. 99–110.

Specht, R. C. (1950). Phosphate Waste Studies. Bull. 42, Florida Eng. and Ind. Exp. Sta., Univ. of Florida, Gainesville, Feb.
Specht, R. C. (1960). Disposal of wastes from the phosphate industry. *J. Water Poll. Control Fed., 32* (9), 963–974.
UNIDO (1974). Minimizing Pollution from Fertilizer Plants. Rept. Expert Group Meeting, Helsinki, ID/WG 175/19.
Wakefield, J. W. (1952). Semi-tropical industrial waste problems. In *Proceedings of the 7th Industrial Waste Conference*, Purdue Univ., Lafayette, Ind., Vol. 7, pp. 495–508.
Wang, L. K. and Wang, M. H. (1990). Decontamination of groundwater and hazardous industrial effluents by high-rate air flotation process. Presented at Proc. Great Lakes '90 Conf., Hazardous Materials Control Res. Inst., Silver Springs, Md., Sept.

Index

Acid, 32, 33, 35
Acid cleaner, 128
Acid pickling, 293
Acid precipitation, 89
Aeration, 304, 306
Agricultural industry, 93
Air emission, 36
Air-SO_3 sulfation/sulfonation, 253, 255
Alkaline chlorination, 155, 156
Alkaline cleaner, 128
Ammonia nitrogen, 189
Ammonium phosphate, 341, 370
   fertilizer, 370
Animal farming, 94
Anodizing, 129, 132
   chromic acid, 132
   hard, 132
   sulfuric acid, 132
Anthraquinone disulfonic acid (ADA), 44, 45, 47
Arsenic removal, 112

Atmospheric contaminants, 88
Automotive assembly plant, 49
Auto painting, 49
Azo dye, 311

Bath dump reduction, 144
BEST process, 43
Biodegradation, 232, 234
Biological treatment, 108, 191, 288, 283, 318
Bleach, 178–179, 188, 203, 204, 308
   fixes, 119
   regeneration, 213
   reuse, 215
Blend fertilizer, 343
Boiler blowdown waste, 348

Cadmium, 150, 188, 293
Calcium phosphate, 334

385

Index

Capital investment, 26
Carbon adsorption, 110, 275, 281, 317
Centrifuging, 201
Chelate and complex breaker, 149
Chemical, 110, 197, 200, 216
   oxidation-reduction, 197
   precipitation, 200
   recovery, 216
Chlorination, 112, 155, 156
Chromate conversion coating, 132, 133
Chrome-anodizing bath regeneration, 168
Chromic acid, 32, 35
Chromium, 33, 34, 148, 150, 187, 218, 293
Chromium precipitation, 218
Cleaner, 128–130
   acid, 128, 130
   alkaline, 128, 130
   electro, 129, 131
Cleaning, 128, 130–131
   acid, 128, 130
   alkaline, 128, 130
Coagulation, 110, 281, 282, 317
Coatings, 132–133
Color index, 309
Coloring, 132–133
   metal, 133
Combined sewer overflow, 62
Concentration, 104
Condenser waste, 348
Conservation, 13, 219, 222
   washwater, 222
   water, 13
Contaminated stormwater, 102
Cooling water, 347, 357
Cost, 25–29
   capital, 25, 26
   operating, 27, 28
Cotton, 308
Cross connection, 99
Crystallization, 296

Cyanide destruction, 155, 156
Cyano complexes, 188

Defluorinated phosphate rock, 333, 354, 358
Defluorinated phosphoric acid, 335, 355
Degreaser, 128, 130
Depression storage, 65
Desizing, 308
Detergent, 109, 229, 232, 237, 241–253, 256–280
   bar, 266
   categorization, 241
   dry, 262
   formulation, 256
   liquid, 262, 263, 278
   manufacture, 241, 253, 264, 265
   spray-dry, 256, 261
   toxicity, 237
   wastewater, 268–273, 280
Developer, 175
   color, 212
   regeneration, 212
Dissolved air flotation, 107, 109, 281, 317
Distillation, 37
Drag, 143
Drainage, 90–92, 119
   base mental mining, 119
   mining, 92
   roadway, 91
   site, 90
Dye, 38, 311
Dyeing, 308

Economic evaluation, 31
Economic information, 19
Effluent limitation, 353–356
   defluorinated phosphate rock, 354
   defluorinated phosphoric acid, 355

Index

[Effluent limitation]
  mixed and blend fertilizer, 353
  phosphate fertilizer, 352
  sodium phosphate, 356
Electric furnace, 348, 364
Electrochemical reduction, 148
Electrocleaner, 129
Electrodialytic system, 167–168
Electroless plating, 135, 138
  bath, 138
Electrolytic silver recovery, 205
Electronic, 32
Electropolishing, 132, 133
  aluminum, 132
  nickel, 132
Electrostatic paint, 51
Electrowinning, 164–167
Engine shop, 47
Environmental information, 19
Equalization, 282
Erosion, 82
Evaporation, 145, 146, 199

Facility assessment, 17
Facility information, 19
Farm pond discharge, 117
Fatty acid, 243, 245, 246
  manufacture, 245
  neutralization, 243, 246
Ferrocyanide, 217
Fertilizer, 343, 370, 375, 376
  ammonium phosphate, 370
  blend, 343
  mixed, 343
  N-P-K, 375, 376
  production waste treatment, 375
Filtration, 107, 201, 281, 282, 317
Finishing, 307
Fixer, 177, 215
  reuse, 215
Flammable and explosive materials, 190

Flotation, 107, 109, 273, 281, 317
Flow diagram, 20
Flowing rinse, 137, 139
Flow through rinsing tank, 143
Fluid catalytic cracker, 40
Fluoride, 365, 380
Foam fractionation, 273
Freeze drying, 200
Freezing, 200
Freon-113, 33, 34, 36
Fume/exhaust scrubbing, 140

Galvanized pipe manufacturing, 299
Glycerine, 246
  recovery, 246, 248
Granular activated carbon, 317

Hazardous material spill, 113
Hexacyanoferrate, 188
Holding tank, 220
Housecleaning, 90
Hydrocracking, 40
Hydrogen fluoride, 33, 35
Hydrogen sulfide, 43, 44, 46
Hydrograph, 66
Hydroquinone, 189
Hydrologic consideration, 64
Hydroxide precipitation, 148, 149, 150, 151
Hyperfiltration, 313

Impact
  on biodegradation, 234
  on drinking water, 237
  on phosphate industry, 349
  on public health, 234
  on river, 233
  on soil and groundwater, 237
  on wastewater treatment, 235
Incineration, 4, 198

Infiltration, 65
Information, 19–21
  design, 19
  economic, 19
  environmental, 19
  material balance, 21
  production, 19
  raw material, 19
Ion exchange, 112, 159–163, 207, 281, 282
  exclusion, 281
  regeneration, 163
Iron, 150, 187, 305

Lag factor, 72, 73
Lagoon, 113, 116
Landfill, 55, 57, 58, 96
Landfill runoff, 96
Land treatment, 320
Land use, 70, 71
Leachate, 96
Liability, 30
Lime, 303
LO-CAT process, 46

Material balance, 20, 21
Mean annual suspended sediment load, 86
Mercerization, 308
Metal, 40, 127, 134, 147, 157, 186, 293, 305
  acid pickling, 293
  finishing, 127
  heavy, 147, 186, 188
  plating, 127, 134
  sludge, 157
Methyl chloroform, 33, 36
Microchips, 32, 47
Microelectronic circuits, 32, 47
Microstrainer, 108
Minewater, 119

Mining drainage, 92
Mixed fertilizer, 343, 353

Neutralization, 191, 243, 256, 260, 295, 304–306
  fatty acid, 243, 247
  sulfonic acid, 256, 260
  sulfuric acid ester, 256, 260
Nitrate, 189
Nonhazardous solid waste, 53
Nonpoint sources, 95
Normal superphosphate, 341

Oleum sulfonation/sulfation, 253, 254
On-site treatment, 315
Overland flow routing, 68
Overland-flow treatment, 114
Overland travel time, 69
Oxidation-reduction, 191, 196, 197
  chemical, 197
  electro, 197
Ozonation, 156, 317
Ozone, 196
Ozone-depleting chemical, 33, 35

Paint, 38, 39
Painting, 49–52
Paint shop, 37–39, 50
Paint sludge, 52
Pathway for contamination, 87, 88
Peak runoff flow, 73–74
Petroleum refinery, 40, 47, 115
Phenolic compound, 188
Phosphate, 189–372
  ammonium, 341
  fertilizer, 338, 340, 351, 352, 379
  fertilizer plant, 379
  industry, 323
  manufacture, 351, 357–362

Index

[Phosphate]
  mining, 367, 372
  ore, 325
Phosphate, 189–372
  removal, 365
  rock, 324, 327, 333, 341
  processing, 372
  production, 328–330
  super, 341
Phosphoric acid, 341, 360, 366, 370
  plant, 370, 375, 376
Phosphorus production, 329, 331, 348, 373
Phosphorus removal, 380
Phossy water, 374
Photographic process, 173–185
  black-and-white, 177, 185
  bleach, 178
  color, 177, 184
  development, 174
  exposure, 173
  stop bath, 176
  wash, 180
Photoprocessing effluent, 182–190
Pickling, 129–131, 293–295
  aluminum, 131
  copper, 131
  steel, 131
Plating, 127, 134–139
  bath, 136, 139
  electro, 134, 136
  electroless, 135
  immersion, 134
  metal, 127
Pollution prevention, 7
Polyethylene, 55
POTW, 314
Precipitation, 149, 152, 200, 209–217
  chromium, 218
  ferrocyanide, 217
  hydroxide, 149
  sulfide, 152, 209, 210
Pretreatment, 315

Primer distillation, 37
Probable maximum rainfall, 64
Product change, 6, 8
Production information, 19
Profitability analysis, 29

Rainfall, 64–67
  depth, 64
  duration, 64
  evaportranspiration, 65
  frequency, 64
  interception, 65
  probable maximum, 64
Rate reduction, 97
Rational formula, 73
Raw material, 19, 90
Raw material stockpiles, 90
Recovery 217–218, 374
  color developing agent, 217
  coupler, 217
  ferrocyanide, 217
  phosphate, 218
  phosphorus, 374
Recycling, 4, 5, 6, 7, 11, 12, 55
Regeneration, 211–213
  bleach, 213
  color developer, 212
  solution, 211
Removal, 380
  fluoride, 380
  phosphorus, 380
Replenishment, 219
Retention pond, 347
Reverse osmosis, 201, 313
Rinse, 137, 139, 140
  cascade, 140
  flowing, 137, 139
  on-demand, 139
  spray, 139
  stagnant, 139
Rinsewater, 135, 139, 143
Rinsing ratio, 143

Index

Risk, 30
Roadway drainage, 91
Rotating biological contactor, 112
Runoff, 66-77, 93, 96, 114, 115
   coefficient, 74
   construction site, 93
   feedlot, 114
   landfill, 96
   peak, 73
   petroleum refinery, 115
   surface, 66
   urban, 67

Scouring, 82, 308
SCOT process, 46
Screening, 103
Scrubbing, 140, 141, 142
   packed-bed, 142
   spray chamber, 141
Seasoning, 181
Secure land disposal, 4
Sedimentation, 82, 103, 201, 282, 295
Sediment, 83–86
   delivery, 83, 85
   rating curve, 86
Sodium phosphate, 338, 339, 356, 358
Seepage control, 364
Separation, 191, 199, 281
   API, 281
   oil and grease, 281, 282
Selectivity, 160
Settling and decanting, 103, 201, 282
Silver, 187, 194, 195, 202, 205, 211
Site, 23, 90, 97
   drainage, 90
   inspection, 23
   planning, 97
Skimming, 102
Sludge solidification, 157
Sludge stabilization, 157

Snyder's equation, 66
Soap, 241–276
   bar, 249, 251
   categorization, 241
   liquid, 249, 252, 276
   manufacture, 241, 243, 244, 250
   neutralization, 247
   production, 249
Sodium hydroxide, 303,
Software, 31
Soil loss, 87
Solidification, 157, 158
Solid waste, 53–58
Solution
   carryover, 191
   regeneration, 211
Solvent, 37, 38, 40
Source reduction, 4, 5, 6, 7, 8, 41, 42, 45
Source control, 6, 8
Spent material stockpiles, 91
Spill, 99, 113
Spray painting, 50
Squeegee, 181, 219
Stabilizer, 180
Stabilization, 157
Steel mill plant, 378
Stop bath, 217
Storage volume, 79
Stormwater, 78–98, 100–102
   pollutant loading, 80
   retention basin, 78
   storage, 98
   treatment, 100–102
Straining, 107
Stripping, 132, 134
   cadmium, 132
   copper, 132
   plated metals, 134
Sulfation, 253–259
   air-SO_3, 253, 255
   chlorosulfonic acid, 256, 259
   oleum, 253, 254

Index

[Sulfation]
  sulfamic acid, 256, 258
  sulfonation, 253–255
Sulfide precipitation, 152, 153, 154, 209, 210
Sulfonation, 253–277
  air-SO$_3$ sulfation, 253–255
  effluent limits, 277
  oleum sulfation, 253, 254, 277
  SO$_3$ solvent and vacuum, 256, 257
Sulfur, 43–45, 47
Sulfur dioxide, 46
Sulfur dioxide chrome reduction, 148
Sulfuric acid, 341, 342
Superphosphate, 341
Surface active agent, 229–232
Surface runoff, 66, 118, 122
  combined, 122
  pesticide load, 118
Surface treatment, 129–134, 136
  bath, 136
Surfactant, 229–231, 253, 288–289
  amphoteric, 231
  anionic, 230, 289
  cationic, 231, 289
  manufacture, 253
  nonionic, 231, 288
  waste stream, 253
Suspended sediment, 83
Synthetic fiber, 309

Textile waste, 307, 310
Thiocyanate, 189
Time of concentration, 67
Topcoat distillation, 37
Toxicity characteristic leaching procedure (TCLP), 157
Toxicity reduction, 14
Treatment, 315–318
  on-site, 315–316
  physical-chemical, 316
  pre-, 315

Trickling, 111
1,1,1-Trichloroethane, 333
Trichlorotrifluoroethane, 33
Triple superphosphate, 341

Ultrafiltration, 202, 313
Unit hydrograph, 76
Universal soil loss equation, 86
Urban, 77, 95
  industrial area, 95
  runoff model, 77
Urban runoff model, 77

Volatiles, 190
Volume reduction, 12, 13, 43, 47, 53, 57, 97

Waste, 2, 13, 18, 32, 44–53, 115, 221, 368–375
  concentration, 13, 221
  disposal, 368
  elimination, 32
  fertilizer production, 375
  furnace, 373
  generation, 2, 44, 372
  management, 18, 47, 53, 115, 368
Waste minimization, 1–19, 22–25, 142
  assessments, 19
  barriers, 15, 16
  economics, 25
  options, 22
  practice, 14
Waste pickle liquor, 294–298
Waste reduction, 47–52
Waste segregation, 10, 13, 159
Waste treatment, 18, 21, 352–385
  phosphate fertilizer, 352
  phosphate industry, 368

[Waste treatment]
  phosphoric acid, 375
Water conservation, 13, 222
Water quality model, 81
Water quality parameters, 82

Wet process phosphoric acid, 341
Wool, 308

Zinc, 150, 153, 188, 293, 304